# An
# Invitation to
# Astrophysics

# WORLD SCIENTIFIC SERIES IN ASTRONOMY AND ASTROPHYSICS

Editor: Jayant V. Narlikar
Inter-University Centre for Astronomy and Astrophysics, Pune, India

*Publication cancelled.

World Scientific Series in Astronomy and Astrophysics – Vol. 8

# An Invitation to Astrophysics

## Thanu Padmanabhan

*Inter-University Centre for Astronomy and Astrophysics, India*

**World Scientific**

NEW JERSEY · LONDON · SINGAPORE · BEIJING · SHANGHAI · HONG KONG · TAIPEI · CHENNAI

*Published by*

World Scientific Publishing Co. Pte. Ltd.

5 Toh Tuck Link, Singapore 596224

*USA office:* 27 Warren Street, Suite 401-402, Hackensack, NJ 07601

*UK office:* 57 Shelton Street, Covent Garden, London WC2H 9HE

**British Library Cataloguing-in-Publication Data**
A catalogue record for this book is available from the British Library.

The background picture used in the cover design was adapted from the figure kindly provided by
J. S. Bagla. The N-Body simulations, from which the picture was developed, were performed at the
Cluster Computing Facility in the Harish-Chandra Research Institute.

ISBN  981-256-638-4
ISBN  981-256-687-2 (pbk)

Printed in Singapore by World Scientific Printers (S) Pte Ltd

In memory of the late Shri Neelakanda Sarma, who introduced me
to the fun of doing Science.

# Preface

This book is exactly what its title claims to be. It is an *invitation* to students and researchers in different areas of physics to share the fun and excitement of astrophysics.

During the last two decades, technological advances have allowed us to observe the cosmos with unprecedented accuracy, making astrophysics and cosmology branches of science driven forward by observations. This excitement and growth has attracted more people to enter this area and, given the current enthusiasm, it is inevitable that astrophysics will continue to be a leading branch of science in the early part of this millennium.

This subject has also fascinated the layman all through the ages, resulting in a proliferation of popular articles and books. Every branch of astrophysics, of course, has excellent text books authored by the experts. There are also a few comprehensive text books at an advanced level intended for serious researchers in the field. In spite of all these, currently available literature leaves a gap which the present book is designed to fill. This book will be useful, for example, to a theoretical physicist who wants to have a quick overview of the entire subject of astrophysics — and appreciate the excitement — but without getting lost in the details. In a way, it is expected to serve as 'a guide to astrophysics for an intelligent outsider'; maybe a researcher in theoretical physics, a student of physics or even an observational astronomer who wants to dabble a bit with the theory.

I have designed the book keeping the above perspective in mind. To begin with, it is by no means expected to be comprehensive. For that, there are other books (including the three volume course of astrophysics, running to about 1800 pages, authored by me and published by Cambridge University Press). The choice of topics here stresses both the foundations and the frontiers, moving lightly over anything in between.

Second, I have made sure that the book is self contained as far as astronomy and astrophysics are concerned. For example, astrophysics, like every other subject, has a collection of jargon specific to it and I do *not* expect the reader to have any familiarity with this jargon. Astronomical jargon is explained in the language of the physicist so that the book is accessible to a wider community. The same criterion is used in the choice of theoretical foundation that is laid in the first three chapters. These three chapters deal with those areas of theoretical physics which are widely used in astrophysics but might require some amount of brushing up for an average student of theoretical physics. For example, bremsstrahlung or ionization equilibrium or the propagation of a shock wave may not be a part of every researcher's arsenal even though they might have learnt basic electrodynamics or fluid mechanics. The first three chapters gather together several such topics in order to make this book self contained. At the same time, I have *not* tried to teach the basic areas of theoretical physics; I expect the reader to know the basics of classical and quantum mechanics, special relativity, electrodynamics, *etc.* (But not general relativity, which is introduced briefly in Chapter 1.)

Third, I have added a fair number of exercises at the end of each chapter which will provide some amount of practice for the physicist in applying the concepts to astrophysical contexts. In particular, there are quite a few problems titled 'Real Life Astronomy' which take the reader directly to published literature.

Fourth, I have kept the algebra to the minimum commensurate with clarity. Whenever the derivation of an equation is straightforward (even though lengthy), I have merely indicated how the result can be obtained rather than hold the reader by hand and take him/her through the algebra. I think this allows for a crisper pace and those who are interested in the key concepts and results can always take algebra for granted.

Finally, I have tried to keep the discussion in the later chapters fairly topical. The emphasis is on research areas which are currently popular rather than on more traditional topics. This has the disadvantage that some of the ideas might change a bit over the next decade or so but the advantage of conveying the current excitement in the frontier areas more than compensates for this.

The cognoscenti will find that many of the derivations of even standard results (Kepler problem, Compton scattering, Kompaneets equation, Schwarzschild metric, gravitational lensing deflection ...) are different from the ones usually found in textbooks. Hopefully, even the experts find something new and enjoyable in this book. I have also not shied away from discussing some of the topics usually considered to be rather advanced but the physical content of which is quite simple (for example, the concept of black hole temperature).

The idea for this book originated from several colleagues who suggested to me that an introductory text, to supplement my three volume course in astrophysics, will be a useful addition to the literature. I was initially hesitant since such a book will necessarily overlap to a certain extent with the existing textbooks in the market, including mine. However, many people convinced me that there is really a need for such a book and some amount of overlap will do no harm. I am thankful to J.V. Narlikar, who is the editor of this series published by World Scientific, for an enthusiastic and positive response. I thank L. Sriramkumar for reading the manuscript and giving extensive comments. I have also benefited from my discussions with K. Subramanian, R. Nityananda, D. Lynden–Bell and the members of the Astronomy Division of Caltech regarding several aspects of pedagogy related to astrophysics. In particular, K. Subramanian tried out parts of the material in the book in an introductory astrophysics course he was teaching at Pune University. I am also grateful to R. Gupta, H.M. Antia, C.C. Steidel and Jim Kneller for providing the data which were used in generating some of the figures. As I have already mentioned, it is impossible to avoid certain amount of overlap in the discussion with other existing text books including my own Theoretical Astrophysics Vol I–III published by Cambridge University Press. I have referred to this and other text books wherever appropriate in the Notes and References at the end of each chapter.

It was a pleasure working with Ms Magdalene Ng and her World Scientific team in this project and I thank them for excellent processing of the book.

I thank Vasanthi Padmanabhan for dedicated support and for taking care of the entire latexing, production of figures and for formatting the text. It is a pleasure to acknowledge IUCAA for the research facilities and ambience it has provided.

T. Padmanabhan

# Prologue

Astrophysics involves the application of laws of physics to understand the cosmic structures which, broadly speaking, include anything bigger than the planets. The purpose of this prologue is to give you a bird's eye view of these structures and introduce some essential jargon. The scales of physical variables, like mass, length, time, temperature, *etc.* which we encounter in astrophysics are summarised in the Appendix at the end of the book.

The description of astrophysical structures could begin either from small scales and proceed to the large scales or the other way around. We will proceed from the largest scale downwards. You would certainly know — even if you are a rank outsider to astronomy — that our universe is expanding. The current rate of expansion (called *Hubble constant*) is measured to be about $H_0 \equiv 0.3 \times 10^{-17} h$ s$^{-1}$ where $h \approx 0.72$ is a dimensionless factor of order unity that parameterizes the actual rate of expansion. This gives you a characteristic time scale, $t_{\text{univ}} \equiv H_0^{-1} \approx 10^{10} h^{-1}$ years, that can be associated with our universe. This is typically the largest time scale over which cosmic evolution takes place. Multiplying it by the speed of light, one can construct a length scale $ct_{\text{univ}}$ associated with the universe which is roughly the size of the visible universe today. Numerically, $ct_{\text{univ}} \approx 10^{28} h^{-1}$ cm and cosmologists usually write this as $ct_{\text{univ}} \approx 3000 h^{-1}$ Mpc where 1 *Mpc* $\approx 3 \times 10^{24}$ cm is a convenient unit for extragalactic astronomy. (We will also use 1 kpc $\approx 3 \times 10^{21}$ cm and 1 pc $\approx 3 \times 10^{18}$ cm in describing galactic and sub-galactic scales.)

The rate of expansion $H_0$ has the dimension of frequency. You know that, if $\rho$ is a mass density, $\sqrt{G\rho}$ also has the dimensions of frequency. Therefore, by introducing the Newtonian gravitational constant $G$, you can define a characteristic density, $\rho_c$, from the rate of expansion $H_0$. It is conventional (and useful) to define it with an extra numerical factor, as

$\rho_c \equiv 3H_0^2/(8\pi G) = 1.88h^2 \times 10^{-29}$ gm cm$^{-3}$. Incredibly enough, the *actual density* of our universe is very close to this value.

An expanding universe would have been smaller, denser and hotter in the past and — at sufficiently early moments — matter would have existed as a primordial soup of elementary particles. Observations show that our universe is embedded in a thermal radiation — called *Cosmic Microwave Background Radiation* (CMBR) — at a temperature of about 2.73 K (which corresponds to an energy scale of $k_B T \approx 2 \times 10^{-4}$ eV using $k_B = 0.86 \times 10^{-4}$ eV K$^{-1}$). In the past, the temperature would have been higher, increasing essentially as the inverse power of the size of the universe. When $k_B T$ is higher than the energy scale $E_{\rm pl} \equiv (\hbar c^5/G)^{1/2} \approx 10^{19}$ GeV — which can be obtained from the three fundamental constants — quantum gravitational effects would have been important in the universe. We have no clue about this phase but, fortunately, this ignorance doesn't seem to affect our ability to predict what happens in the universe as it cools to more moderate energies.

Two events in the later life of the universe are of deep significance (see Chapter 6). First, when the universe was about a minute old, it was cool enough to allow the protons and neutrons to come together and synthesize some amount of deuterium and helium (but not significant amounts of heavier elements). From the observed abundance of these primordial elements, one can put a bound on the amount of baryonic matter (that is, protons and neutrons) that can exist in the universe. This turns out to be a fraction $\Omega_B \approx 0.05$ of the critical density $\rho_c$ introduced earlier. Since other observations show that the mean density of the universe is actually about $\rho_c$, most of the matter in the universe must be made of non baryonic matter. In fact, observations suggest that nearly 30 per cent of the matter is made of *Weakly Interacting Massive Particles* (WIMPs), usually called *dark matter*, while the remaining constituent is in a more exotic form of a fluid that exerts negative pressure (called *dark energy*). Understanding the nature of this dark energy is the greatest theoretical challenge in cosmology today.

Second, when the size of the universe was about $10^3$ times smaller than the current size, it was cool enough for the electrons to combine with protons and helium nuclei to form a neutral gaseous mixture of hydrogen and helium. The formation of the neutral atoms allowed the radiation to decouple from matter and flow freely without significant scattering or absorption till today. It is this relic radiation that we observe as the CMBR mentioned earlier. Clearly, CMBR will contain a fossilized record of the state of the universe when it was $10^3$ times smaller. Direct observations of CMBR now tells us that — at that time — the matter in the universe was distributed very smoothly but with tiny fluctuations in energy density amounting to about 1 part in $10^5$. The detection of these tiny fluctuations, in the last decade or so, has been a high point in our understanding of the cosmos.

As the universe evolved further, these tiny fluctuations in the density grew by gravitational instability and eventually formed the different structures which we see in the universe. The dark matter particles cluster gravitationally to form large, virialized, self-gravitating lumps in which baryonic matter gets embedded in the form of, say, *galaxies*. In fact, direct observations of the gravitational potential surrounding the visible part of the galaxies show that galaxies are embedded in large dark matter halos which are probably at least ten times as massive. The normal matter located within these dark matter halos radiate energy, cool and sink to the central regions of the halos finally reaching a size of about 15 kpc in radius (see Chapters 7 and 8). Incidentally, one can obtain this mass scale of the galaxy entirely in terms of fundamental constants:

$$ M_g \simeq \alpha_G^{-2} \alpha^5 \left( \frac{m_p}{m_e} \right)^{1/2} m_p \simeq 3 \times 10^{44} \text{gm} $$

where $\alpha_G \equiv (Gm_p^2/\hbar c) \simeq 10^{-38}$ and $\alpha = (e^2/\hbar c) \simeq 10^{-2}$ are dimensionless constants. Clearly, the scale of galaxies is no accident but is determined by physics.

The formation of galaxy like structures took place from the time when the universe was about a tenth of the present size. In fact, even larger structures — which are gravitationally bound — exist in the universe. *Groups of galaxies* may typically consist of about (5–100) galaxies; a system with more than 100 galaxies is conventionally called a *cluster*. The sizes of groups range from a few hundred kiloparsecs to few Megaparsecs. Clusters have a size of typically a few Megaparsecs. The gravitational potential energy is comparable to the kinetic energy of random motion in these systems. The line of sight velocity dispersion in the groups is typically 300 km s$^{-1}$ while that in the clusters can be nearly 1000 km s$^{-1}$. About ten percent of all galaxies are members of large clusters. In addition to galaxies, clusters also contain very hot intra-cluster gas at temperatures $(10^7 - 10^8)$K.

The lumps of gas within the galaxies will further undergo gravitational contraction and when the temperature at the central region of such a contracting gas cloud is high enough to ignite nuclear reaction, it will reach a stable configuration in the form of a *star* (see Chapter 4). The typical mass of a star is 1 $M_\odot \approx 2 \times 10^{33}$ gm (which is the mass of the Sun) and the radius is about $R_\odot \approx 7 \times 10^{10}$ cm. Again, this scale is determined by fundamental constants:

$$M_* \cong \left(\frac{m_p}{m_e}\right)^{3/4} \left(\frac{\alpha}{\alpha_G}\right)^{3/2} m_p \simeq 10^{32} \text{ gm}.$$

For comparison, the mass and radius of the Earth are $M_\oplus = 6 \times 10^{27}$ gm and $R_\oplus = 6 \times 10^8$ cm, respectively. (Astronomers use the subscript $\odot$ for matters pertaining to Sun and $\oplus$ for matter pertaining to Earth.) From the previous estimate, it is clear that a typical galaxy contains about $10^{11}$ to $10^{12}$ stars. The mean distance between galaxies in the universe is about 1 Mpc while the mean distance between the stars in the galaxy is 1 pc.

The first generation of stars (which are called *Population II* or *Pop II* stars) will essentially be made from a cloud of hydrogen and helium. Nucleosynthesis in the hot stellar furnace leads to the formation of heavier elements inside the star. When the nuclear fuel is exhausted, the stars die and become *stellar remnants* which could take different forms (see Chapter 5). For example, high mass stars end their life in spectacular explosions (called *supernova*) which are the brightest optical events in the sky. The supernova explosion spews out the heavier elements synthesized inside the stars on to the *interstellar medium* of the galaxy from which the next generation of stars (usually called *Pop I*) form. What is left behind in a supernova explosion will be a rotating *neutron star* usually called a *pulsar*. Less massive stars may end their life leaving behind a remnant called *white dwarf*. Both white dwarfs and neutron stars are quantum mechanical objects in the sense that the gravitational attraction is balanced by quantum degeneracy pressure of electrons or neutrons in these objects. It is one of the many examples in which microphysics dictates the form of large scale structures in the universe. And the really high mass stars leave behind a *black hole* as the remnant.

Since stars form in collapsing gas clouds, they very often form in pairs or larger groups. In fact, some of the oldest structures in the universe are structures known as *globular clusters* which are conglomerations of about $10^6$ stars. Our galaxy, for example, contains several of them distributed in a roughly spherical manner. When the stars form in pairs (*binaries*, as they are called) it is quite possible for one of them to evolve faster and leave behind a remnant (white dwarf, neutron star or a black hole) which will acrete matter from the second star leading to high energy radiation.

Several such systems, often consisting of a star with an invisible companion, have been observed.

Accretion of matter, allowing gravitational potential energy to be converted into radiation, is an extraordinarily efficient process and works in another class of systems, called *active galactic nuclei* (see Chapter 9). These are galaxies, which harbor massive ($> 10^7 M_\odot$) black holes in their centers. Matter accreting to them produces luminosity exceeding that of an entire galaxy from a region smaller in size than the solar system! Many such systems are known and they have existed even when the universe was about five time smaller in size. The farther we can see, the earlier we can probe, because of the finite time taken by the light to traverse the distance; AGN can be seen from very large distances and hence can be used to probe the universe when it was considerably younger.

The formation of galaxies by gravitational instability cannot, of course, be one hundred per cent efficient and one would expect fair amount of gas to be left in between the galaxies in the form of an *intergalactic medium*. Observations suggest that the intergalactic medium is ionized to a high degree, possibly from the radiation emitted by the first structures, which formed in the universe when it was smaller.

The above summary illustrates the diversity of structures we encounter in astrophysics. The density varies from $10^{-28}$ g cm$^{-3}$ in the intergalactic medium to $10^{14}$ g cm$^{-3}$ in a neutron star; the sizes vary from a $10^8$ cm for the radius of a planet to $10^{25}$ cm for a large galaxy cluster; the distance scales range from a few parsecs in the interstellar medium to a few thousand Megaparsecs in the size of the universe; the masses range from $10^{33}$ gm for a typical star to about $10^{47}$ gm for a cluster of galaxies. Consequent of this variety is the fact that matter exists in widely different forms in the astrophysical systems: ionized plasma, neutral gas, solid of moderate density and material with nuclear densities. Clearly, one requires a wide variety of input physics to understand these phenomena.

In this book we will attempt to cover all these phenomena (and more) from the point of view of a curious physicist. The first three chapters provide the theoretical background describing a key processes in gravity, theory of radiation, fluid mechanics *etc.* which are extensively used in astrophysics. Chapters 4 and 5 take up the physics of stars and stellar remnants which is the corner stone of astrophysics; star formation, stellar evolution, white dwarfs, neutron stars, supernova and supernova remnants, binary stars ..... all of them find their place here. In Chapters 6 and 7 we move to the other end of the scale. Chapter 6 discusses the current ideas in cosmology and Chapter 7 builds this further to describe the formation of galaxies. The actual description of galaxies as physical systems is presented in Chapter 8. The last chapter covers the more exotic type of galaxies, viz., the active galactic nuclei.

# Contents

# Chapter 1

# Gravitation

Given the fact that gravity plays the key role in astrophysics — at all scales from planets to the universe — it makes sense to begin our discussion with a study of gravitational physics. Though the correct description of gravity requires general relativity, you can get by, most of the time, with the simpler Newtonian approximation based on three assumptions:

(i) Gravity obeys the *principle of superposition*: If particle 1 and particle 2 individually produce gravitational fields $\mathbf{g}_1$ and $\mathbf{g}_2$, then the gravitational field produced in the presence of both the particles is just the sum $\mathbf{g}_1 + \mathbf{g}_2$. So if we know the gravitational field $d\mathbf{g}$ at a location $\mathbf{x}$ due to an infinitesimal amount of matter $dM = \rho(t, \mathbf{x}') \, d^3x'$ located around $\mathbf{x}'$, we can find the gravitational field produced by any mass distribution by integrating over the volume.

(ii) So all we need to know is the gravitational field at $\mathbf{x}$, produced by an infinitesimal amount of matter $dM$ at $\mathbf{x}'$. Following the footsteps of Newton, we will take this to be $d\mathbf{g} = -G(dM)(\hat{\mathbf{r}}/r^2) = -G\rho(t, \mathbf{x}') \, d^3x'(\hat{\mathbf{r}}/r^2)$ where $\mathbf{r} = \mathbf{x} - \mathbf{x}'$, which is just the inverse-square law for the gravitational force. The total field due to any mass distribution can then be found by the integral:

$$\mathbf{g} = -\int d^3x' \, G\rho(t, \mathbf{x}')\frac{\mathbf{x} - \mathbf{x}'}{|\mathbf{x} - \mathbf{x}'|^3}. \tag{1.1}$$

This force can be expressed as a gradient of a potential: $\mathbf{g} \equiv -\nabla_{\mathbf{x}}\phi$ where the gravitational potential due to the density distribution $\rho(t, \mathbf{x})$ is taken to be

$$\phi(t, \mathbf{x}) \equiv -\int d^3x' \, \frac{G\rho(t, \mathbf{x}')}{|\mathbf{x} - \mathbf{x}'|}. \tag{1.2}$$

The equation $\mathbf{g} \equiv -\nabla_{\mathbf{x}}\phi$ by itself does not fix the normalization for $\phi$

1

and you can add any constant to it without changing $\mathbf{g}$. But the explicit definition in Eq. (1.2) fixes it: with this definition, $\phi \to 0$ as $|\mathbf{x}| \to \infty$ if the density distribution falls faster than $|\mathbf{x}|^{-3}$ at large distances. Using the identity $\nabla^2(1/|\mathbf{x}|) = -4\pi\delta_D(\mathbf{x})$, where $\delta_D$ is the Dirac delta function (which is most easily proved by integrating both sides over a spherical region of radius $r$) we see that Eq. (1.2) also implies the *Poisson's equation*: $\nabla^2\phi = 4\pi G\rho$.

(iii) To complete the story, we need to know how particles move in a given gravitational field. We will assume that a gravitational field $\mathbf{g}$ exerts a force $\mathbf{f} = m\mathbf{g}$ on a particle of mass $m$.

These simple and familiar rules of Newtonian gravity hide several non-trivial features. First, note that particles produce (and respond to) gravitational field with a strength proportional to their masses. There is no *a priori* reason why they should! For example, particles produce (and respond to) electric field with a strength proportional to electric charge $q$ so that the equation of motion for a charged particle in an electric field is $m\mathbf{a} = q\mathbf{E}$. Here $m$ is the inertial mass of the particle which occurs in Newton's second law $\mathbf{F} = m\mathbf{a}$ in which the acceleration $\mathbf{a}$ could have been produced by *any* force $\mathbf{F}$. One can easily imagine the existence of a "gravitational charge" $q_g$ so that the gravitational force scales in proportion to $q_g$ with the equation of motion for a particle in a gravitational field being $m\mathbf{a} = q_g\mathbf{g}$. But this is not the case. The mass $m$ which occurs in the equation for the gravitational force, $\mathbf{f} = m\mathbf{g}$, is the same mass $m$ which occurs in the Newton's second law. An immediate consequence is that the acceleration experienced by particles with different values of $m$ is the *same* in a given gravitational field. (This fact — which is one version of *principle of equivalence* — is a key ingredient in Einstein's general relativity which describes gravity as curvature of space time; see Sec. 1.5.)

Second, the gravitational potential $\phi(t, \mathbf{x})$ (and the force) at time $t$ depend on the density distribution $\rho(t, \mathbf{x})$ at the *same* time $t$; that is, Newtonian gravity acts instantaneously. If the Sun disappears, its effect on Earth will vanish at the same instant. Obviously, this will not be true in a relativistic theory in which signals cannot propagate at speeds greater than that of light. If the Sun disappears, it should take at least about 8 minutes for the effect to be felt on Earth. General relativity takes care of this but, as we said before, we can ignore this feature in the nonrelativistic limit.

And, finally, the principle of superposition — something you might have thought is obvious — is *not* true in the correct description of gravity based on general relativity. This is because relativity demands that all energies contribute to gravity including the energy of the gravitational field itself; this makes the theory non linear. So, even if you know the field produced by two particles individually, you cannot find the field when both are present by just adding them up. You need to go and solve Einstein's equations — which is the analogue of Poisson's equation — again. This is one reason why we have only a handful of known exact solutions in general relativity.

What is the necessary condition for the validity of the non relativistic description of gravity? The gravitational potential in Eq. (1.2) has the dimension of square of velocity and we can use Newtonian gravity as long as $|\phi| \ll c^2$. In the case of a particle of mass $M$ located at the origin, the gravitational potential at a distance $r$ is $|\phi| = GM/r = v^2$ where $v$ is the speed of a particle in a circular orbit at the radius $r$. The quantity, $|\phi|/c^2 = GM/(c^2r) \approx (v/c)^2$ is, therefore, a measure of relativistic effects in the system. Newtonian gravity can be trusted for $v^2 \ll c^2$, that is at distances $r \gg r_S \equiv 2GM/c^2$. The length scale $r_S$ is called the *Schwarzschild radius* for a particle of mass $M$ and the factor 2 is included in order to agree with exact definition in general relativity (which we will discuss in Sec. 1.5). Numerically, $r_S \approx 3$ km $(M/M_\odot)$ where $M = M_\odot \approx 2 \times 10^{33}$ gm is the mass of the Sun. Clearly, $r_S$ is tiny compared to the size of the Solar system, or for that matter, the size of the Sun itself. Hence, Newtonian gravity works reasonably well in the Solar system.

## 1.1 Orbits in Newtonian gravity

The simplest problem in Newtonian gravity is that of a one-body problem: say, that of a planet moving around the Sun, with the Sun assumed to stay fixed at the origin. It turns out that the trajectory of a particle, moving under the attractive inverse square law force, is a circle (or part of a circle) in the *velocity space*. The proof is quite straightforward. Start with the text book result that, for particles moving under any central force $f(r)\hat{\mathbf{r}}$, the angular momentum $\mathbf{J} = \mathbf{r} \times \mathbf{p}$ is conserved. (Here $\mathbf{r}$ is the position vector, $\mathbf{p}$ is the linear momentum and $\hat{\mathbf{r}}$ is the unit vector in the direction of $\mathbf{r}$. The conservation law arises because in $d\mathbf{J}/dt$, the $d\mathbf{r}/dt$ is in the direction of $\mathbf{p}$ and the $d\mathbf{p}/dt$ is in the direction of $\mathbf{r}$ for the central force). The constancy of $\mathbf{J}$ implies, among other things, that the motion is confined to plane perpendicular to $\mathbf{J}$. Let us introduce in this plane the polar coordinates $(r, \theta)$ and the Cartesian coordinates $(x, y)$. The conservation law for $\mathbf{J}$

implies that $r^2\dot{\theta}$ is a constant giving

$$\dot{\theta} = \text{constant}/r^2 \equiv hr^{-2}. \tag{1.3}$$

(Incidentally, this is equivalent to Kepler's second law, since $(r^2\dot{\theta}/2) = h/2$ is the area swept by the radius vector in unit time.) In a small interval of time $\delta t$, the magnitude of the velocity changes by $\Delta v = (GM/r^2)\delta t$ according to Newton's law for the inverse square law force. The angle changes by $\Delta\theta = (h/r^2)\delta t$ from the conservation of angular momentum expressed as Eq. (1.3). Dividing the two relations, we get

$$\frac{\Delta v}{\Delta\theta} = \frac{GM}{h} = \text{constant}. \tag{1.4}$$

But, in the velocity space, $\Delta v$ is the arc length of the curve traced by the moving particle and $\Delta\theta$ is the angle of turn. If the ratio between the two is a constant, then the curve must be (part of) a circle.

An algebraic derivation of the same result is equally easy: Newton's laws of motion give $m\dot{v}_x = f(r)\cos\theta; m\dot{v}_y = f(r)\sin\theta$ for any central force. Dividing these equations by Eq. (1.3) we get

$$m\frac{dv_x}{d\theta} = \frac{f(r)r^2}{h}\cos\theta; \quad m\frac{dv_y}{d\theta} = \frac{f(r)r^2}{h}\sin\theta. \tag{1.5}$$

The miracle is now in sight for inverse square law force, for which $f(r)r^2$ is a constant. For the planetary motion we have $f(r)r^2 = -GMm$, leading to

$$\frac{dv_x}{d\theta} = -\frac{GM}{h}\cos\theta; \quad \frac{dv_y}{d\theta} = -\frac{GM}{h}\sin\theta. \tag{1.6}$$

Integrating these equations, with the initial conditions $v_x(\theta = 0) = 0; v_y(\theta = 0) = u$, squaring and adding, we get the equation to the trajectory in the velocity space, (called *hodograph*, if you like to know that sort of things):

$$v_x^2 + (v_y - u + GM/h)^2 = (GM/h)^2 \tag{1.7}$$

which is a circle with center at $(0, u - GM/h)$ and radius $GM/h$. So you see, planets actually move in circles!

A little thought shows that the detailed structure depends crucially on the ratio between the initial velocity $u$ and $GM/h$, motivating us to introduce a dimensionless quantity $e$ by writing $(u - GM/h) \equiv e(GM/h)$. The geometrical meaning of $e$ is clear from Fig. 1.1. If $e = 0$, i.e, if we had chosen the initial conditions such that $u = GM/h$, then the center

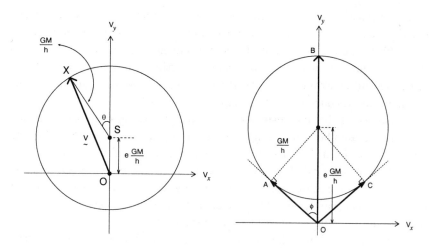

Fig. 1.1 Trajectory of a particle moving under $r^{-2}$ force in the velocity space with (a) $e < 1$ (left) and (b) $e > 1$ (right).

of the hodograph is at the origin of the velocity space and the magnitude of the velocity remains constant. Since $h = ur_i$, where $r_i$ is the initial position, we get $u^2 = GM/r_i^2$ leading to a circular orbit in the real space as well. When $0 < e < 1$, the origin of the velocity space is inside the hodograph but is located off-center. As the particle moves the magnitude of the velocity changes between a maximum of $(1+e)(GM/h)$ and a minimum of $(1 - e)(GM/h)$. When $e = 1$, the origin of velocity space is at the circumference of the hodograph and the magnitude of the velocity vanishes at this point. In this case, the particle goes from a finite distance of closest approach to infinity, reaching infinity with zero speed. Clearly, $e = 1$ implies $u^2 = 2GM/r_i$ which is just the text book condition for escape velocity.

When $e > 1$, the origin of velocity space is outside the hodograph and Fig. 1.1(b) shows the behaviour in this case. The maximum velocity achieved by the particle is $OB$ when the particle is at the point of closest approach in real space. The asymptotic velocities of the particle are $OA$ and $OC$ obtained by drawing the tangents from $O$ to the circle. From the figure it is clear that $\sin \phi = e^{-1}$. During the *unbound motion* of the particle, the velocity vector traverses the part ABC.

The structure of the orbits can also be understood from the point of view of dimensional analysis. The trajectory of a planet is independent of the mass of the planet and hence is determined essentially by the energy (per unit mass) and the angular momentum (per unit mass) of the planet. Let the energy per unit mass, with the dimension of square of the velocity, be $(1/2)u^2$ where $u$ is the initial velocity. Given $GM$ and $u^2$, we can construct one quantity with the dimension of length: $GM/u^2$. The availability of the second constant $h$ — with the dimension of velocity times distance — allows us to construct a *dimensionless* ratio $GM/uh \equiv (1 + e)^{-1}$. It is the existence of this dimensionless ratio that determines the orbital structure. If the angular momentum is small in a gravitating system, then its size is essentially determined by a single parameter $GM/u^2$ where $u$ is the characteristic velocity of the system. This result has an astonishing range of validity from satellites orbiting Earth to galaxies in the expanding universe, which is (partly) due to the fact that the relevant eccentricities play a sub-dominant role in all astrophysical systems.

So much for life in velocity space. To do anything practical, you need the trajectories in *real* space which — fortunately — turns out to be as easy: Consider the tangential velocity in the real space, $v_T$, at any instant. This is clearly the component perpendicular to the instantaneous radius vector in real space. But in the central force problem, the velocity change $\Delta v$ is parallel to the radius vector. So $v_T$ is also perpendicular to $\Delta v$. This result, of course, is true in the velocity space as well in which $\Delta v$ is tangent to the circle. The $v_T$, which is perpendicular to $\Delta v$, must be the component of velocity parallel to the radius of the circle in the *velocity* space. From Fig. 1.1a, it is just the radius XS plus the projection of OS on XS (which is OS $\cos \theta$). So

$$v_T = \frac{GM}{h} + \frac{GM}{h} e \cos\theta = \frac{GM}{h}(1 + e\cos\theta) = \frac{h}{r} \qquad (1.8)$$

with the last relation coming from the definition of angular momentum, viz. $h = rv_T$. You will recognize that the trajectory $r(\theta)$ in Eq. (1.8) is a conic section with a latus rectum of $l = h^2/GM$ and eccentricity of $e$.

This derivation, however, leaves one question unanswered. A particle moving in 3 space dimensions has a phase space which is 6 dimensional. For any time independent central force, we have constancy of energy $E$ and angular momentum $\mathbf{J}$. Conservation of these four quantities $(E, J_x, J_y, J_z)$ confines the motion to a region of $6 - 4 = 2$ dimensions. The projection of this phase space trajectory on to xy-plane will, in general, fill a two dimensional region of space. So you would expect the orbit to fill a finite two dimensional region of this plane, if there are no other conserved quantities. But we know that the motion is on a conic section in the xy-plane, which is an *one* dimensional curve. So, there must exist yet another conserved quantity for the inverse square law force which keeps the planet stick to one dimension rather than wander off in a two dimensional region. And indeed there is, which provides another cute way of solving the Kepler problem.

To discover the last constant, consider the time derivative of the quantity $(\mathbf{p} \times \mathbf{J})$ in any central force $f(r)\hat{\mathbf{r}}$. We have

$$\frac{d}{dt}(\mathbf{p} \times \mathbf{J}) = \dot{\mathbf{p}} \times \mathbf{J} = \frac{f(r)}{r}\mathbf{r} \times (\mathbf{r} \times m\dot{\mathbf{r}})$$

$$= \frac{mf(r)}{r}[\mathbf{r}(\mathbf{r} \cdot \dot{\mathbf{r}}) - \dot{\mathbf{r}}r^2] = -mf(r)r^2\frac{d}{dt}\left(\frac{\mathbf{r}}{r}\right). \qquad (1.9)$$

That is,

$$\frac{d}{dt}(\mathbf{p} \times \mathbf{J}) = -mf(r)r^2\frac{d\hat{\mathbf{r}}}{dt}. \qquad (1.10)$$

The miracle of inverse square force is again in sight: When $f(r)r^2 = $ constant$= -GMm$, we find that the vector (called *Runge-Lenz vector* though several others are also connected with its discovery):

$$\mathbf{A} \equiv \mathbf{p} \times \mathbf{J} - \frac{GMm^2}{r}\mathbf{r} \qquad (1.11)$$

is conserved. But we needed only one constant of motion while we now have got 3 components of $\mathbf{A}$ which will prevent the particle from moving at all! Such an overkill is avoided because $\mathbf{A}$ satisfies the following two, easily verified, relations:

$$A^2 = 2mJ^2E + (GMm^2)^2; \quad \mathbf{A}.\mathbf{J} = 0 \qquad (1.12)$$

where $E = p^2/2m - GMm/r$ is the conserved energy for the motion. The first relation tells you that the magnitude of $\mathbf{A}$ is fixed in terms of other constants of motion and the second one shows that $\mathbf{A}$ lies in the orbital plane. These two constraints reduce the number of independent constants in $\mathbf{A}$ from 3 to 1, exactly what we needed. It is this extra constant that keeps the planet on a closed orbit. To find that orbit, we only have to take the dot product of Eq.(1.11) with the radius vector $\mathbf{r}$ and use the identity $\mathbf{r}.(\mathbf{p} \times \mathbf{J}) = \mathbf{J}.(\mathbf{r} \times \mathbf{p}) = J^2$. This gives

$$\mathbf{A}.\mathbf{r} = Ar\cos\theta = J^2 - GMm^2r \qquad (1.13)$$

or, in more familiar form, the conic section:

$$\frac{1}{r} = \frac{GMm^2}{J^2}\left(1 + \frac{A}{GMm^2}\cos\theta\right). \qquad (1.14)$$

(Compare with Eq. (1.8).) As a bonus we see that $\mathbf{A}$ is in the direction of the major axis of the ellipse. One can also verify that the offset of the center in the hodograph, $(GM/h)e$, is equal to $(A/hm^2)$. Thus $\mathbf{A}$ also has a geometrical interpretation in the velocity space. It all goes to show how special the inverse square law force is!

If you add a $1/r^3$ or $1/r^4$ component to the force, (which can arise, for example, from the general relativistic corrections to Newton's law of gravitation or because the Sun is not spherical and has a small quadrupole moment) $\mathbf{J}$ and $E$ are still conserved but not $\mathbf{A}$. If the perturbation is small, it will make the direction of $\mathbf{A}$ slowly change in space and we will get a "precessing" ellipse. As an explicit example, consider the motion in a potential $V = -(GMm/r) - (\beta/r^2)$ where the second term is treated as a small perturbation. From Eq. (1.10), we find that the rate of change of Runge-Lenz vector is now given by $\dot{\mathbf{A}} = (2\beta m/r)(d/dt)(\mathbf{r}/r)$. The change $\Delta\mathbf{A}$ per orbit is obtained by integrating $\dot{\mathbf{A}}dt$ over the range $(0, T)$ where $T$ is the period of the original orbit. Doing one integration by parts and changing the variable of integration from $t$ to the polar angle $\theta$, we get $\Delta\mathbf{A}$ per orbit to be

$$\Delta\mathbf{A}\big|_{\text{orbit}} = 2\beta m \int_0^{2\pi} \frac{\mathbf{r}}{r^3}\frac{dr}{d\theta}d\theta. \tag{1.15}$$

Let us take the coordinate system such that the unperturbed orbit originally had $\mathbf{A}$ pointing along the $x-$axis. After one orbit, a $\Delta A_y$ component will be generated and the major axis of the ellipse would have precessed by an amount $\Delta\theta = \Delta A_y/A$. The $\Delta A_y$ can be easily obtained from Eq. (1.15) by using $y = r\sin\theta$, converting the dependent variable from $r$ to $u = (1/r)$ and substituting $(du/d\theta) = -(A/J^2)\sin\theta$ [which comes from Eq. (1.14)]. This gives the angle of precession per orbit to be

$$\Delta\theta = \frac{\Delta A_y}{A} = -\frac{2\beta m}{A}\int_0^{2\pi}\sin\theta\frac{du}{d\theta}d\theta = \frac{2\pi\beta m}{J^2}. \tag{1.16}$$

In this particular case, the problem can be solved exactly (see Ex. 1.5) but the above trick will work for any general perturbation to the original $(1/r)$ term.

The next logical step from planetary motion is the gravitational two-body problem which, fortunately, turns out to be as simple as the one-body problem we have just solved. If we take two bodies with masses $m_1$ and $m_2$ and positions $\mathbf{r}_1$ and $\mathbf{r}_2$, then the total momentum of the system $\mathbf{p}_{\text{tot}} = m_1\dot{\mathbf{r}}_1 + m_2\dot{\mathbf{r}}_2$ can be written as $(m_1 + m_2)\dot{\mathbf{R}}$ where $\mathbf{R} =$

$[(m_1\mathbf{r}_1 + m_2\mathbf{r}_2)/(m_1 + m_2)]$ is the position of the center of mass. For the closed system, $\mathbf{p}_{\text{tot}}$ is conserved and the center of mass moves with uniform velocity. By going to another inertial frame which moves with the center of mass, we can treat $\mathbf{R}$ to be a constant and by shifting the coordinates, make the center of mass, the origin. If we further define the relative distance between the particles as $\mathbf{r} = (\mathbf{r}_1 - \mathbf{r}_2)$, then it is easy to see that $\ddot{\mathbf{r}} = -G(m_1 + m_2)(\mathbf{r}/r^3)$; that is, the relative separation obeys the same equation as in the one-body problem. In the coordinate system with the center of mass at the origin, the two bodies move around the center of mass.

Incidentally, this fact also shows that, in any planetary system, the planets *and the central star* are moving around the common center of mass. Hence, the motion of the planet will lead to a wobble of the central star thereby providing one possible way of detecting extra-solar planets — which is a prelude to close encounters of the appropriate kind! Since the orbital velocity of Earth (with mass $M_\oplus$) around Sun (with mass $M_\odot$) is about $v = 29.7$ km s$^{-1}$, the 'reflex' velocity of a Sun-like star due to an earth-like planet orbiting at the Earth-Sun distance 1 AU $= 1.5 \times 10^{13}$ cm (called the *astronomical unit*), will be $V \approx (M_\oplus/M_\odot)v \approx 9$ cm s$^{-1}$ which, however, is quite tiny. If, instead, we had a planet like Jupiter (about 320 times more massive than Earth), orbiting at 1 AU from the star, then we can easily detect the wobble with current technology. The current detection limit of about 3 ms$^{-1}$ will make it possible to catch a planet with a mass of about 33 $M_\oplus$ orbiting a 1 $M_\odot$ star at 1 AU under best orbital inclination. (See exercise 1.7 for the effect of orbital plane inclination).

An interesting special case of astrophysical significance arises when the two bodies move around the common center of mass in *circular* orbits. The angular velocity is then given by $\Omega = [G(m_1 + m_2)/a^3]^{1/2}$ with $a$ denoting the separation between the objects. If we transform to a coordinate system rotating with this angular velocity $\Omega$, then the two bodies will be stationary in this frame. But now the *effective* potential $\phi$ at any point $\mathbf{r}$ in the $x - y$ plane (which we take to be the plane of orbit) will have a contribution from the centrifugal force as well as the normal gravitational field. This net (effective) gravitational potential is given by

$$\phi = -\frac{Gm_1}{|\mathbf{r} - \mathbf{r}_1|} - \frac{Gm_2}{|\mathbf{r} - \mathbf{r}_2|} - \frac{1}{2}\Omega^2 r^2. \tag{1.17}$$

The first two terms are the standard gravitational potential while the gradient of the last term gives the centrifugal force. The equilibrium points (that is, points on the $x - y$ plane at which the acceleration, $\mathbf{g} = -\nabla\phi$, vanishes) in this potential are of considerable interest. This is because, if you keep a test particle at such a point, it will not move in this rotating frame. (In the original, inertial, frame the particle will co-rotate with the two bodies.) The condition that the acceleration should vanish at some point $\mathbf{r}$ is given by:

$$-\frac{Gm_1}{|\mathbf{r} - \mathbf{r}_1|^3}(\mathbf{r} - \mathbf{r}_1) - \frac{Gm_2}{|\mathbf{r} - \mathbf{r}_2|^3}(\mathbf{r} - \mathbf{r}_2) + \Omega^2\mathbf{r} = 0. \qquad (1.18)$$

It is fairly easy to show that there will be 3 such points (called the *Lagrange points* $L_1, L_2$ and $L_3$) along the line joining the masses. But you can easily verify that equilibria at these points are unstable. Unfortunately, there is no simple formula for the location of $L_1, L_2, L_3$. (If you write down the relevant equations, you will find that it is fifth order which has no ready made formula for its solution.) One can, however, produce an approximate solution when, say, $m_2 \ll m_1$. Since $m_1$ is then the primary source of gravity, $L_1$ must be quite close to $m_2$ for the two gravitational forces to be comparable. If we take the location of $L_1$ as $x = a_2 - R$ with $R \ll a$, you can solve for $R$ to first order and obtain $R \approx (m_2/3m_1)^{1/3}a$. This radius is called *Roche radius* or radius of *Hill sphere* (depending on whether you are talking to a stellar astronomer or a planetary person). It is obvious that, if a body of mass $m_2$ is held together by self gravity, then it will be torn apart (by the tidal force of mass $m_1$) if its size is bigger than $R$. For example, consider a binary star system made of a normal star and a compact companion star. If the normal star expands beyond $R$ some of its mass will flow on to the compact companion which is a process of great importance in several astrophysical sources (see Chapter 5).

There are also two equilibrium points $(L_4, L_5)$ that are *off axis* which are of greater interest. To find such solutions, take the dot product and cross product of Eq. (1.18) with $\mathbf{r}$ and use the facts that $\mathbf{r}_1 = a_1\hat{\mathbf{x}}$, $\mathbf{r}_2 = -a_2\hat{\mathbf{x}}$ and $m_1a_1 = m_2a_2$. The cross product gives the relation

$$|\mathbf{r} - \mathbf{r}_1| = |\mathbf{r} - \mathbf{r}_2| \equiv d \qquad (1.19)$$

which shows that $L_4$ and $L_5$ must be on the line bisecting the line joining the two masses. If either of these points are at a distance $y$ from the line joining the masses, then $d = [y^2 + (a^2/4)]^{1/2}$. Next, taking the dot product of Eq. (1.18) with $\mathbf{r}$, dividing by $r^2$ and re-arranging the terms gives

$$\frac{G(m_1 + m_2)}{a^3} = \frac{Gm_1}{d^3}\left(1 + \frac{a_1}{r}\right) + \frac{Gm_2}{d^3}\left(1 - \frac{a_2}{r}\right). \qquad (1.20)$$

Since $m_1 a_1 = m_2 a_2$, those two terms cancel out giving the result $d = a$. That is, we conclude

$$|\mathbf{r} - \mathbf{r}_1| = |\mathbf{r} - \mathbf{r}_2| = a. \qquad (1.21)$$

This remarkable result shows that the two off axis equilibrium points form an equilateral triangle of side $a$ with respect to $m_1$ and $m_2$! This result is independent of the ratio $(m_1/m_2)$. It can be verified that while $L_1, L_2, L_3$ are unstable equilibrium points, $L_4, L_5$ are indeed stable equilibrium points provided the mass ratio is less than about $(1/25)$. (It is not obvious since these points are actually local *maxima* of the potential. The reason is subtle and has to do with the Coriolis force.)

In our solar system, taking Sun and the Jupiter as the two dominant masses, one can determine the location of the two points $L_4$ and $L_5$, which are called the *Trojan points*. The asteroids close to these points (called *Trojan asteroids*) rotate around in the solar system fairly stably, forming an equilateral triangle with Sun and Jupiter.

Finally, we mention that even the unstable Lagrange points have their uses. Since the net acceleration is zero at, for example, $L_2$, a spacecraft which is put there needs comparatively less course correction. This is one of the reasons why the satellites with Next Generation Space Telescope (NGST) or the Laser Interferometer Space Antenna (LISA, a satellite based gravity wave detector) are planned to be put on the Earth-Sun $L_2$ point.

## 1.2 Precession and tides

So far we have treated all the bodies involved in the gravitational interactions as point particles. Let us next look at some "real life" effects that occur in solar system in which we have to recognize the fact that planets and satellites are deformable bodies of finite size and not structure-less point particles.

To the lowest order of approximation, Earth can be treated as a flattened spheroid with the deviation from the spherical shape attributed, in large part, to the rotation. To estimate the flattening, let us note that the acceleration due to gravity at the poles is $g_p \approx -GM_\oplus/R_p^2$ where $R_p$ is the polar radius while the acceleration at the equator is $g_e \approx -(GM_\oplus/R_e^2)$. (You will recall that the subscript $\oplus$ stands for values pertaining to Earth.) The difference in these two accelerations will be of the order of centrifugal acceleration $\Omega^2 R_e$ where $\Omega$ is the angular velocity of rotation of earth and

$R_e$ is the equatorial radius. Assuming that $R_e - R_p \equiv \delta R \ll R_\oplus$ (where $R_\oplus$ is the mean radius), we get $(GM_\oplus/R_\oplus^2)(\delta R/R_\oplus) \approx \Omega^2 R_\oplus$. Such a deviation is characterized by an eccentricity of

$$e \equiv \frac{\delta R}{R} \approx \frac{\Omega^2 R_\oplus^3}{GM_\oplus} \approx \frac{1}{300}. \qquad (1.22)$$

To make the numerical estimate, remember that $2\pi(R_\oplus^3/GM_\oplus)^{1/2} = 1.41$ hours while $2\pi/\Omega = 24$ hours. (The former time scale occurs in different contexts; it is (i) the gravitational collapse time scale for matter with the density of earth as well as (ii) the orbital period of a satellite hovering just above earth. It is a good number to remember.) Of course, once the earth is taken to be non-spherical, the gravitational acceleration need to be worked out again in a self-consistent manner. This can be done and the $e$ changes only by a factor $(5/4)$.

This flattening at the poles makes earth a solid body with two moments of inertia (about the principle axis) being nearly equal and the third one a little different from the other two: $I_x = I_y \neq I_z$ with $\delta I/I = \delta R/R = 1/300$. This deviation in the shape of the earth from a perfect sphere leads to some interesting effects.

To begin with, since Earth rotates with an angular velocity $\omega_z = \Omega = 7.3 \times 10^{-5}$ per second (corresponding to a period of 24 hours), its axis of rotation will undergo (what is called) a *free precession*. To obtain this result, we need to recall some basic facts from rotational dynamics. In an inertial frame, the angular momentum of the body changes in accordance with the law $(d\mathbf{J}/dt) = \mathbf{K}$ where $\mathbf{K}$ is the external torque. The time derivative of any vector $\mathbf{A}$ in the inertial and rotating frames differ by the term $\boldsymbol{\Omega} \times \mathbf{A}$. This is most easily seen by noting that if the vector $\mathbf{A}$ does not change in the moving frame, then its rate of change in the fixed frame is only due to rotation and we must have $\delta\mathbf{A} = \delta\theta \times \mathbf{A} = (\boldsymbol{\Omega} \times \mathbf{A})dt$. Hence this term represents the difference in $(d\mathbf{A}/dt)$ as measured in the rotating and inertial frames. Therefore, the equation of motion for the angular momentum in the rotating frame is given by

$$\frac{d\mathbf{J}}{dt} + \boldsymbol{\Omega} \times \mathbf{J} = \mathbf{K}. \qquad (1.23)$$

Taking the components of this equation along the principle axis of the system, we get

$$I_1 \frac{d\Omega_1}{dt} + (I_3 - I_2)\Omega_2\Omega_3 = K_1,$$

$$I_2 \frac{d\Omega_2}{dt} + (I_1 - I_3)\Omega_3\Omega_1 = K_2,$$

$$I_3 \frac{d\Omega_3}{dt} + (I_2 - I_1)\Omega_1\Omega_2 = K_3. \tag{1.24}$$

These are the *Euler's equations* which govern the rotational motion of a rigid body. Consider the special case of free rotation (i.e. no external torque, **K** = 0) of a body with axial symmetry, for which $I_1 = I_2 \neq I_3$ and $\Omega_3 =$ constant (as in the case of earth rotating on its axis). Then we see from Eq. (1.24) that $\Omega_1$ and $\Omega_2$ satisfy the equations $\dot{\Omega}_1 = -\omega_p\Omega_2, \dot{\Omega}_2 = \omega_p\Omega_1$ where $\omega_p = \Omega_3(I_3 - I_1)/I_1$ is a constant. Multiplying one of the equations by $i$ and adding to the second equation, one can easily solve the set and obtain

$$\Omega_1 = A\cos\omega_p t; \qquad \Omega_2 = A\sin\omega_p t \tag{1.25}$$

where $A = (\Omega_1^2 + \Omega_2^2)^{1/2}$ is a constant. The motion, therefore, consists of the vector $\boldsymbol{\Omega}$ precessing with an angular velocity $\omega_p$ around the symmetry axis of the body, while remaining constant in magnitude. In fact, the component along axis-3 as well as the magnitude of the projection of $\Omega$ on to 1-2 plane remain separately constant during such precession. For earth,

$$\omega_p = \frac{I_z - I_x}{I_x}\Omega \simeq \frac{\Omega}{300} \simeq 2.4 \times 10^{-7}. \tag{1.26}$$

This precession corresponds to a period of about 300 days which turns out to be about 50 per cent smaller than the observed value. It is believed that the discrepancy arises due to complicating factors like tidal distortion of the Earth.

A more interesting phenomenon is the *forced precession* of the axis of rotation of the Earth, which arises due to the torque exerted by Sun (and Moon; we will concentrate on the Sun) on Earth. This precession rate can be estimated fairly easily. Since the precession rate is small — something which we will verify at the end of the calculation — we can adopt a geocentric view and think of Sun as going around the Earth, once every year. For time scales much larger than a year, this is equivalent to the mass of the Sun, $M_\odot$, being distributed around earth along a circular ring of radius $d_\odot$ (equal to the mean earth-sun distance) as far as computation of the torque causing the precession is concerned. In this case, $\langle \dot{\mathbf{J}} \rangle = 0$ and Eq. (1.23) gives, $\mathbf{\Omega} \times \mathbf{J} = \mathbf{K}$ where $\mathbf{K}$ is the external torque. The rotation axis of earth is inclined at an angle $\theta \approx 23.5$ degrees to the normal to the plane containing the ring (see Fig. 1.2). A perfectly spherical earth, of course, will feel no torque from the ring; but since the earth has bulge around equator with equatorial and polar radii differing by $\delta R/R \approx 1/300$, there will be a torque, as if earth is a small dumbbell with masses $\delta M \approx M(\delta R/R)$ located at the ends of a rod of length $2R$. This net force $F$ is due to the *difference* in the gravitational force of the sun (treated as a ring around earth) on the two masses of the dumbbell; so $F \approx (GM_\odot \delta M/d_\odot^3)(2R\cos\theta)$ and the corresponding torque is $K \approx F(2R\sin\theta)$. The rate of precession will be $\Omega \approx K/J\sin\theta$ (since $\mathbf{\Omega} \times \mathbf{J} \simeq \mathbf{K}$ from Eq. (1.23)) where $J \approx (MR^2)\omega$ is the angular momentum due to daily rotation of earth. Combining all these and using $(GM_\odot/d_\odot^3) = \nu^2$ where $\nu$ is the frequency of revolution of earth around the sun, we get

$$\Omega \approx \frac{4\nu^2}{\omega}\frac{\Delta R}{R}\cos\theta; \qquad (1.27)$$

$$T_{\text{pre}} = \frac{2\pi}{\Omega} = T_{orbit}\left(\frac{T_{orbit}}{4T_{spin}}\right)\left(\frac{\Delta R}{R}\right)^{-1}\sec\theta \simeq 3\times 10^4 \text{ years}. \qquad (1.28)$$

There is a comparable contribution from the moon going around the earth. The gravitational force of Sun on Earth is, of course, significantly larger than that of Moon on Earth. But the *tidal force* due to Sun and the Moon has the ratio

$$\frac{(M_\odot/d_\odot^3)}{(M_{\text{moon}}/d_{\text{moon}}^3)} \simeq \frac{(M_\odot/R_\odot^3)(R_\odot/d_\odot)^3}{(M_{\text{moon}}/R_{\text{moon}}^3)(R_{\text{moon}}/d_{\text{moon}})^3}$$

$$= \frac{\rho_\odot}{\rho_{\text{moon}}}\left(\frac{\theta_\odot}{\theta_{\text{moon}}}\right)^3 \simeq \frac{\rho_\odot}{\rho_{\text{moon}}} \simeq \frac{1.4}{3.3} \simeq 0.42 \quad (1.29)$$

where $\theta \equiv (R/d)$ is the angle subtended by Sun or Moon at Earth and we have used the fact $\theta_\odot \simeq \theta_{\text{moon}}$. So adding the effects of both sun and moon reduces the period to about $2\times 10^4$ years which is close to what is observed. This means that the axis of rotation of earth will slowly precess and what we call the pole star, for example, will be different over few thousands of years! The direction of the equinoxes, with respect to fixed stars, will also shift slowly at the same rate. If the axis of rotation precesses by 360 degrees in $2\times 10^4$ years, it implies a shift of about 1 arcminute per year, which even the ancient civilizations could take note of. Obviously, astronomical observations, which routinely work at fraction of arcsec precision needs to

corrected for this and one usually specifies the coordinates of the objects
in the sky at a fixed reference date (used to be 1950A.D and is changing to
2000A.D).

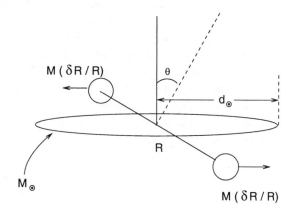

Fig. 1.2   Simplified picture to derive the precession of the equinoxes (definitely not to
scale !).

The same effect will try to align the orbital plane of the moon (which
is inclined at about 5 degrees) with the ecliptic. Similar analysis shows
that the line of intersection of these two planes precesses with a period of
about 18 years. Since the intersection of the the orbital plane of the moon
with the ecliptic gives the directions near which eclipses occur, this effect
is of considerable practical importance. The precession of the moon's orbit
plus its inclination implies that the torque it exerts *on* earth has small
fluctuations. Consequently, the precessing path of earth's axis has a tiny
wobbling motion called *nutation*. Once again, this was known for centuries!

Another phenomenon which arises because bodies are deformable is the
tide raised on earth, which we shall consider next. Let us begin with the
tidal effects due to the Moon. The action of Moon's gravitational force
deforms the surface of the Earth periodically thereby leading to *tidal bulges*
on the surface of the Earth. These tidal bulges will be located directly
under the Moon if: (i) there is no other force (like friction) acting on the
system and (ii) the rotational period of earth is identical to the orbital
period of moon around earth. In reality the surface deformation of earth is
affected by fairly strong frictional forces and the rotation of earth is quite
fast compared to the orbital rate of the moon around earth. Hence the tidal

bulge will not be aligned along the direction of the Moon. The fast rotation of the planet will drag the axis of the bulge ahead of the earth-moon line leading to several interesting consequences:

This causes a torque between the Moon and the earth tending to align the axis of the bulge along the earth-moon line; an equal and opposite torque acts on the moon increasing its orbital angular momentum. The net effect is to decrease the angular velocity of earth (thereby lengthening the day) and increase the orbital angular momentum of moon which in turn, makes the earth-moon distance to increase (thereby lengthening the month). In fact study of fossilized corals show that the length of the month was shorter in the past; earth has indeed slowed down and moon has moved away. You can see that, eventually, earth will rotate once in its axis exactly in the time it takes for moon to complete one orbit around earth. Also, since circular orbits have the lowest energy for a given angular momentum, this process also leads to the circularisation of orbits.

The final length of the day resulting from the tidal interaction (which will be the same as the final orbital period of Moon) can be easily estimated from the conservation of angular momentum. Let $M_1, M_2$ be the masses of earth and moon, $\omega_1$ is the rotational speed of Earth, $\omega_2$ is the orbital speed of Moon, $\omega_f$ is the final values for these two in synchronous rotation, $R$ the distance between earth and moon and $R_f$ is the corresponding final distance then the conservation of angular momentum gives

$$I_1\,\omega_1 + I_2\,\omega_2 + M_2\,\omega_2\,R^2 = I_1\,\omega_f + I_2\,\omega_f + M_2\,\omega_f\,R_f^2. \qquad (1.30)$$

In arriving at this equation, we have ignored the spin angular momentum of Moon at the *present* epoch and the spin angular momenta of Earth and Moon in the *final* state; you can easily verify that these are ignorable. Further, orbital dynamics requires $(R_f/R) = (\omega_2/\omega_1)^{2/3}$. Taking $I_1 = (2/5)M_1 R_1^2$ etc. we get

$$\frac{\omega_2}{\omega_f} = \left[ 1 + \frac{2}{5}\left(\frac{M_1}{M_2}\right)\left(\frac{R_1}{R}\right)^2\left(\frac{\omega_1}{\omega_2}\right) \right]^3 = 1.99. \qquad (1.31)$$

Hence the final day (*and* month) will be about 54 days and the final Moon's orbital radius will be about $5.9 \times 10^5$ km.

At present, the fractional increase in the distance to the moon is $1.3 \times 10^{-10}$ per year and the lengthening of the day is by $7 \times 10^{-4}$ sec per century. Precise calculation of the tidal friction, described above, suggests that — while it could be the dominant effect — there could be other contributions to the lengthening of day *etc.* A completely satisfactory model, accounting for all the observations is not yet available.

A corresponding effect, of course, occurs due to the tides produced *on* the moon *by* earth. If the initial position of the moon was closer to earth and its orbital period was different from the rotational period about its own axis, then the tides raised by earth on moon will have the effect of synchronizing the orbital and rotational periods of moon. Being a smaller body, this effect on moon occurs at a shorter time scale and has already caused the moon to be in such a synchronous, tidally locked orbit with respect to earth. In fact, most of the satellites in solar system have this synchronous feature.

Let us consider some details of the tidal interaction between earth and moon using a fairly simple model. We will assume that earth-moon system goes around the common center of mass in circular orbits and work in a coordinate system that is co-rotating with them. Suppose that a point near the surface of earth has the coordinates $(l, \theta)$ where $l$ measures the radial distance from the center of the Earth and $\theta$ measures the angle from the instantaneous earth-moon line. If $M_1$ and $M_2$ are the masses of earth and the moon, $R$ is the distance between their centers, we can compute the *effective* gravitational potential at any given point with coordinates $(r, \theta)$ using Eq. (1.17). (Note that in this equation the $\mathbf{r}$ refers to the position of a particle with respect to the center of mass of the Earth-Moon system and *not* with respect to the center of the Earth.) The potential due to the Moon is given by

$$
\begin{aligned}
-\frac{GM_2}{|\mathbf{r} - \mathbf{r}_2|} &= -\frac{GM_2}{R} \left( 1 + \frac{l^2}{R^2} - 2\frac{l}{R}\cos\theta \right)^{-1/2} \\
&\cong -\frac{GM_2}{R} \left[ 1 + \frac{l}{R}\cos\theta + \frac{1}{2}(3\cos^2\theta - 1)\frac{l^2}{R^2} + \mathcal{O}\left(\frac{l}{R}\right)^3 \right]
\end{aligned}
\tag{1.32}
$$

where we have done a Taylor expansion in $l/R$. (If you know your Legendre polynomials, you will see this immediately.) The term $(l/R)\cos\theta$ varies linearly along the $z = l\cos\theta$ direction from $M_1$ to $M_2$ so that its gradient gives a constant force. This is precisely canceled by a corresponding term in the centrifugal force; which is, of course, necessary for the centrifugal force to keep the bodies in circular orbits. Noting that the center of mass (from which the vector $\mathbf{r}$ is measured) is at a distance $M_2R/(M_1 + M_2)$, we can compute the centrifugal contribution to be

$$\frac{1}{2}\omega^2 r^2 = \frac{1}{2}\omega^2 \left[ \left( \frac{M_2}{M_1 + M_2} R \right)^2 + l^2 - 2 \left( \frac{M_2}{M_1 + M_2} R \right) a \cos\theta \right]. \quad (1.33)$$

Using $\omega^2 = G(M_1 + M_2)/R^3$ we see that the term linear in $\cos\theta$ indeed cancels. Adding the contribution due to Earth, the net potential near the surface of the Earth varies as

$$\Phi(l,\theta) = -\frac{GM_1}{l} - \frac{3}{2}\frac{GM_2 a^2}{R^3} \cos^2\theta + \text{constant}. \quad (1.34)$$

The equi-potential surfaces, given by $\Phi(l,\theta) = \text{constant}$, will provide the lowest order deformation of the spherical shape of the earth due to tidal forces. To get a better feeling for this shape, one can set $l = R_\oplus + h$ (where $R_\oplus$ is the radius of the Earth) and expand $\Phi(l,\theta)$ in a Taylor series in $h$. A simple calculation now gives, to the lowest non trivial order in $h$, the expression

$$\Phi(h,\theta) \approx g \left[ h - \frac{3}{2} \left( \frac{M_2}{M_1} \right) \left( \frac{R_\oplus}{R} \right)^3 R_\oplus \cos^2\theta \right] + \text{constant} \quad (1.35)$$

where $g$ is the acceleration due to gravity on the surface of the earth and higher order terms in $(R_\oplus/R)$ and $(h/R)$ are ignored. Setting $\Phi$ to constant shows that the deformation of the surface, $h$, is given by

$$h = \frac{3}{2} \left( \frac{M_2}{M_1} \right) \left( \frac{R_\oplus}{R} \right)^3 R_\oplus \cos^2\theta. \quad (1.36)$$

For $(M_2/M_1) \approx (1/80)$; $R_\oplus \approx 6400$ km and $R = 3.8 \times 10^5$ km, the difference between $h(\theta = 0)$ and $h(\theta = \pi/2)$ is about 54 cm. The corresponding calculation for Sun's gravitational field on earth produces a shift of about 25 cm. It should be noted that the tidal force has a $\cos^2\theta = (1/2)(\cos 2\theta + 1)$ dependence where $\theta$ is the angle measured from the location of the moon. This leads to a maximum at both $\theta = 0$ and $\theta = \pi$ which is characteristic of a tidal force.

The same formalism also allows you to obtain another interesting result. Consider a *small* body of mass $m$ which is orbiting close to a large body of mass $M$. It is intuitively obvious that, if the small body gets too close to the large one, tidal force can rip it apart. How close can it get and still survive ? To determine this, align the z-axis along the line joining the centers of the bodies and consider the gravitational acceleration along the z-axis. This is given by the gradient of the potential in Eq. (1.34), which we will now write, using $z = l\cos\theta$ and dropping the constant term, as:

$$\Phi(l,\theta) = -\frac{Gm}{l} - \frac{3}{2}\frac{GMl^2}{R^3}\cos^2\theta = -\frac{Gm}{|z|} - \frac{3}{2}\frac{GMz^2}{R^3} \qquad (1.37)$$

for points along the z-axis. It is easy to see that the gradient gives $g_z = -d\phi/dz = Gz[(3M/R^3) - (m/|z|^3)]$. The force in the first term is trying to rip the body apart while the second term is the restoring force due to self gravity. The stability of the body against tidal disruption requires $(3M/R^3) < m/|z|^3$. This allows us to define a critical radius $R_{\text{roche}} = (3M/m)^{1/3}|z|$ (which is exactly the result we obtained earlier from the study of Lagrange points; see the discussion after Eq. (1.18)). Writing $|z|$ as $r$ and the mass of the satellite as $m = 4\pi\rho r^3/3$, the condition for stability is $R > R_{\text{roche}}$ where

$$R_{\text{roche}} = \left(\frac{9M}{4\pi\rho}\right)^{1/3} \approx 1.44\left(\frac{\rho_*}{\rho_1}\right)^{1/3} R_* \qquad (1.38)$$

is the Roche radius and $M = 4\pi\rho_* R_*^3/3$ where $\rho_*$ and $R_*$ are the density and radius of the central object. (A more precise calculation changes the factor 1.44 to about 2.9.) Of course, this limit really makes sense only when $R_{\text{roche}}$ is larger than the radius $R_*$ of the massive object. As an example, consider a small body which gets close to the sun. The condition $R_{roche} > R_\odot$ leads to the constraint $\rho < (9/4\pi)(M_\odot/R_\odot^3) \approx 4.2$ gm cm$^{-3}$, which holds for most objects in solar system. So any of them will be disrupted tidally if they get too close to the sun.

## 1.3 Virial theorem

When we move from 2, 3 or few body problem to the genuine many body problem in gravity, life becomes rather difficult. Usually, such a many body limit of a system can be understood using statistical physics. But conventional ideas of statistical mechanics do not work with gravity because conventional statistical mechanics assumes that energy is extensive. If you divide the molecules of a gas into two parts, the total energy is (to a very good approximation) the sum of the internal energies of the two individual parts. But if you divide a star cluster or galaxy into two parts, the total energy is not the sum of two parts; the gravitational interaction energy between the two parts makes a significant contribution. Molecules in a gas interact through short range forces and hence the interaction between molecules in one bulk region with molecules in another region is essentially confined to those which are near the surface separating the two regions. Hence the ratio of interaction energy to bulk energy scales as the ratio of surface area to the volume, $L^2/L^3 \simeq L^{-1}$ where $L$ is a typical size of the region. In any large volume of sufficiently dilute gas, we can ignore the interaction energy and treat energy as an extensive variable. This does not work in the presence of gravity which is a long range interaction.

There is, however, one beautiful result — called *virial theorem* — which can be proved in gravitational N-body problem. Consider a collection of particles with masses $m_i$ and positions $\mathbf{r}_i$ interacting through gravity. Starting from the equation of motion for the $i$-th particle

$$m_i\ddot{\mathbf{r}}_i = -\sum_{j\neq i} \frac{Gm_jm_i(\mathbf{r}_i - \mathbf{r}_j)}{|\mathbf{r}_i - \mathbf{r}_j|^3} \qquad (1.39)$$

we take the dot product of both sides with $\mathbf{r}_i$ and sum over all $i$ to get

$$\sum_i m_i\mathbf{r}_i \cdot \ddot{\mathbf{r}}_i = -\sum_i\sum_{j\neq i} \frac{Gm_jm_i\mathbf{r}_i \cdot (\mathbf{r}_i - \mathbf{r}_j)}{|\mathbf{r}_i - \mathbf{r}_j|^3}. \qquad (1.40)$$

Writing $\mathbf{r}_i = (1/2)(\mathbf{r}_i - \mathbf{r}_j) + (1/2)(\mathbf{r}_i + \mathbf{r}_j)$ and using the antisymmetry of $(\mathbf{r}_i - \mathbf{r}_j)$ under pair exchange, one can write this as

$$\sum_i m_i\mathbf{r}_i \cdot \ddot{\mathbf{r}}_i = -\frac{1}{2}\sum_{j\neq i} \frac{Gm_jm_i}{|\mathbf{r}_i - \mathbf{r}_j|} = U \qquad (1.41)$$

where $U$ is the gravitational potential energy. On the other hand, we can relate the left hand side of Eq. (1.41) to the second time derivative of the quantity $I = \sum m_i r_i^2$. Direct algebra gives

$$\frac{1}{2}\frac{d^2I}{dt^2} = \sum_i m_i\ddot{\mathbf{r}}_i \cdot \mathbf{r}_i + \sum_i m_i\dot{\mathbf{r}}_i \cdot \dot{\mathbf{r}}_i = \sum_i m_i\ddot{\mathbf{r}}_i \cdot \mathbf{r}_i + 2K \qquad (1.42)$$

where $K$ is the kinetic energy of the system. This result, when combined with Eq. (1.41) leads to the virial theorem:

$$\frac{1}{2}\frac{d^2I}{dt^2} = U + 2K. \qquad (1.43)$$

(Historically, the quantity $\sum \mathbf{F}_i \cdot \mathbf{r}_i$ was called virial and hence the name). You might think this is rather useless since $\ddot{I}$ is not a simple quantity to determine. The utility of virial theorem comes from the fact that we can handle it in two important cases. First, when the system is (at least approximately) in steady state we can set $\ddot{I} = 0$ thereby obtaining a relation between kinetic, potential and total energies:

$$2K + U = 0; \qquad E = -K = \frac{1}{2}U. \tag{1.44}$$

Second, suppose the system is (reasonably) bound so that the time average $< I >$ is at least quasi-periodic making its time derivative vanish. Then, taking the time average of both sides of Eq. (1.43) we get $2\langle K \rangle + \langle U \rangle = 0$. In either context, we have an estimate of the relevant energies.

One of the key applications of virial theorem is in the determination of masses of large groups and clusters of galaxies. For a group of $N$ galaxies, treated as point masses with masses $m_i$, positions $\mathbf{r}_i$ and velocities $\mathbf{v}_i$, the virial theorem can be written as

$$\sum_{i=1}^{N} m_i v_i^2 = \sum_{i \neq j}^{N} \frac{G m_i m_j}{|\mathbf{r}_i - \mathbf{r}_j|}. \tag{1.45}$$

Treating the quantities appearing in the above equation as approximately equal to the time-averages, we can use this relation to determine the total mass of the system. Observationally, we can only determine the line-of-sight velocity dispersion $\sigma^2$ and the two dimensional projection $\mathbf{R}_i$ of $\mathbf{r}_i$ on to the plane of the sky. On the average, we expect $v_i^2 = 3\sigma^2$; the projected inverse separation can be related to the true inverse separation by averaging over all possible orientations of the group. It is easy to show that

$$\langle (|\mathbf{R}_i - \mathbf{R}_j|)^{-1} \rangle = \frac{\pi}{2} \left( |\mathbf{r}_i - \mathbf{r}_j| \right)^{-1}. \tag{1.46}$$

Finally, we will also assume that the *mass-to-light ratio* $Q$ of all the galaxies is the same so that $m_i = QL_i$ where $L_i$ is the luminosity of the $i-$th galaxy. Substituting $m_i = QL_i$ in Eq. (1.45) and using Eq. (1.46) we get

$$Q = \frac{3\pi}{2G} \frac{\sum_{i=1}^{N} L_i \sigma_i^2}{\sum_{i \neq j}^{N} L_i L_j \left( |\mathbf{R}_i - \mathbf{R}_j|^{-1} \right)}. \tag{1.47}$$

Since all the quantities on the right hand side can be determined from the observations, we can find $Q$ for the galaxies in the cluster. This procedure will give reasonably reliable results for groups with sufficiently large number of galaxies. Several groups have been studied using the above equation and observations suggest a median value of $Q \approx 260 \ Q_{\odot}$. This is one of the key evidences for the existence of significant quantities of *dark matter* in virialized clusters and groups of galaxies.

One curious consequence of the virial theorem is that gravitating systems have negative specific heat. Recall that we can associate a temperature $T$ to the kinetic energy of random motion $K$ by $K \approx (3/2)Nk_BT$. The specific heat is $C_V = (dE/dT)$; since $T \propto K$, the sign of $C_V$ is same as that of $(dE/dK) = -1$ from virial theorem, making $C_V < 0$. This behaviour, of course, is well known even in the case of gravitational one body problem. A satellite in a circular orbit around Earth, when it loses energy, will go to an orbit with lower radius but *higher* orbital velocity. Loss of energy heats up a self gravitating system.

The importance of this result arises from the fact that *no* system in canonical ensemble can have $C_V < 0$. To see why, recall that the mean energy of a system in canonical ensemble is given by

$$E \equiv \langle \epsilon \rangle = \frac{1}{Z} \sum_i \epsilon_i e^{-\beta \epsilon_i} = -(\ln Z)'; \quad Z \equiv \sum_i e^{-\beta \epsilon_i} \qquad (1.48)$$

where $\beta = (k_BT)^{-1}$ and the prime denotes the derivative with respect to $\beta$. It follows that

$$\frac{C_V}{k_B} = \frac{dE}{d(k_BT)} = -\beta^2 E' = \beta^2 (\ln Z)'' = \beta^2 \left( \frac{Z''}{Z} - \left( \frac{Z'}{Z} \right)^2 \right). \qquad (1.49)$$

But the $\langle \epsilon^2 \rangle = (Z''/Z)$ and $\langle \epsilon \rangle^2 = (Z'/Z)^2$ giving

$$\frac{C_V}{k_B} = \beta^2(\langle \epsilon^2 \rangle - \langle \epsilon \rangle^2) = \beta^2(\Delta \epsilon)^2 > 0. \qquad (1.50)$$

How does one reconcile this result with the fact that the specific heat of a self gravitating system is negative? The contradiction implies that the statistical description of a self gravitating system can never be based on canonical ensemble. It turns out that one can handle the situation using microcanonical ensemble for which $C_V$ can be negative. The proof for the equivalence of the two ensembles — given in any text book on statistical mechanics — uses the extensivity of energy which, as we noted before, no longer holds.

In many astrophysical contexts, one has to deal with gaseous systems described by standard thermodynamics rather than point particles. To obtain the virial theorem for such systems, let us first recall some basic results for ideal gases with constant specific heats. For these systems, the definition $C_V = dU_{int}/dT$ integrates to $U_{int} = C_VT$. For example, if the gas has some $f$ internal degrees of freedom — vibrational, rotational *etc.* — in addition to the 3 degrees of freedom for translational motion, then $U_{int} = (3/2)Nk_BT[1+(f/3)]$ and $C_V = (3/2)Nk_B[1+(f/3)]$. On the other hand, using

$$dQ = dU_{int}+PdV = d(U_{int}+PV)-VdP = C_V dT+Nk_B dT-VdP \quad (1.51)$$

we find that the specific heat at constant pressure, $C_P \equiv (dQ/dT)_P = C_V + Nk_B$. The ratio of specific heats, $\gamma$, then becomes

$$\gamma = \frac{C_P}{C_V} = 1 + \frac{Nk_B}{C_V} = 1 + \frac{Nk_BT}{U_{int}} \quad (1.52)$$

allowing us to write

$$U_{int} = (\gamma - 1)^{-1}Nk_BT = (\gamma - 1)^{-1}PV; \qquad P = (\gamma - 1)\epsilon \quad (1.53)$$

where $\epsilon$ is the energy per unit volume.

For such a system, we need to take into account the force due to pressure gradients while deriving the virial theorem. In this case the virial will pick up an additional contribution which is the sum of $\mathbf{r}_i \cdot \mathbf{F}_i = -\mathbf{r}_i \cdot \nabla P$ where $P$ is the pressure. Writing

$$\mathbf{r}_i \cdot \mathbf{F}_i = -\mathbf{r}_i \cdot \nabla P = -\nabla \cdot (\mathbf{r}_i P) + P\nabla \cdot \mathbf{r}_i = -\nabla \cdot (\mathbf{r}_i P) + 3P \quad (1.54)$$

the summation over all particles can be thought of as an integration over the volume. The first term gives a surface contribution and we will take the surface to be one at constant pressure $P_s$. Noting that the integral of $\mathbf{r} \cdot \mathbf{n}$ over the surface is just the volume integral of $\nabla \cdot \mathbf{r}$, this term gives $-3P_sV$ where $V$ is the volume of the system. In the second term we can use Eq. (1.53) to write $P = (\gamma - 1)\epsilon$. The integral of $\epsilon$ over the volume will give $U_{int}$ so that this term becomes $3(\gamma - 1)U_{int}$. Then the virial theorem reads:

$$\frac{1}{2}\frac{d^2I}{dt^2} = 2K + U_{gr} + 3(\gamma - 1)U_{int} - 3P_sV. \quad (1.55)$$

Now $K$ denotes the kinetic energy of *bulk* motion of gas (if any) while the kinetic energy due to random, thermal, motion is incorporated into $U_{int}$. Let us look at some consequences of this relation.

First, consider the case in which $P_s = 0$ and $K = 0$ (no bulk motion) which will represent a gaseous, self-gravitating system. Writing the total energy of the system as $E = U_{\text{gr}} + U_{\text{int}}$, the virial theorem becomes

$$\frac{1}{2}\frac{d^2 I}{dt^2} = U_{\text{gr}} + 3(\gamma - 1)U_{\text{int}} = 3(\gamma - 1)E - (3\gamma - 4)U_{\text{gr}}. \tag{1.56}$$

In steady state, the time derivative must vanish, implying

$$E = \frac{3\gamma - 4}{3(\gamma - 1)}U_{\text{gr}} \tag{1.57}$$

which gives a simple relation between the total energy and the gravitational potential energy $U_{\text{gr}}$. (Of course, Eq. (1.57) reduces to Eq. (1.44) when $\gamma = 5/3$ which corresponds to an ideal monatomic gas.) For a bound system, we require $E < 0$; since $U_{\text{gr}} < 0$ it is necessary that $\gamma > (4/3)$. The system will be unstable when $\gamma < (4/3)$. Some insight into the nature of instability can be obtained by rewriting the original Eq. (1.56) in the form

$$E = \frac{\gamma - (4/3)}{\gamma - 1} U_{\text{gr}} + \frac{(d^2 I/dt^2)}{6(\gamma - 1)}. \tag{1.58}$$

When $\gamma > (4/3)$ and $E < 0$, the last term can vanish and the system can be in equilibrium. If some physical process now drives $\gamma$ towards $(4/3)$ with $E$ still negative, the last term cannot vanish and we must have $2E \approx (d^2 I/dt^2)$ when $\gamma \approx (4/3)$. Since $E < 0$, this implies that $(d^2 I/dt^2) < 0$. Usually such a condition would imply a collapse of the system.

As a second example, consider an ideal monatomic gas, (with $U_{int} = (3/2)k_B T, \gamma = 5/3$ in the absence of bulk motion $(K = 0)$ and in steady state $(\ddot{I} = 0)$. The virial theorem now gives $3k_B T + U_{gr} = 3P_s V$. Let us apply this relation to a (nearly) spherical configuration of fluid with mass $M$, radius $R$ and temperature $T$ in equilibrium with the gravity balanced by the pressure. The gravitational potential energy of such a configuration will be $U_{\text{gr}} = -\alpha(GM^2/R)$ where $\alpha$ is a numerical constant possibly of order unity. In terms of the one dimensional velocity dispersion $u^2 = (k_B T/m)$ the virial theorem becomes

$$P_s = \frac{1}{4\pi}\left(\frac{3Mu^2}{R^3} - \alpha\frac{GM^2}{R^4}\right). \tag{1.59}$$

In the absence of gravity $(G = 0)$, this equation gives $P_s V = Mu^2$ which is merely the ideal gas law. In the presence of gravitational field, such a cloud will have a tendency to collapse under self gravity; so we expect $P_s$ to be reduced. Equation (1.59) shows that this is precisely what happens. As we increase the value of $M$, a point is reached when $P_s$ vanishes. For larger values of $M$, the pressure $P_s$ will become negative indicating an instability in the system. The critical value of mass is given by

$$M_{\text{crit}} = \frac{3}{\alpha} \left( \frac{u^2 R}{G} \right) = \frac{3}{\alpha} \left( \frac{k_B T R}{mG} \right). \tag{1.60}$$

Clouds of a given size $R$ and temperature $T$ will collapse under self gravity if $M > M_{\text{crit}}(T, R)$. This critical mass is usually called the *Jean's mass*. The instability of systems with $M > M_{\text{crit}}$ is the key to understanding both star formation and galaxy formation, as we shall see in later chapters.

## 1.4   Gravitational collisions and relaxation

There is another significant aspect in which self-gravitating systems differ from more familiar gaseous systems. In a gas, at normal temperatures and pressures, collisions are very frequent. That is, the mean free path due to molecular collisions is small compared to the characteristic size of the system. The situation is very different for self gravitating systems. The exchange of energy between particles due to gravitational scattering off each other is a very slow process. Hence this process cannot relax the system, in contrast to molecular scattering in a normal gas. Let us estimate the time scale for this effect.

Consider a gravitational encounter between two bodies in an N-body system (say, a cluster of stars) each of mass $m$ and relative velocity $v$. If the impact parameter is $b$, then the typical transverse velocity induced by the encounter is $\delta v_\perp \simeq \left( Gm/b^2 \right) (2b/v) = (2Gm/bv)$. The deflection of the stars can be significant if $(\delta v_\perp/v) \gtrsim 1$, which occurs for collisions with impact parameter $b \lesssim b_c \simeq \left( Gm/v^2 \right)$. Thus the effective cross section for collision with significant change of momentum is $\sigma \simeq b_c^2 \simeq \left( G^2 m^2/v^4 \right)$. The collisional relaxation time-scale will be

$$t_{gc} \simeq \frac{1}{(n\sigma v)} \simeq \frac{R^3 v^3}{N \left( G^2 m^2 \right)} \simeq \frac{N R^3 v^3}{G^2 M^2}. \tag{1.61}$$

If the system is virialized, then $(GM/R) \simeq v^2$ and we can write

$$t_{gc} \simeq N \left( \frac{R}{v} \right) \left( \frac{R^2 v^4}{G^2 M^2} \right) \simeq N \left( \frac{R}{v} \right). \tag{1.62}$$

Thus the collisional time-scale is $N$ times the "crossing time" $(R/v)$ for the system. For most gravitational systems, $(R/v)$ will be comparable to the dynamical time scale while $N$ will be a large number making $t_{gc}$ much larger than the time scales of interest.

There is, however, another effect which operates at a slightly shorter time-scale. When the impact parameter $b$ is far larger than $b_c$, the critical impact parameter $b_c \simeq (Gm/v^2)$ obtained above, individual collisions still impart a small transverse velocity $(\delta v_\perp) \simeq (Gm/bv)$ to a given star. The effect of a large number of such collisions is to make the star perform a random walk in the velocity space. The net mean-square-velocity induced by collisions with impact parameters in the range $(b, b + db)$ in a time interval $\Delta t$ will be the product of mean number of scatterings in time $(\Delta t)$ and $(\delta v_\perp)^2$. The former is given by the number of scatterers in the volume $(2\pi b\, db)(v\Delta t)$. Hence

$$\langle(\delta v_\perp)^2\rangle = (2\pi bdb)\,(v\Delta t)\,n\left(\frac{Gm}{bv}\right)^2 \qquad (1.63)$$

where $n$ is the number density of scatterers. The total mean-square transverse velocity due to all stars is found by integrating over $b$ within some range $(b_1, b_2)$:

$$\langle(\delta v_\perp)^2\rangle_{\text{total}} \simeq \Delta t \int_{b_1}^{b_2} (2\pi bdb)\,(vn)\left(\frac{G^2m^2}{b^2v^2}\right)$$
$$= \frac{2\pi nG^2m^2}{v}\Delta t \ln\left(\frac{b_2}{b_1}\right). \qquad (1.64)$$

The logarithmic factor shows that we cannot take $b_1 = 0, b_2 = \infty$ and one needs to use some physical criteria to fix $b_1$ and $b_2$. It is reasonable to take $b_2 \simeq R$, the size of the system; As regards $b_1$, notice that the effect discussed earlier in Eq. (1.62) becomes important for $b < b_c$. So we may take $b_1 \simeq b_c \simeq (Gm/v^2)$. Then $(b_2/b_1) \simeq (Rv^2/Gm) = N\,(Rv^2/GM) \simeq N$ for a system in virial equilibrium. This effect is important over time-scales $(\Delta t)$ which is long enough to make $\langle(\delta v_1)^2\rangle_{\text{total}} \simeq v^2$. Using this condition and solving for $(\Delta t)$ we get:

$$(\Delta t)_{gc} \simeq \frac{v^3}{2\pi G^2m^2n\ln N} \simeq \left(\frac{N}{8\ln N}\right)\left(\frac{R}{v}\right) \simeq \left(\frac{t_{gc}}{8\ln N}\right) \qquad (1.65)$$

which is a shorter time-scale compared to $t_{gc}$ obtained in Eq. (1.62). The numerical factor arises from a more precise calculation and the above expression gives the correct order of magnitude for the gravitational relaxation time of several systems.

For most astrophysical systems this time scale is very large compared to, say, the age of the universe. For example, a typical galaxy has $R \approx 10$ kpc, $N \approx 10^{11}, v \approx 300$ km s$^{-1}$, giving $t_{gc} \approx 3 \times 10^{18}$ yrs and $(\Delta t)_{gc} \simeq 1.2 \times 10^{17}$ yrs which are much longer than the age of the universe $t_U \approx 10^{10}$ yrs. Among the systems of astrophysical interest, only globular clusters — which are gravitationally bound systems of about $10^6$ stars — have a relaxation time (of about 0.4 Gyr) which is about a tenth of their age; one therefore expects to see the effects of gravitational relaxation in these systems. For all other systems like galaxies, clusters of galaxies *etc.*, the relaxation time is much larger than the age of the system, thereby making it difficult to observe any effects of relaxation.

Incidentally, the above result depends crucially both on the dimension of the space and the nature of the force law. If the analysis is repeated in a $D$-dimensional space with a force law which varies as $r^{-n}$ then we will need to evaluate the integral of $b^{D-2n}$ over all impact parameters in the range $b_{min} < b < b_{max}$. In 3-dimension ($D = 3$), the integral converges on the upper limit for $n > 2$ and gives a logarithmic divergence for $n = 2$. This again shows that distant encounters are important for gravity ($n = 2$) but not for short range forces ($n \geq 3$). On the other hand, a strictly 2-dimensional system (like a disk of self gravitating matter) interacting via inverse square gravitational force ($D = 2, n = 2$) will have a relaxation time ($R/v$) which is the same as crossing time. Thus strictly two dimensional mass distributions can relax through gravity within a few crossing times.

Our analysis also allows us to estimate the time scale for another closely related effect. Consider a test body of mass $M_t$ moving through a cluster of stars, say, where each "field" star has a mass $m_f$. The energy exchanged between the test mass and the field stars is again governed by the same time scale as the one estimated above in Eq. (1.65), with $Gm^2$ replaced by $GM_t m_f$. If the field stars were moving with typical random velocities larger than that of the test mass, then the net effect of the stars on the test mass will be to exert a drag force on the mass. This process — called dynamical friction — has the time scale

$$
\begin{aligned}
\tau_{\text{dyn.fri}} &\simeq \frac{v^3}{2\pi G^2 M_t (m_f n_f) \ln N} \\
&\approx \frac{10^{10}\text{yr}}{(\ln N)} \left( \frac{v}{10\text{km s}^{-1}} \right)^3 \left( \frac{M_t}{M_\odot} \right)^{-1} \left( \frac{\rho_f}{10^3 \, M_\odot \text{pc}^{-3}} \right)^{-1} . \quad (1.66)
\end{aligned}
$$

Note that the result only depends on the background mass density $\rho_f =$

$nm_f$ rather than on the mass $m_f$. A body moving through a dark matter halo of density $\rho$ will have the same $\tau_{\rm dyn.fri}$ irrespective of whether the halo is made of elementary particles (like neutrinos) or planet like objects with the mass of Jupiter. This process can be of importance in many astrophysical contexts since the time scale is now smaller by the factor $(m_f/M)$ compared to $t_{\rm gc}$. The motion of stars in the central regions of galaxies, the motion of clusters *etc.* are all affected by this phenomenon (see eg., Exercises (8.1), (8.14)).

Finally, we must mention another peculiar phenomena which takes place in a system containing a large number of particles interacting through gravitational forces. Such particles can relax in a time scale which is typically a few times the dynamical time scale by a process known as *violent relaxation*. This works by exchange of energy in a time dependent potential as well as through mixing of groups of particles in phase space. The details of this process are fairly complex but there is no doubt that this is the most rapid process which works among self gravitating system of particles allowing them to relax to a configuration which is approximately Maxwellian in the velocity distribution. The importance of this process lies in the fact that it operates at a time scale which is just a factor of few larger than the dynamical time scale.

## 1.5   Relativistic gravity

So far, we have discussed gravity in the Newtonian framework. There are several astrophysical applications in which one needs to go beyond this and use concepts from general relativity, which we will briefly introduce in this section.

We developed the Newtonian gravity starting with the notion of gravitational potential $\phi$ related to the mass density $\rho$ by Eq. (1.1). With suitable boundary conditions, this equation is equivalent to $\nabla^2\phi(t,\mathbf{x}) = 4\pi G\rho(t,\mathbf{x})$. This equation shows that if you move a mass near the origin slightly, its gravitational field at a distant point should change instantaneously. This, of course, does not agree with the basic notion in special relativity which requires signals to propagate at most with the speed of light. It is clear that — at the very least — we need to tinker with this equation in order to get a theory of gravity which is consistent with special relativity. Actually, the situation is much more complicated than that, as we shall see.

The peculiar features arise from the principle of equivalence which we mentioned in page 2. Consider a region of space (say, near the surface of Earth), in which we can take the gravitational acceleration $\mathbf{g} = g\hat{x}$ to be uniform and constant. A non relativistic particle moving in this gravitational field along the x-axis will have a trajectory $x = (1/2)gt^2$ if it starts at rest from the origin at $t = 0$. This trajectory is completely independent of any of the properties of the particle, in particular its mass. Two particles of masses $m$ and $m'$ would have followed the same trajectory if the initial conditions were the same.

Consider now a different situation. Suppose there are two particles of masses $m$ and $m'$ at rest in some frame of reference $S$. Let us now view these particles from another frame of reference $(\bar{S})$ — say, a moving car — which is moving with an acceleration $(-g)$ along the x-axis. The coordinates in $\bar{S}$ and $S$ are related by $\bar{x} = x + (1/2)gt^2$. Hence, viewed from the moving frame, a particle which is located at the origin of $S$ will be given the trajectories $\bar{x} = (1/2)gt^2$. Once again, both the masses $m$ and $m'$ will have the same trajectory because the "motion" of the particle arises merely due to coordinate transformation between $S$ and $S'$. This simple example shows that the motion of particles under Newtonian gravity can be mimicked by the motion viewed from an accelerated frame, as long as we can treat $\mathbf{g}$ as a constant.

Let us now push this analogy further. Consider a clock which is attached to the rim of a disk of radius $r$ which is rotating with an angular velocity $\omega$ about an axis perpendicular to the disk through the center. The speed of the rotating clock, $v = \omega r$, remains constant as it goes around. If we now compare the reading of this clock, $T(r)$, with that of another clock located at the center of the disk, $T(0)$, we will find that they differ due to special relativistic time dilation:

$$T(r) = T(0)\sqrt{1 - (v^2/c^2)} = T(0)\sqrt{1 - (\omega^2 r^2/c^2)}. \qquad (1.67)$$

The rotating clock also experiences an acceleration $g = \omega^2 r$ which is independent of any property of the clock. If we enclose the clock in a small box, then any mass inside the box will experience the same acceleration. Since the acceleration is independent of the mass of the particles, one could attribute this acceleration to a gravitational field generated by a potential $\phi(r)$ such that $-\partial\phi/\partial r = \omega^2 r$. This gives $\phi(r) = -(1/2)\omega^2 r^2$, allowing us to write

$$T(r) = T(0)\sqrt{1 - (\omega^2 r^2/c^2)} = T(0)\sqrt{1 + (2\phi(r)/c^2)}. \qquad (1.68)$$

This suggests that the rate of flow of time as measured by clocks is affected by the presence of gravitational potential! In special relativity, two clocks which are at rest with respect to each other will run at the same rate. If our result is correct, then a clock kept on a table (in a gravitational field) will run at a rate different from a clock kept on the ground since the gravitational potential felt by the two clocks are different.

Another argument leading to the same dramatic result, based on special relativistic dynamics, can be provided along the following lines. The Lagrangian for a non relativistic free particle is given by $L = (1/2)mv^2$ while that for a particle in a gravitational potential $\phi$ will have an additional piece $(-m\phi)$ incorporating the contribution from the potential energy. In special relativity, the corresponding action for a *free* particle is given by

$$A = -mc^2 \int dt \sqrt{1 - \frac{v^2}{c^2}} = -mc \int ds \qquad (1.69)$$

where

$$ds^2 = c^2 dt^2 - d\mathbf{x}^2 \qquad (1.70)$$

is the spacetime interval. Since $ds/c$ denotes the proper time for the particle, one can introduce the gravitational potential into special relativity by adding an extra term to the above action, leading to:

$$A = -mc \int ds - \frac{m}{c} \int \phi\, ds = -mc \int ds(1 + \frac{\phi}{c^2}). \qquad (1.71)$$

This expression is Lorentz invariant if we treat $\phi$ as a scalar. Further, $m$ appears as an overall multiplicative constant which does not affect the equations of motion for a particle, thereby ensuring the validity of principle of equivalence. (Of course, this action also has the correct non relativistic limit.) Writing $ds = cdt \sqrt{1 - (v^2/c^2)}$ and using

$$c^2 dt^2 \left[1 - \frac{v^2}{c^2}\right] \left(1 + \frac{\phi}{c^2}\right)^2 \approx c^2 dt^2 - v^2 dt^2 + 2\phi dt^2 + \mathcal{O}(c^{-2}) \qquad (1.72)$$

we notice that our action in Eq. (1.71) can be written in a much more meaningful way for $\phi \ll c^2$:

$$A = -(mc) \int (cdt) \left[ \sqrt{1 - \frac{v^2}{c^2} \left(1 + \frac{\phi}{c^2}\right)} \right]$$

$$\cong -(mc) \int \left[ \left(1 + \frac{2\phi}{c^2}\right) c^2 dt^2 - d\mathbf{x}^2 \right]^{1/2}. \qquad (1.73)$$

This form suggests that one possible effect of gravity (when $\phi \ll c^2$) is to modify the line interval in the spacetime to the form

$$ds^2 = \left(1 + \frac{2\phi}{c^2}\right) c^2 dt^2 - d\mathbf{x}^2 = g_{ik} dx^i dx^k. \qquad (1.74)$$

The second equation defines the symmetric *metric tensor* $g_{ab}$ with $g_{00} = (1 + 2\phi/c^2)$ and $g_{\alpha\beta} = -\delta_{\alpha\beta}$ and we have assumed the Einstein summation convention (one of the most important contributions Einstein made to science!): Any index repeated in an expression is summed over. Here $i, k$ are summed over $0, 1, 2, 3$. Given this modification for the form of $ds^2$, the action for the particle can still be taken to be that of a "free" particle in Eq. (1.69) with a different form for $ds$.

This is a truly profound conclusion for several reasons. To begin with, it ties up nicely with the result of our clock experiment. Suppose a clock is at rest ($d\mathbf{x} = 0$) in some coordinate system which has a gravitational potential $\phi$. The proper time recorded by the clock is now given by the Eq. (1.74) to be $ds = dt\sqrt{1 + 2\phi/c^2}$ which is exactly what we found earlier [see Eq. (1.68)]. Gravity does affect the rate of flow of clocks in precisely the manner we originally asserted. Second, the above modification of the line interval from that in Eq. (1.70) to the one in Eq. (1.74) shows that the particle still moves from event to event extremizing the spacetime interval. That is, the trajectory of a particle is a curve of extremum length (called *geodesic*) in the spacetime even in the presence of gravitational field. All that has happened is that the underlying distances in spacetimes are modified due to gravity. Third, the modification of the line interval into a form like Eq. (1.74) makes the spacetime curved. This opens up a completely new way of interpreting gravity — not as a force but as curvature of spacetime. All this is possible mainly because of principle of equivalence which allows one to state that the effect of gravity on a particle is independent of the properties of the particle.

The above result also blurs the distinction between different coordinate systems which one might use in describing physics. In special relativity, one gives preference to inertial frames of reference in which the spacetime interval has the form in Eq. (1.70). The Lorentz transformations preserve the form of this line interval and the class of all coordinate systems connected to each other by Lorentz transformations enjoy a special status in special relativity. But since gravity is locally indistinguishable from an accelerated frame, we no longer need to confine our attention to frames of reference which are moving with uniform velocity with respect to each other. Once we introduce coordinate transformations connecting frames which are accelerated with respect to each other, the line interval no longer retains the simple form in Eq. (1.70). We now need to study more general line intervals in which the coefficients are functions of space and time. The expression for the line interval in Eq. (1.74) emphasizes the same point. In general, both the existence of genuine gravitational field as well as the use of accelerated coordinate systems in an inertial frames will lead to modification of the line element and one no longer retains a special status to any particular class of coordinate systems.

Another general result which follows from our discussion is that the frequency of radiation will change when it propagates from event to event in a curved spacetime, since the rate of clocks at different points are different. In the case of a *static* metric, the proper rate of flow of clock at a point $\mathbf{x}$ is given by $\sqrt{g_{00}}dt$. Since the frequency is inversely related to the period, it follows that

$$\frac{\omega(\mathbf{x})}{\omega(\mathbf{x}')} = \frac{\sqrt{g_{00}(\mathbf{x}')}}{\sqrt{g_{00}(\mathbf{x})}}. \qquad (1.75)$$

If we assume that $g_{00} \approx 1$ at very large distances from a mass distribution, then the frequency of radiation measured by an observer at infinity $(\omega_\infty)$ will be related to the frequency of radiation emitted at some point $\mathbf{x}$ by $\omega_\infty = \omega(\mathbf{x})\sqrt{g_{00}(\mathbf{x})}$.

So far we have assumed that $\phi \ll c^2$. In this limit, only the coefficient of $dt^2$ gets modified due to the presence of gravity and one can still describe the gravitational field in terms of just one potential $\phi$. In general, all the coefficients in the quadratic expression for $ds^2$ will get modified and the gravitational field will be described by the symmetric metric tensor $g_{ab}(= g_{ba})$ which has 10 independent components. One of the simplest situations arises in the case of spacetime outside a spherically symmetric distribution of matter with total mass $M$. The metric for this case can be obtained in a heuristic manner, which we will now describe.

Let us consider a point $P$ at a distance $r$ from the origin where the gravitational potential is $\phi_N = -GM/r$ according to Newtonian theory. If we consider a small box around $P$ which is freely falling towards the origin, then the metric in the coordinates used by a freely falling observer in the box will be just that of special relativity:

$$ds^2 = c^2 dt_{\text{in}}^2 - d\mathbf{r}_{\text{in}}^2. \qquad (1.76)$$

This is because, in the freely falling frame, the observer is weight-less and there is no effective gravity. (The subscript 'in' is for inertial frame.) Let us now transform the coordinates from the inertial frame to a frame $(T, \mathbf{r})$ which will be used by observers who are at rest around the point $P$. Since the freely falling frame will be moving with a radial velocity $\mathbf{v}(r) = -\hat{\mathbf{r}}\sqrt{2GM/r}$ around $P$ (if it starts at rest at infinity), the required (non-relativistic!) transformations are: $dt_{\text{in}} = dT, dr_{\text{in}} = dr - \mathbf{v}dT$. Substituting these in Eq. (1.76), we find the metric in the new coordinates to be

$$ds^2 = \left[ 1 - \frac{2GM}{c^2 r} \right] c^2 dT^2 - 2\sqrt{(2GM/r)} dr dT - d\mathbf{r}^2. \qquad (1.77)$$

Incredibly enough, this turns out to be the correct metric describing the spacetime around a spherically symmetric mass distribution of total mass $M$! It will be nicer to have the metric in diagonal form. This can be done by making a coordinate transformation of the time coordinate (from $T$ to $t$) in order to eliminate the off-diagonal term. We look for a transformation of the form $T = t + J(r)$ with some function $J(r)$. This is equivalent to taking $dT = dt + K(r)dr$ with $K = dJ/dr$. Substituting for $dT$ in Eq. (1.77) we find that the off-diagonal term is eliminated if we choose $K(r) = \sqrt{2GM/c^4 r}(1 - 2GM/c^2 r)^{-1}$. In this case, the new time coordinate is:

$$\begin{aligned} ct &= c \int dT + \frac{1}{c^2} \int dr \frac{\sqrt{(2GM/r)}}{(1 - \frac{2GM}{c^2 r})} \\ &= cT - \left[ \sqrt{\frac{8GMr}{c^2}} - \frac{4GM}{c^2} \tanh^{-1} \sqrt{\frac{2GM}{c^2 r}} \right]. \end{aligned} \qquad (1.78)$$

This leads to the line element

$$ds^2 = \left( 1 - \frac{2GM}{c^2 r} \right) c^2 dt^2 - \left( 1 - \frac{2GM}{c^2 r} \right)^{-1} dr^2 - r^2 \left( d\theta^2 + \sin^2\theta d\phi^2 \right) \qquad (1.79)$$

called the *Schwarzschild metric*, which is used extensively both in the study of general relativistic corrections to motion in the solar system and in black hole physics. It is clear from the form of Eq. (1.79) that something peculiar is happening at the radius $r_S \equiv (2GM/c^2)$ called the Schwarzschild radius. It turns out that no signal can propagate from $r < r_S$ to $r > r_S$ — which earns the surface $r = r_S$ the name *event horizon*. We shall say more about this in chapter 5, Sec. 5.4 while discussing black holes. Of course, if the body has a radius $R$ greater than $r_S$, then the metric in Eq. (1.79) is valid only for $r > R$. Such a metric could still describe the geometry around, say, Sun.

## 1.6   Gravitational lensing

Another key feature of general relativity is that it affects the trajectory of light rays. Once again, the basic idea can be understood from the equivalence principle. Consider a cubical box of size $l$ which is being accelerated upwards with an acceleration $g$. Let a light source on one wall of the box emit a pulse of light towards the opposite wall. It will take a time of about $t \approx l/c$ to reach the opposite wall during which time the wall would have moved up by an amount $x = (1/2)gt^2 = (1/2)(gl^2/c^2)$. An observer inside the box will see that the light beam has been deflected downwards by an angle $\theta = x/l = gl/2c^2$. The observer will interpret this deflection as due to the effect of gravity on the light beam and will estimate it as

$$\theta \simeq \int dl\, \frac{\nabla \phi}{c^2}. \tag{1.80}$$

If one uses this formula for a light beam which is grazing the surface of Sun, the total deflection it would suffer would be about $GM_\odot/c^2 R_\odot$. (In general, we expect a bending angle of the order of $(\phi/c^2)$ in a gravitational potential $\phi$.) For a sun like star, $GM_\odot/c^2 R_\odot \simeq R_S/R_\odot \simeq$ few arc sec. This deflection is *huge*, in spite of general relativists sometimes claiming it is small; if you don't believe that, ask any radio astronomer who works with milliarcsec accuracy for sources in the sky.

To get the details and the numerical factors right, one can proceed as follows. We first recall that when the light propagates in a media with a spatially varying refractive index $n(\mathbf{x})$, its trajectory can be determined from the Fermat's principle, which extremizes the time of travel

$$\tau = \int \frac{n(\mathbf{x})}{c}\, dl = \frac{1}{c} \int n(\mathbf{x})(g_{\mu\nu}dx^\mu dx^\nu)^{1/2} \tag{1.81}$$

where $dl$ is the *spatial* line element and we follow the convention that the *Greek* indices run over just $1, 2, 3$. Extremizing this integral gives the equation of motion

$$\frac{d}{dl}\left( n\frac{dx^\alpha}{dl} \right) = \frac{\partial n}{\partial x^\alpha}. \tag{1.82}$$

On using Eq. (1.81) again in the form, $dl = cd\tau/n$, this equation becomes

$$\frac{1}{c^2}\frac{d}{d\tau}\left(n^2\frac{dx^\alpha}{d\tau}\right) = \frac{\partial \ln n}{\partial x^\alpha}. \tag{1.83}$$

Our aim now is to approximate a weak gravitational field as a medium with a spatially varying refractive index. To do this, we note that, in the metric in Eq. (1.79), the proper time is $d\tau = dt(1 + 2\phi/c^2)^{1/2}$, while the proper length is $dl = dr(1 + 2\phi/c^2)^{-1/2}$ giving

$$\frac{dl}{d\tau} = \frac{dr}{dt}(1 + 2\phi/c^2)^{-1} \approx \frac{dr}{dt}(1 - 2\phi/c^2). \tag{1.84}$$

Assuming that light always travels with the speed $c$ in proper coordinates we have $dl/d\tau = c$ and taking $v = (dr/dt)$, one gets the effective refractive index to be $n = c/v \simeq (1 - 2\phi/c^2)$. This can be interpreted as the effective refractive index of a spacetime with the gravitational potential $\phi$. When this is used in Eq. (1.83) to the lowest order in $\phi/c^2$, we have the equation of motion for the light ray to be

$$\frac{d^2x^\alpha}{dt^2} \approx -2\frac{\partial\phi}{\partial x^\alpha}. \tag{1.85}$$

Thus the net force acting on a light beam in a gravitational field is *twice* the Newtonian force. Though we used the form of the metric in Eq. (1.79) to get this result, it happens to be of general validity to the lowest order in $\phi/c^2$.

Let us estimate the deflection of light $\mathbf{d}$ due to a bounded density distribution $\rho(\mathbf{x})$ producing a gravitational potential $\phi(\mathbf{x})$. If a light ray is moving along the $z-$axis, it will experience a total transverse angular deflection by the amount

$$\boldsymbol{\theta} = 2\int_{-\infty}^{\infty} dz \left[\frac{^{(2)}\nabla\phi}{c^2}\right] \tag{1.86}$$

where $^{(2)}\nabla \equiv (\partial/\partial x, \partial/\partial y)$ is the gradient in the $x - y$ plane. Such a deflection, of course, can change the apparent position of the source in the sky, distort the shape of an extended object, produce multiple images of the same source *etc.*. All these effects are collectively called *gravitational lensing* and each of them is a powerful tool in observational astronomy. We will now discuss some key features of gravitational lensing.

To do this, we shall assume that all the deflection takes place on a single plane which we take to be the plane of the deflecting mass — usually called the lens — and project the image position and the source position on to the lens plane. The distances between source and observer, source and lens, and observer and lens are give by $D_{OS}, D_{LS}, D_{OL}$, respectively. From Fig. 1.3, it is clear that $\theta_s = \theta_i - (D_{LS}/D_{OS})\theta$. Let $\mathbf{s}$ and $\mathbf{i}$ denote the two dimensional vectors giving the source and image positions projected on the two dimensional lens plane; and let $\mathbf{d(i)}$ be the (vectorial) deflection produced by the lens corresponding to the angular deflection $\boldsymbol{\theta}$. (See Fig.1.3). Then, the relation $\theta_s = \theta_i - (D_{LS}/D_{OS})\theta$ leads to the vector relation:

$$\mathbf{s} = \mathbf{i} - \frac{D_{LS}}{D_{OS}}\mathbf{d(i)}. \tag{1.87}$$

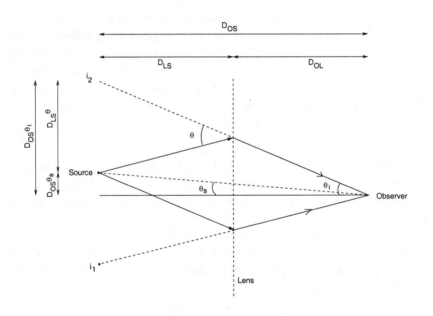

Fig. 1.3   The geometry for gravitational lensing.

Clearly, all the physics is in the deflection vector $\mathbf{d(i)}$. Consider now the 2−dimensional divergence $^{(2)}\nabla \cdot \mathbf{d}$ of this vector. Using

$$^{(2)}\nabla^2\phi = \nabla^2\phi - \frac{\partial^2\phi}{\partial z^2} = 4\pi G\rho(\mathbf{x}) - \frac{\partial^2\phi}{\partial z^2} \tag{1.88}$$

and Eq. (1.86), we get

$$^{(2)}\nabla\cdot\mathbf{d} = \frac{2D_{OL}}{c^2}\int_{-\infty}^{\infty} dz \left[4\pi G\rho(x,y,z) - \frac{\partial^2\phi}{\partial z^2}\right] = \frac{8\pi G D_{OL}}{c^2}\Sigma(x,y) \tag{1.89}$$

where $\Sigma(x, y)$ is the surface mass density corresponding to $\rho(\mathbf{x})$ obtained by integrating $\rho(x, y, z)$ along the $z-$axis. [The $\partial^2\phi/\partial z^2$ term vanishes on integration]. This two-dimensional Poisson equation has the standard solution:

$$\mathbf{d}(\mathbf{i}) = \frac{4GD_{\mathrm{OL}}}{c^2} \int d^2x \Sigma(\mathbf{x}) \frac{(\mathbf{i} - \mathbf{x})}{|\mathbf{i} - \mathbf{x}|^2} \qquad (1.90)$$

which gives the deflection $\mathbf{d}$ in terms of the surface density $\Sigma(\mathbf{x})$. This completely solves the problem.

As a simple application, consider the lensing due to a point mass, for which $\Sigma(x, y) = M\delta(x)\delta(y)$; Eq. (1.90) then gives the deflection $\mathbf{d}(\mathbf{i}) = (4GMD_{\mathrm{OL}}/c^2)(\mathbf{i}/i^2)$. Substituting into Eq. (1.87), we find

$$\mathbf{s} = \mathbf{i}\left[1 - \frac{L^2}{i^2}\right]; \qquad L^2 = \frac{4GMD_{\mathrm{LS}}D_{\mathrm{OL}}}{c^2 D_{\mathrm{OS}}}. \qquad (1.91)$$

This result shows that $\mathbf{s}$ and $\mathbf{i}$ are in the same direction and their magnitudes are related by $s = i - (L^2/i)$. Solving this quadratic equation, we find the *two* solutions for the image position $i$ to be

$$i_{\pm} = \frac{1}{2}\left(s \pm \sqrt{s^2 + 4L^2}\right). \qquad (1.92)$$

Thus any source produces two images when lensed by a point mass with the two images located on either side of the source. You can also see that one image is inside a circular ring of radius $L$ (called *Einstein ring*) and the other outside. As the source moves away from the lens (ie., as $s$ increases) one of the images approaches the lens (and becomes fainter, as we shall see below) while the other approaches the true position of the source (and tends towards a magnification of unity).

In the cosmological context in which lensing often occurs, a foreground galaxy (or a cluster) will act as a lens for a background quasar (or a galaxy). In this case, all relevant distances can be taken to be $D \approx 3000$ Mpc (see Chapter 6; 1 Mpc $\simeq 3 \times 10^{24}$ cm). Then the angular size of this Einstein ring will be

$$\theta_L \equiv \frac{L}{D} = \left(\frac{4GM}{c^2 D}\right)^{1/2} \approx 2 \text{ arc sec} \left(\frac{M}{10^{12}M_{\odot}}\right)^{1/2} \left(\frac{D}{3000\,\mathrm{Mpc}}\right)^{-1/2}. \qquad (1.93)$$

One can also think of $\theta_L$ as the typical angular separation between images produced by lens of mass $M$ located at cosmological distances. For galaxy with $M \simeq 10^{12} M_\odot$, we have $\theta \simeq 2$ arcsec, while for a cluster $M \approx 10^{15} \, M_\odot$ and $\theta_L \approx 1$ arc arcmin.

Even when the lensing action does not produce multiple images, it can significantly distort the image of an extended object. (It is usual to call lensing action which produces multiple images as *strong* lensing and the one which merely distorts the background image as *weak* lensing.) Consider, for example, the lensing action of a foreground cluster of galaxies on several distant background galaxies. The cluster will distort the background images of the galaxies and will shear them in the tangential direction. Mapping this distortion allows you — among other things — to determine the mass of the foreground cluster which is acting as a lens. Since all the gravitating matter contributes to the lensing action, this approach provides a direct measurement of the total gravitating matter in the cluster including the dark matter. Such observations have indicated that there is nearly few hundred times more dark matter in clusters of galaxies than visible matter.

This lensing action can also lead to magnification of the source. This magnification can be calculated by considering the change in the image position $\delta \mathbf{i}$ for a small change in the source position $\delta \mathbf{s}$. Since all the photons from a small area element $\Delta s$ around $\mathbf{s}$ reach another small area element $\Delta i$ around $\mathbf{i}$, the magnification $A = (\Delta i / \Delta s)$ will be the determinant of the matrix of the transformation between the two:

$$A = \det \left| \frac{\partial i_a}{\partial s_b} \right| = \left[ \det \left| \frac{\partial s_b}{\partial i_a} \right| \right]^{-1}. \tag{1.94}$$

The amplification of the two images in Eq. (1.92), for example, are given by

$$A_\pm = \left[ 1 - \left( \frac{L}{i_\pm} \right)^4 \right]^{-1} = \frac{u^2 + 2}{2u\sqrt{u^2 + 4}} \pm \frac{1}{2} \tag{1.95}$$

where $u = s/L$. It is clear that the magnification of the image inside the Einstein ring is negative. The total magnification of the flux obtained by adding the absolute value of the two magnifications is given by

$$A_{\text{tot}} = A_+ + A_- = \frac{u^2 + 2}{u\sqrt{u^2 + 4}}. \tag{1.96}$$

An important application of the lensing by a point mass is in a phenomenon called *microlensing*. If, for example, a compact object in our galaxy (say a Jupiter like non luminous object) moves across the line of sight to a distant star in, say, the Large Magellanic Cloud (LMC, which is a small satellite galaxy of Milky Way), it will lead to a magnification of the light from the star under optimal conditions. The peak magnification in Eq. (1.96) is at $u = 1$ is $A_{\max} = 1.34$ which should be easily detectable. This is an indirect way of detecting massive, compact halo objects (called MACHOS) in the halo of our galaxy. The typical time scale for a microlensing event is

$$t = \frac{D_{\mathrm{OL}}\theta_L}{v} = 0.2\,\mathrm{yr} \left(\frac{M}{M_{\odot}}\right)^{1/2} \left(\frac{D_{\mathrm{OL}}}{10\,\mathrm{kpc}}\right)^{1/2} \left(\frac{D_{\mathrm{LS}}}{D_{\mathrm{OS}}}\right)^{1/2} \left(\frac{v}{200\,\mathrm{km\ s^{-1}}}\right)^{-1}.$$
$$(1.97)$$

In the case of MACHOS located in our galactic halo acting as lenses for sources in the LMC, we need to sample the light curve of stars in LMC for between about an hour and a year to detect lenses with the mass range of $(10^{-6} - 10^2)M_{\odot}$. The probability for such lensing action, of course, is low and hence it is necessary to monitor a large number of stars for a long period of time in order to produce observable effects. If the number density of lenses is $n$ and if we take the typical cross section of each lens to be $\pi\theta_L^2$ (see Eq. (1.93)) then the probability of lensing over a distance $D$ will be $P \simeq (nD^3)(\theta_L^2) \simeq G\rho D^2/c^2$. (Incidentally, this depends only on the mass density of the lenses and not on the actual masses.) For lensing of stars in LMC by compact objects in the halo of our galaxy, we may assume that all the dark matter density in the galactic halo is contributed by these lensing objects. Then we can use $\rho = 10^{-26}\mathrm{g\ cm^{-3}}$, $D = 500$ kpc; this gives a rather small value of $P \simeq 10^{-6}$. You are saved by the fact that there are about $5 \times 10^9$ stars in the LMC and a large fraction of them can be studied systematically.

## Exercises

### Exercise 1.1    *Shapes don't matter:*

Compute the gravitational potential energy of a sphere of radius $R$ and mass $M$, if it has a constant density. Now suppose that the same mass $M$ is flattened to a thin disk of radius $R$ with constant surface density. Compute the the potential energy of the new configuration. What is the moral of the story? [Hint: For a sphere $U = (3/5)GM^2/R$ while for a disk it is $(8/3\pi)GM^2/R$. Under extreme flattening, the numerical coefficient changes only by a factor 1.41.]

**Exercise 1.2**   *Going full circle:*

Can you find a physical situation in which the minor arc AC of Fig.1.1 could be meaningful? [Hint: Repulsive thoughts might help.]

**Exercise 1.3**   *Dynamic ranges for physical parameters:*

Compute the characteristic speeds [using the $v^2 \approx GM/R$] for (a) planets in solar system; (b) stars in a galaxy and (c) galaxies in the universe. What is the dynamic range in the variation of mass, radius and density in these systems compared to the dynamic range in speed ? What do you conclude ?

**Exercise 1.4**   *Estimates for solar system:*

(a) Where is the center of mass of the solar system? (b) Where is the center of mass of the Earth-Moon system? (c) What is Jupiter's orbital angular momentum? (d) What is the rotational angular momentum of the Sun (which is rotating about its axis with a period of 25 days)? (e) What should be the speed with which an object should be launched to escape the gravitational force of the solar system? What if we want to escape the Milky Way galaxy?

**Exercise 1.5**   *Exact solution with $(1/r^2)$ term:*

Solve the central force problem with the potential $V = -(GMm/r) - (\beta/r^2)$ exactly and show that the bound orbits are of the form $(1/r) = C_1 + C_2 \cos[(1 - (2\beta m/J^2))^{1/2}\theta]$ where $C_1, C_2$ are constants. Hence show that when the $\beta/r^2$ term is a *small* perturbation to the $(1/r)$ potential, the orbit precesses at the rate given by Eq. (1.16). [Hint: In any central force problem, the angular momentum gives a term proportional to $J^2/r^2$; hence the $(1/r^2)$ correction to potential only "renormalizes" the $J^2$ term.]

**Exercise 1.6**   *Basic measurements:*

Some great moments in early astronomy occurred when different civilizations managed to measure the astronomical distances of nearby objects. This exercise explores three of them: (a) Ancient astronomers knew that lunar eclipse occurs when the shadow of the Earth falls on the Moon. Assume that Sun is very far away from the Earth. Show how one can determine the ratio of the diameter of the Moon to the diameter of the Earth from the duration of the lunar eclipse. Given the diameter of the Moon and its angular size in the sky, we can determine the Earth-Moon distance in terms of the Earth diameter. (b) At noon on the summer solstice, the Sun is seen overhead at Syene, a city almost directly south of Alexandria. (This is known from the fact that at that time, buildings and poles cast virtually no shadows.) At the same moment, in Alexandria, a vertical rod AB casts a shadow AC on the ground such that angle ABC is 7.2 degrees. Show how this measurement can be used to determine the diameter of the Earth given the distance between Syene and Alexandria. (This distance was recorded to be 5000 stadia; there is some confusion as to how much is 1 stadia but if it is 157.2 m, as some scholars believe, you get the correct result.) Using the result in (a), one can now determine the Earth-Moon distance. (c) A half Moon occurs when the Sun(S), Moon(M) and the Earth(E) are at the vertices of a right angled triangle SEM. By measuring angle SEM (which is about 89.85 degrees!) show how one can determine distance between Earth and Sun.

**Exercise 1.7**  *Kepler motion of a binary system:*

Consider two stars of masses $m_1$ and $m_2$ orbiting around the common center of mass (in an elliptical path) under their mutual gravitational attraction. The plane of the orbit is inclined at an angle $i$ to the line of sight connecting the observer to the focus. From the Doppler shift of the radiation emitted by the star 1, it is possible to determine the velocity of that star along the line of sight. Show how this information will help one to determine the combination

$$C = \frac{m_2^3 \sin^3 i}{(m_1 + m_2)^2}. \tag{1.98}$$

This procedure is of importance in determining the mass of an invisible companion in a binary stellar system.

**Exercise 1.8**  *Effect of tides:*

Give simple physical arguments to show that (a) the tide raised *on the satellite by the planet* has the following effects: (i) It de-spins the satellite towards synchronous rotation; (ii) it acts to damp the satellite's orbital eccentricity and (iii) it can also tidally heat up the satellite. (b) Also show that the tide raised *on the planet by the satellite* has the following effects. (i) It increases the satellite's semi major axis if $\omega_{\text{rot}} > \omega_{\text{rev}}$ and $\omega_{\text{rev}} > 0$ where $\omega_{\text{rot}}$ is the angular velocity of rotation of the planet and $\omega_{\text{rev}}$ is the angular velocity of the satellite around the planet. If $\omega_{\text{rot}} < \omega_{\text{rev}}$ (as in the case of the satellite Phobos of Mars), the semi major axis will decrease and the satellite can eventually crash on to the planet. (ii) It decreases the planet's spin rate if $\omega_{\text{rot}} < \omega_{\text{rev}}$ and $\omega_{\text{rev}} > 0$.

**Exercise 1.9**  *Standard error in textbooks:*

Some textbooks give the following argument as regards the Schwarzschild radius. "The escape velocity from a body of mass $M$ is $v_{\text{esc}} = \sqrt{2GM/R}$. If the radius of the body is less than $2GM/c^2$, then $v_{\text{esc}} > c$ and not even light can escape from the body." Even assuming that Newtonian considerations are admissible, explain why this argument is completely wrong.

**Exercise 1.10**  *Seeing the Oort cloud:*

Can the comets in Oort cloud lens the background stars in our galaxy? Give reasons for your conclusion; look up the necessary parameters of the Oort cloud.

**Exercise 1.11**   *Bounds on the mass of the dark matter particle:*

In the conventional cosmological models (which we will discuss in chapter 6) one invokes dark matter made of weakly interacting massive particles which form large halos around galaxies and in clusters. For the sake of this problem, assume that the dark matter is made of massive neutrinos (this is *not* true!). During the early epochs of the evolution of the universe, the *maximum* phase space density of neutrinos will be $(dN_\nu/d^3x\,d^3p)_{\mathrm{max}} \approx 2/(2\pi\hbar)^3$. (Explain why.) During the collisionless evolution and clustering, eventually leading to a gravitationally bound structure of mass $M$ and radius $R$, regions of high phase space density will be mixed with regions of low phase space density. In this process, the maximum phase space density can only decrease. Therefore the mean final phase space density has to be less than the original maximum value. Show that this allows one to put a constraint on the mass of the neutrino as

$$m_\nu \gtrsim 30 \text{ eV } \left(\frac{\sigma}{220 \text{ km s}^{-1}}\right)^{-1/4} \left(\frac{R}{10 \text{ kpc}}\right)^{-1/2} \tag{1.99}$$

where $\sigma$ is the velocity dispersion of the dark matter halo.

## Notes and References

More material on gravitation, directly relevant to astrophysics, can be found in chapters 2 and 10 of Padmanabhan (2000). General relativistic description of gravity is discussed in e.g., Landau and Lifshitz (1975a), Misner Thorne and Wheeler (1973) and chapter 11 of Padmanabhan (2000).

1. *Section 1.1:* (a) More details regarding the discovery of extra solar planets can be found, for example, in the website http://cfa-www.harvard.edu/planets. (b) A detailed discussion of reduced three body problem can be found in Henon (1993) and in Marchal (1990).

2. *Section 1.2:* The shape of a rotating fluid mass and related details can be found, for example, in Chandrasekhar (1969) and Binney and Tremaine (1987).

3. *Sections 1.3 and 1.4:* For a detailed discussion of statistical mechanics of gravitating systems, see Padmanabhan (1990). Many of these topics are also covered in Binney and Tremaine (1987). The process of violent relaxation was discovered by Lynden-Bell. See Lynden-Bell (1967).

4. *Section 1.5:* For a discussion of this particular approach to deriving Schwarzschild metric, see Visser (2003) and references cited therein.

# Chapter 2

# Radiative Processes

Electromagnetic radiation is the single most important diagnostic tool in astrophysics. In any given astrophysical context, you need to relate the properties of the radiation received on earth to the properties of the source and identify the signatures of the processes which are involved. So we have to familiarize ourselves with some of the key radiative processes, which is the task we address in this chapter.

Broadly speaking, one can divide the radiation from astrophysical sources into *line* and *continuum* radiation. Line radiation arises when the source makes a transition between two discrete energy levels thereby producing a distinct spectral line. This process is usually quantum mechanical in nature. The continuum radiation, on the other hand, can arise in *both* classical and quantum mechanical contexts. Classically, different kinds of accelerating mechanisms for the charged particle will lead to different radiation patterns. Quantum mechanically, it is possible for a system to make a transition from a discrete to continuum energy level (like when an electron is kicked off a hydrogen atom) thereby leading to continuum radiation. All these processes play their role in astrophysics.

## 2.1 The origin of radiation

We will first describe the origin of radiation in classical and quantum theories and then discuss the specific processes.

### 2.1.1 *Radiation in classical theory*

The Coloumb field of a charged particle varies as $E \propto r^{-2}$. The energy density of this field, which scales as $E^2 \propto r^{-4}$, falls fairly rapidly with distance. Contrast this with the radiation field produced by an accelerated charge for which $E \propto (1/r)$ and energy density varying as $r^{-2}$. How does this come about (and the caterpillar becomes the butterfly, as Feynman calls it) just because the charge is accelerated?

To understand this, let us begin by noting that a charge $q$ moving with uniform velocity $\mathbf{v}$ is equivalent to a current $\mathbf{j} = q\mathbf{v}$. This current will now produce a magnetic field (in addition to the electric field) which scales in proportion to $\mathbf{j}$. Uniform velocities, of course, cannot produce any qualitatively different phenomena compared to those due to a charge at rest, because you could always view it from another Lorentz frame in which the charge is at rest. The situation becomes interesting when we consider a charge moving with an acceleration. If $\mathbf{a} = \dot{\mathbf{v}} \neq 0$, it will produce a non zero $(\partial \mathbf{j}/\partial t)$ and hence a non zero $(\partial \mathbf{B}/\partial t)$. Through Faraday's law, the $(\partial \mathbf{B}/\partial t)$ will induce an electric field which scales as $(\partial \mathbf{j}/\partial t)$. (That is, if $(\partial \mathbf{j}/\partial t)$ changes by factor 2, the electric field will change by factor 2.) It follows that an accelerated charge will produce an electric field which is linear in $(\partial \mathbf{j}/\partial t) = q\mathbf{a}$. [This, of course, is in addition to the usual Coulomb term which is independent of $\mathbf{a}$ and falls as $r^{-2}$.] Let us consider this electric field in the *instantaneous rest frame* of the charge. It has to be constructed from $q, a, c$ and $r$ and has to be linear in $qa$. Hence it *must* have the general form

$$E = C\left(\theta\right) \frac{qa}{c^n r^m} = C\left(\theta\right) \left(\frac{q}{r^2}\right) \left(\frac{a}{c^n r^{m-2}}\right) \tag{2.1}$$

where $C$ is a dimensionless factor, depending only on the angle $\theta$ between $\mathbf{r}$ and $\mathbf{a}$, and $n$ and $m$ need to be determined. (Since $\mathbf{v} = 0$ in the instantaneous rest frame, the field cannot depend on the velocity.) From dimensional analysis, it immediately follows that $n = 2, m = 1$. Hence

$$E = C\left(\theta\right) \frac{qa}{c^2 r}. \tag{2.2}$$

Thus, dimensional analysis plus the fact that $\mathbf{E}$ must be linear in $q$ and $a$, implies the $r^{-1}$ dependence for the radiation term.

There are several ways of determining the factor $C(\theta)$. The most rigorous method is to consider the field of a charged particle which has just begun to accelerate. We will, however, use a more heuristic approach. At very large distances from the charged particle, the electric field is $\mathbf{E} = \mathbf{E}_{\mathrm{coul}} + \mathbf{E}_{\mathrm{rad}}$ in the instantaneous global rest frame of the charged particle. Here $\mathbf{E}_{\mathrm{coul}}$ is the standard Coulomb field which is along the radial direction and $\mathbf{E}_{\mathrm{rad}}$ is the radiation field which has the *magnitude* given by

Eq. (2.2). We can determine $C(\theta)$ if we can determine the direction of $\mathbf{E}_{\text{rad}}$. To do this, we only need to note that one of the Maxwell's equations, $\nabla \cdot \mathbf{E} = 4\pi\rho$ continues to be valid even in the case of an accelerated charge. Hence the flux of the electric field through a spherical surface at a large distance should be equal to the total charge enclosed by the surface. Since the Coulomb field already satisfies this condition, the total flux of $\mathbf{E}_{\text{rad}}$ over such a surface must vanish. If this has to happen without violating the symmetries of the problem, then $\mathbf{E}_{\text{rad}}$ must be normal to the radius vector which requires $C(\theta) = \sin\theta$.

Hence the two components of the electric field are given by: (i) the radial Coulomb field $E_{\parallel} = E_r = \left(q/r^2\right)$ and (ii) the transverse radiation field (with $a_{\perp} \equiv a\sin\theta$):

$$E_{\perp} = a_{\perp}\left(\frac{r}{c^2}\right)\cdot\frac{q}{r^2} = \frac{q}{c^2}\left(\frac{a_{\perp}}{r}\right). \tag{2.3}$$

It is also obvious from special relativistic considerations that the acceleration of a charge located at $\mathbf{x}'$ at time $t'$ can only affect the behaviour of the field at a location $\mathbf{x}$ at a time $t$ where $t = t' + |\mathbf{x}' - \mathbf{x}|/c$. Thus to determine the field at $(t, \mathbf{x})$, we need to use the state of motion of the particle at a *retarded time* $t' = t - |\mathbf{x}' - \mathbf{x}|/c$. This is the radiation field located in a shell at $r = ct$ and is propagating outward with a velocity $c$. One can express it concisely in the vector notation as

$$\mathbf{E}_{\text{rad}}\left(t,\mathbf{r}\right) = \frac{q}{c^2}\left[\frac{1}{r}\hat{\mathbf{n}} \times (\hat{\mathbf{n}} \times \mathbf{a})\right]_{\text{ret}} \tag{2.4}$$

where $\mathbf{n} = (\mathbf{r}/r)$ and the subscript "ret" implies that the expression in square brackets should be evaluated at $t' = t - r/c$. The full electric field *in the frame in which the charge is instantaneously at rest*, is $\mathbf{E} = \mathbf{E}_{\text{coul}} + \mathbf{E}_{\text{rad}}$. The velocity dependence of this field can be obtained by making a Lorentz transformation. You will recall that, if a primed reference frame moves with a velocity $\mathbf{v}$ with respect to an unprimed reference frame, then we have:

$$\mathbf{E}'_{\perp} = \gamma\left(\mathbf{E} + \frac{\mathbf{v}}{c} \times \mathbf{B}\right)_{\perp}, \qquad \mathbf{B}'_{\perp} = \gamma\left(\mathbf{B} - \frac{\mathbf{v}}{c} \times \mathbf{E}\right)_{\perp} \tag{2.5}$$

where $\mathbf{E}_{\perp}$ and $\mathbf{B}_{\perp}$ are the components of electric and magnetic fields perpendicular to $\mathbf{v}$; the components $\mathbf{E}_{\parallel}$, $\mathbf{B}_{\parallel}$ parallel to $\mathbf{v}$ do not change under the Lorentz transformation.

There are two crucial ways in which the radiation field differs from the Coulomb like field. First, the radiation field is *transverse* to the direction of propagation in contrast to the Coulomb field, which is radial. Second, it falls only as $(1/r)$ in contrast to the Coulomb field which falls as $r^{-2}$. It is also clear that each spherical shell containing the radiation field moves forward with the speed of light $c$. Since the energy density in the electromagnetic field is $U = (E^2 + B^2)/8\pi$, the amount of energy crossing a sphere of radius $r$ centered at the charge is

$$\frac{d\mathcal{E}}{dt} = c(4\pi r^2) \cdot 2 \cdot \frac{\langle E_{rad}^2 \rangle}{8\pi}. \tag{2.6}$$

Here the factor 2 takes into account an equal contribution from magnetic field and the symbol $\langle \cdots \rangle$ denotes an averaging over the solid angle. Using $\langle \sin^2 \theta \rangle = 1 - \langle \cos^2 \theta \rangle = 1 - (\langle z^2 \rangle / r^2) = 2/3$ (where the result $\langle z^2 \rangle / r^2 = 1/3$ follows from the symmetry among the three axes) we find that the total amount of energy radiated per second in all directions by a particle with charge $q$ moving with acceleration $a$ is given by

$$\frac{d\mathcal{E}}{dt} = \frac{2}{3} \frac{q^2}{c^3} a^2. \tag{2.7}$$

provided the acceleration is measured in the frame in which the particle is instantaneously at rest. The rate of energy emission, of course, is independent of the frame used to define it because both $d\mathcal{E}$ and $dt$ transform in the identical manner under Lorentz transformation in this case. This result in Eq. (2.7) is called *Larmor's formula* and can be used to understand a host of classical electromagnetic phenomena.

Since $\mathbf{d} = (q\mathbf{x})$ is the "dipole" moment related to a charge located at position $\mathbf{x}$, this formula shows that the total power radiated is proportional to the square of $\ddot{\mathbf{d}}$. In bounded motion, if $\mathbf{d}$ varies at frequency $\omega$ (so that $\ddot{\mathbf{d}} = -\omega^2 \mathbf{d}$), then the energy radiated is given by

$$\frac{d\mathcal{E}}{dt} = \frac{2}{3} \frac{d^2}{c^3} \omega^4. \tag{2.8}$$

This turns out to be the correct formula for radiation from an oscillating dipole. Different physical phenomena are essentially characterized by different sources of acceleration in Eq. (2.7) for the charged particle.

### 2.1.2   *Radiation in quantum theory*

Let us next consider the origin of radiation in quantum theory. The radiation field in quantum theory is described in terms of photons and the emission or absorption of photons arises when a system makes a transition from one energy level to another. Consider an initial state with an atom, say, in ground state $|G\rangle$ and $n$ photons present. Absorbing a photon, the atom makes the transition to the excited state $|E\rangle$, leaving behind a $(n-1)$ photon state. Let the probability for this process be

$$P\left(|G\rangle|n\rangle \rightarrow |E\rangle|n-1\rangle\right) \propto n \equiv Qn. \qquad (2.9)$$

The fact that this absorption probability $P$ is proportional to $n$ seems intuitively acceptable. Consider now the probability $P'$ for the 'time reversed' process $(|E\rangle|n-1\rangle \rightarrow |G\rangle|n\rangle)$. By principle of microscopic reversibility, we expect $P' = P$ giving $P' \propto n \equiv Qn$. Calling $n-1 = m$, we get

$$P'\left(|E\rangle|m\rangle \rightarrow |G\rangle|m+1\rangle\right) = Qn = Q(m+1). \qquad (2.10)$$

You now see that $P'$ is non-zero even for $m = 0$; clearly, $P'\left(|E\rangle|0\rangle \rightarrow |G\rangle|1\rangle\right) = Q$ which gives the probability for a process conventionally called *spontaneous emission*. The term $Qm$ gives the corresponding probability for *stimulated emission*. Thus, the fact that absorption probabilities are proportional to $n$ while emission probabilities are proportional to $(n + 1)$ originates most directly from the principle of microscopic reversibility. Between any two levels, we can write the two probabilities as

$$R_{abs} = Qn; \qquad R_{em} = Q(n+1). \qquad (2.11)$$

Historically, however, radiation was described in terms of a quantity called *intensity* before we knew about the existence of photons. To connect the two concepts, let us consider the number of photons with momentum in the interval $(\mathbf{p}, \mathbf{p} + d^3\mathbf{p})$ which is given by $dN = 2n[Vd^3p/(2\pi\hbar)^3]$. Here the factor in the square brackets gives the number of quantum states in the phase volume $Vd^3p$ and the factor 2 takes into account the two spin states for each photon. The corresponding energy $dE$ flowing through the volume element $d^3x = dA(cdt)$ will be $dE = h\nu dN = 2nh\nu \, dA[cdt][d^3p/h^3]$. Using $|\mathbf{p}| = h\nu/c$ and writing $d^3p = p^2dpd\Omega = (h/c)^3\nu^2d\nu d\Omega$ where $d\Omega$ is the solid angle in the direction of $\mathbf{p}$, we get

$$dE = \frac{2h\nu^3}{c^2} n \, dA \, dt \, d\Omega \, d\nu \qquad (2.12)$$

which suggests the definition of a quantity called *intensity*, which is the energy per unit area per unit time per solid angle per frequency (that is, energy per everything), as

$$\frac{dE}{dAdtd\Omega d\nu} \equiv I_\nu = \frac{2h\nu^3}{c^2}n(\nu). \qquad (2.13)$$

Among other things, this equation shows that: (i) $(I_\nu/\nu^3)$ is Lorentz invariant, being proportional to the number of photons per phase cell; often, this provides a quick way to transform the intensity from one frame to another. (ii) $I_\nu$ measured from a source in the sky is independent of the distance to the source if no photons are lost in the way; in fact $I_\nu$ does not change even on passing through any lossless optical device (like, for example, a lens).

While using intensity, it was conventional to redefine the rates of absorption and emission in Eq. (2.11) by

$$R_{abs} = BI_\nu; \qquad R_{em} = A + BI_\nu \qquad (2.14)$$

where the coefficients $A, B$ are called *Einstein coefficients*. Using $I_\nu = (2h\nu^3/c^2)n$, and comparing with Eq. (2.11) we find that

$$Q = \left(\frac{2h\nu^3}{c^2}\right) B; \qquad \frac{A}{B} = \left(\frac{2h\nu^3}{c^2}\right). \qquad (2.15)$$

In the old fashioned language based on intensity, one feels somewhat mystified as to why the ratio of $A/B$ has this strange value; of course, if you think in terms of photons, the corresponding ratio in Eq. (2.11) is just unity.

To gain a feel for the relative importance of the spontaneous emission (governed by $A$) and stimulated emission, governed by $B$, let us compute the intensity $I_c$ of radiation at which the stimulated and spontaneous emission rates are equal; $BI_c = A$. The energy flux contained in a small frequency interval $d\omega$ in such a beam will be $I_c d\omega = (A/B)d\omega = (\hbar\omega^3/2\pi^2c^2)d\omega$. For visible light, with $\omega \approx 3 \times 10^{15}$ Hz and $d\omega \approx 2\pi \times 10^{10}$ Hz, we get $I_c d\omega \approx 10$ Watt cm$^{-2}$. Conventional light sources do not have such high intensities while a laser beam does.

The intensity defined in Eq. (2.13) has a $d\Omega$ indicating that it is per unit solid angle where the solid angle is measured in steradians. The definition of 1 steradian is just 1 square radian which is equal to $(360/2\pi)^2 = 3.28 \times 10^3$ square degree. This leads to the conversions

$$1 \text{ st} = 4.25 \times 10^{10} \text{ arc s}^2; \qquad 1 \text{ sq deg} = 3 \times 10^{-4} \text{ st}. \qquad (2.16)$$

The solid angle in entire sphere is $4\pi$ steradians which corresponds to $4.12 \times 10^4$ deg$^2$.

## 2.2 Thermal radiation

As an immediate consequence of the result in Eq. (2.11) we can determine the the energy distribution of photons in equilibrium with matter at temperature $T$. In such a situation, photons will be continuously absorbed and emitted by matter. Consider the rate of absorption (or emission) of photons between any two levels, say, a ground state $|G\rangle$ and an excited state $|E\rangle$. We saw in Eq. (2.11) that the absorption rate per atom is given by $Qn$ while the emission rate is $Q(n + 1)$ where $n$ is the number of photons present and $Q$ is to determined from the quantum theory of radiation. In steady state, the number of upward and downward transitions must match, which requires $N_G Q n = N_E Q(n + 1)$ where $N_G, N_E$ are the number of atoms in ground state and excited state, respectively. Solving for $n$ we get $n = [(N_G/N_E) - 1]^{-1}$. Since matter is in thermal equilibrium at temperature $T$, we must also have $(N_E/N_G) = \exp(-\Delta E/k_B T)$ where $\Delta E$ is the energy difference between the two levels. This relation should hold for all forms of matter with arbitrary energy levels; hence we can take the ground state energy to be zero (i.e, arbitrarily small) and the excited state energy to be $E$, leading to

$$n = \frac{1}{(N_G/N_E) - 1} = \frac{1}{\exp(E/k_B T) - 1}. \tag{2.17}$$

This equation gives the number of photons with energy $E = h\nu$ [or — equivalently — with momentum $p = (E/c)$] in thermal equilibrium with matter. The intensity of this *thermal radiation* obtained from Eq. (2.13) is

$$\frac{dE}{dA\,dt\,d\Omega\,d\nu} \equiv B_\nu = \frac{2h\nu^3}{c^2} n(\nu) = \frac{2h\nu^3}{c^2} \frac{1}{e^{h\nu/k_B T} - 1}. \tag{2.18}$$

This is known as Planck spectrum. For the sake of practical applications, we will give the numerical results as well. The numerical value of $B_\nu$ is

$$B_\nu = 8.7 \times 10^{-7} \left(\frac{T}{1\ \mathrm{K}}\right)^3 f(x)\ \mathrm{erg\ cm^{-2}\ s^{-1}\ Hz^{-1}\ st^{-1}}; \quad f(x) = \frac{x^3}{e^x - 1} \tag{2.19}$$

and $x = h\nu/k_B T$. From Eq. (2.13) it is clear that integrating $I_\nu$ over $d\Omega$ and dividing by $c$ gives the energy density in a frequency band $U_\nu = dE/dA\,(cdt)\,d\nu$. For isotropic radiation, the integration over $d\Omega$ is same as multiplying by $4\pi$ so that $U_\nu = (4\pi/c)I_\nu$ which becomes, for thermal spectrum, $U_\nu = (4\pi/c)B_\nu$. Integrating this energy density over all $\nu$, we get the total energy density of blackbody radiation to be

$$U = \frac{8\pi^5}{15} \left(\frac{k_B T}{hc}\right)^3 k_B T \equiv aT^4 \simeq \left(\frac{T}{3400 \text{ K}}\right)^4 \text{ erg cm}^{-3}. \qquad (2.20)$$

Numerical work with these quantities can be confusing both because of the steradian factor and because of the frequency dependence. Your answer will change if you use $B_\omega$ defined with respect to $\omega = 2\pi\nu$ by a factor $2\pi$; and will change quite a bit more if you use wavelength rather than frequency and work with $B_\lambda$. One sensible way of handling this (which, unfortunately, has not yet been universally adopted), is to work with $\nu B_\nu$ which gives the intensity per logarithmic band in frequency. Quite obviously, $\nu B_\nu$ is the same as $\omega B_\omega$ and more importantly, it is the same as $\lambda B_\lambda$. In logarithmic band, we have the numerical results:

$$\nu U_\nu = \left(\frac{T}{3400 \text{ K}}\right)^4 g(x) \text{ erg cm}^{-3}; \quad g(x) = \frac{15}{\pi^4} \frac{x^4}{e^x - 1}$$

$$\nu B_\nu = 1.8 \left(\frac{T}{100 \text{ K}}\right)^4 g(x) \text{ W m}^{-2} \text{ st}^{-1}. \qquad (2.21)$$

The quantity $\nu B_\nu$ reaches a maximum value around $h\nu \approx 4k_B T$ which translates to the fact that a blackbody at 6000 K will have the maximum for $\nu B_\nu$ at 6000 Å. The maximum intensity is

$$(\nu B_\nu)_{\max} = 1.3 \left(\frac{T}{100 \text{ K}}\right)^4 \text{ Watt m}^{-2} \text{ ster}^{-1}$$

$$= 3 \times 10^{-16} T_K^4 \text{ erg cm}^{-2} \text{ s}^{-1} \text{ arcsec}^{-2}. \qquad (2.22)$$

Finally let us consider the amount of radiation which is leaking out of a small area on the surface of a blackbody (say, a star). If the energy density is $U$ [erg cm$^{-3}$], then the flux [erg cm$^{-2}$ s$^{-1}$] is given by $(dE/dtdA) = (1/4)cU$; a factor $(1/2)$ arises from counting only the flux which is going outward and the other factor $(1/2)$ comes from averaging $\cos\theta$ to get the normal component of the flux in the forward hemisphere. It is usual to write this result as $(dE/dtdA) = \sigma T^4$ where

$$\sigma = \frac{ac}{4} = \left(\frac{2\pi^5}{15}\right)\left(\frac{k_B^4}{h^3 c^2}\right) = 5.67 \times 10^{-8} \text{ W m}^{-2}\text{K}^{-4}. \qquad (2.23)$$

Using the relation $B = (c/4\pi)U$, we find that $B = (ca/4\pi)T^4 = \sigma T^4/\pi$ so that $F = \pi B$ (also see exercise 2.2). It is clear from (2.17) that there are very few photons with momentum greater than $\bar{p} \approx (k_B T/c)$, so that,

$(N/V) \approx (4\pi/3)(\bar{p}/2\pi\hbar)^3 \approx (k_BT/\hbar c)^3$ and the mean energy is $U_{\mathrm{ER}} \approx k_BT(N/V) \approx (k_BT)^4/(\hbar c)^3$.

The expression for $n(\omega) = (e^{\beta\hbar\omega} - 1)^{-1}$, with $\beta \equiv (k_BT)^{-1}$, has simple asymptotic forms for $\hbar\omega \ll k_BT$ and $\hbar\omega \gg k_BT$. The $\hbar\omega \ll k_BT$ limit corresponds to the long wavelength, classical $(\hbar \to 0)$, regime of the radiation. Equipartition of energy suggests that each mode (having two polarization states) should have energy $\epsilon_\omega = 2 \times (k_BT/2) = (k_BT)$ or $n_\omega = (\epsilon_\omega/\hbar\omega) = (k_BT/\hbar\omega)$. The quantity $n(\omega)$ does reduce to $(k_BT/\hbar\omega)$ in this limit. At these low frequencies, the intensity of thermal radiation given by Eq. (2.18) will be $B_\nu \approx (2k_BT/\lambda^2)$. Because of this relation, it is conventional to define a *brightness temperature* for any source with intensity $I_\nu$ as $T_B \equiv (\lambda^2 I_\nu/2k_B)$ which is (in general) a function of frequency.

The $\hbar\omega \gg k_BT$ limit, on the other hand, corresponds to the regime in which photons behave as particles. In such a case we expect $n_\omega = \exp(-\hbar\omega/k_BT)$ based on Boltzmann statistics. In this limit, $n_\omega \ll 1$ and quantum statistical effects are ignorable; hence we get Boltzmann statistics rather than Bose-Einstein statistics. The limiting domains $\hbar\omega \ll k_BT$ and $\hbar\omega \gg k_BT$ are called the *Rayleigh-Jeans* regime and *Wien's* regime respectively.

From our general results Eqs. (1.49) and (1.50), we can also compute the fluctuations in the energy of the photons:

$$-\frac{\partial\bar{E}}{\partial\beta} = \frac{1}{Z}\frac{\partial^2 Z}{\partial\beta^2} - \left(\frac{1}{Z}\frac{\partial Z}{\partial\beta}\right)^2 = \langle E^2 \rangle - \bar{E}^2 = (\Delta E)^2. \qquad (2.24)$$

In the case of a photon gas with $\bar{E} = \hbar\omega \left(e^{\beta\hbar\omega} - 1\right)^{-1}$, direct differentiation gives

$$(\Delta n)^2 \equiv \left(\frac{\Delta E}{\hbar\omega}\right)^2 = \left(\frac{\bar{E}}{\hbar\omega}\right)^2 + \left(\frac{\bar{E}}{\hbar\omega}\right) = n^2 + n. \qquad (2.25)$$

If photons were to be interpreted as particles, then one would expect $(\Delta n)^2 \simeq n$ giving the usual Poisson fluctuations of $(\Delta n/n) \simeq n^{-1/2}$. For this to occur it is necessary that $n \gg n^2$; that is, $n \ll 1$ which requires $\beta\hbar\omega \gg 1$. On the other hand, if $\beta\hbar\omega \ll 1$ then $n \gg 1$ and $(\Delta n)^2 \simeq n^2$, which characterizes the wave-like fluctuations. Thus, you can think of thermal photons as particles when $\hbar\omega \gg k_BT$ and as waves when $\hbar\omega \ll k_BT$.

Thermal radiation arises in astrophysics when a primary source of energy is thermalized due to some physical process, the most important example being stellar radiation. Given the luminosity $L$ and the radius $R$ of a star, its effective surface temperature is determined by the equation $L = (4\pi R^2)(\sigma T_{\text{eff}}^4)$. For the Sun, this gives $T_{\text{eff}} \approx 5500$ K; the Planckian radiation corresponding to this temperature will have peak value for $\nu B_\nu$ around $\lambda \approx 6500$ Å. The flux of radiation from such a star, located at a distance of 10 pc (where 1 pc $\simeq 3 \times 10^{18}$ cm), will be about $F = 3 \times 10^{-10}$ W m$^{-2}$. It is conventional in astronomy to use a unit called the *bolometric magnitude*, $m$, to measure the flux where $F$ and $m$ are related by $\log F \cong -8 - 0.4(m - 1)$ when $F$ is measured in W m$^{-2}$. When the flux changes by one order of magnitude, the magnitude changes by 2.5. (We will say more about magnitude scales in chapter 4.) A Sun-like star at 10 pc will have a magnitude of about 4.6.

Another important example of thermal radiation, which is of cosmological significance, is the *cosmic microwave background radiation* which is a relic from the early hot phase of the universe. It turns out that the major component of the background radiation observed in the microwave band can be fitted very accurately by a thermal spectrum at a temperature of about 2.73 K. It seems reasonable to interpret this radiation as a relic arising from the early, hot, phase of the evolving universe. The $\nu B_\nu$ for this radiation peaks at a wavelength of about 1 mm and has a maximum intensity of $7.3 \times 10^{-7}$ W m$^{-2}$ rad$^{-2}$. The intensity per square arc second of the sky is about $1.7 \times 10^{-17}$ W m$^{-2}$ arcsec$^{-2} \simeq 23$ mag arcsec$^{-2}$. We will discuss the properties of this radiation in detail in Sec. 6.6.

## 2.3   Monochromatic plane wave

Having discussed the production of radiation in the classical and quantum regimes, let us next consider a few important facets of propagation of radiation. The simplest type of classical radiation one could study corresponds to an electromagnetic wave at distances far away from the source of emission — like, for example, the light we receive from cosmic bodies. At large distances from the source, the wavefront is approximately planar and for simplicity, we will assume that the radiation is monochromatic with a frequency $\omega$.

If we take the direction of propagation to be $\mathbf{n}$, then the plane monochromatic wave is characterized by the wave vector $\mathbf{k} = (\omega/c)\mathbf{n}$. The wave vector $\mathbf{k}$ and the frequency $\omega$ form the components of a four vector just as $\mathbf{x}$ and $t$ form the components of a four vector. (You can get this directly from the fact that the phase $(\omega t - \mathbf{k} \cdot \mathbf{x})$ of a wave should be Lorentz invariant. Or, by noting that $E = \hbar\omega$ and $\mathbf{p} = \hbar\mathbf{k}$ are the energy and momentum in the photon picture.) Since the components of this four vector determine both the frequency and direction of propagation of the wave, different observers will see the wave to have different frequencies and directions of propagation.

As a specific example, consider two Lorentz frames $S$ and $S'$, with $S'$ moving along the positive $x$-axis of $S$ with velocity $v$. We will call $S'$ the rest frame and $S$ the lab frame and denote quantities in these frames by subscripts $R$ and $L$. (You might think that we got the nomenclature switched; usually some object will be moving as seen in the lab frame and if we go go to a frame which is moving with the object, it is conventional to call it the "rest" frame!) To bring out the relativistic effects as dramatically as possible, we will assume that $v$ is close to $c$ so that

$$\left(1 - \frac{v}{c}\right) = \frac{\left(1 - v^2/c^2\right)}{(1 + v/c)} \simeq \frac{1}{2\gamma^2} \tag{2.26}$$

with $\gamma \equiv \left(1 - v^2/c^2\right)^{-1/2} \gg 1$.

Let us look at a plane wave with frequency $\omega_L$ traveling along the direction $(\theta_L, \phi_L)$ in the lab frame. The four-vector $k^i$ describing the wave will have components $(\omega_L/c, \mathbf{k}_L)$ in the lab frame and $(\omega_R/c, \mathbf{k}_R)$ in the rest frame. For motion along $x-$axis, only the angle made by $\mathbf{k}$ with $x-$axis will change; the azimuthal angle is clearly invariant due to the symmetry. Writing $ck_L^x = \omega_L \cos\theta_L, ck_R^x = \omega_R \cos\theta_R$ and using the Lorentz transformations for the four vector $k^i$, we find that

$$\omega_R = \gamma\omega_L[1 - (v/c)\cos\theta_L]; \tag{2.27}$$

$$\omega_R \cos\theta_R = \gamma\omega_L[\cos\theta_L - (v/c)]. \tag{2.28}$$

These equations relate the directions of propagation and the frequencies in the two frames.

Using the Eqs. (2.27) and (2.28), we can express $\mu_R \equiv \cos\theta_R$ in terms of $\mu_L \equiv \cos\theta_L$ :

$$\mu_R = \frac{\mu_L - (v/c)}{1 - (v\mu_L/c)}. \tag{2.29}$$

Figure 2.1 shows a plot of $\mu_R$ against $\mu_L$ for two values of $v/c$. When $\mu_L = 1$ we have $\mu_R = 1$ and when $\mu_L = -1$ we have $\mu_R = -1$; that is, waves traveling along the $x-$axis appear to do so in both the frames. (The

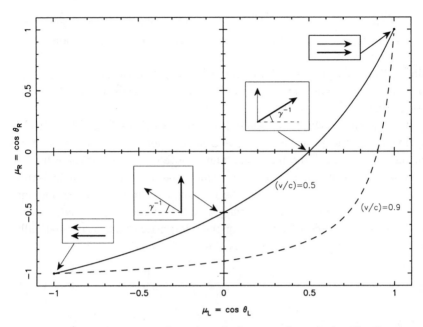

Fig. 2.1   Relation between $\cos\theta_R$ and $\cos\theta_L$ for two values of $v/c$. The directions of two waves in the rest frame and the lab frame are indicated by thin and thick arrows respectively.

directions of two waves in the rest frame and the lab frame are indicated by thin and thick arrows respectively, inside the boxes in the figure) The frequencies, however, are different in the two cases. When $\mu_L = 1$, $\omega_R = \gamma\omega_L\left(1 - (v/c)\right)$, so that $\omega_R < \omega_L$; the radiation is blue shifted as seen in the lab frame. For $\mu_L = -1$, $\omega_R = \gamma\omega_L\left(1 + (v/c)\right)$ and the radiation is red shifted.

When $\mu_L = (v/c)$, we have $\mu_R = 0$. Since $v$ is close to $c$, we expect $\mu_L = \cos\theta_L \simeq \left(1 - \theta_L^2/2\right)$ with $\theta_L \ll 1$. Comparing with Eq. (2.26) we find that $\theta_L \simeq \gamma^{-1} \ll 1$. Hence this direction of propagation is almost along the positive $x-$axis in the lab frame. But since $\mu_R = 0$, it is clear that $\theta_R = \pi/2$ and the wave was propagating *perpendicular* to the $x-$axis in the rest frame. Thus, a wave emitted in the direction perpendicular to the direction of motion will be 'turned around' to (almost) forward direction by the relativistic motion of the source. Similarly, when $\mu_L = 0$, $\mu_R = -(v/c)$. In this case, a wave which has been traveling almost along the negative $x-$axis has been "turned around" to travel orthogonal to the $x-$axis. It is clear from the arrows in the little boxes in Fig. 2.1 that the wave will appear to propagate forward in both frames only if $\theta_L < \gamma^{-1}$. This phenomenon is called relativistic beaming.

In the two cases discussed above, the frequency of the wave changes. When $\mu_L = (v/c)$, $\omega_R = \omega_L \gamma^{-1}$ and the wave is still blue shifted. When $\mu_L = 0, \omega_R = \gamma \omega_L$ and the wave is red shifted. To find the angle of propagation at which there is no frequency change, we can set $\omega_L = \omega_R$ and solve for $\mu$. This gives $\theta_c \simeq \gamma^{-1/2}$. For $\theta < \theta_c$ wave is blue shifted. A wave propagating along $\theta = \theta_c$ appears to make the "same" angle with respect to $x-$axis in both the frames.

The above discussion shows that the motion of a source "drags" the wave forward. Waves originally traveling in the forward hemisphere in the rest frame ($|\theta_R| < \pi/2$) are beamed into a small cone of opening angle $\gamma^{-1}$ around the forward direction. Further almost all the energy flowing out (except those in small cone of opening angle $\gamma^{-1}$ around the backward direction in the rest frame) are flowing into the forward hemisphere in the lab frame. Hence a charged particle, moving relativistically, will "beam" most of the radiation it emits in the forward direction. In exactly same manner, you can show that a charged particle moving relativistically through an isotropic bath of radiation will see most of the radiation as hitting it in the front. These results will be used in the study of several radiative processes later on.

So far we discussed the propagation of the wave in the vacuum. This is also a good place to introduce some aspects related to propagation of radiation *through matter*. In general, this can be described (phenomenologically) using some general parameters called emissivity, opacity *etc.*. Once these quantities are obtained from fundamental physics, the actual application to an astrophysical phenomenon proceeds along the following lines.

If the cross section for some process under consideration (which scatters or absorbs radiation) is given by $\sigma$, and the number density of scatterers is $n$, then the mean free path of a photon is $l = (n\sigma)^{-1}$. In astronomy, however, it is conventional to work with a quantity $\kappa$ called *opacity* (with units $\mathrm{cm}^2\ \mathrm{g}^{-1}$ in c.g.s. units) which is defined as:

$$\kappa = \frac{n\sigma}{\rho} \qquad (2.30)$$

where $\rho$ is the mass density of the scatterers. (The opacity which will play a key role in later chapters.) The *optical depth* of a system of size $R$ is defined to be $\tau = (R/l) \equiv \alpha R$ where $\alpha \equiv n\sigma$. From the standard theory of random walk, we know that a photon will traverse a distance $R$ with $N_c$

collisions where $R = N_c^{1/2}l$; that is, the number of collisions is given by $N_c = (R/l)^2 = \tau^2$ provided $\tau \gg 1$. (When $\tau \ll 1$, we can think of $\tau$ as the probability for the process to take place over a distance $R$.) In general $\sigma, \alpha$ etc will all depend on the frequency and we will indicate it as $\sigma_\nu, \alpha_\nu$ etc. with a subscript $\nu$ when required.

To give a flavour of the formalism based on $\alpha_\nu$, let us consider a simple representation of the propagation of radiation in a medium which is emitting and absorbing radiation. We shall ignore all the scattering effects and also any spatial dependence of the processes. Then, when the radiation travels a distance $dl$, the intensity $I_\nu$ will increase by the amount $j_\nu dl$ due to emission of photons where $j_\nu$ is called *emissivity* of the medium. The absorption of the photons will cause the intensity to decrease by the net amount $(-\alpha_\nu I_\nu dl)$ during the same time. Hence the rate of change of intensity can be described by a differential equation of the form

$$\frac{dI_\nu}{dl} = j_\nu - \alpha_\nu I_\nu \qquad (2.31)$$

if scattering is ignored. This has the solution

$$I_\nu = I_\nu(0)e^{-\alpha_\nu l} + \frac{j_\nu}{\alpha_\nu}\left[1 - e^{-\alpha_\nu l}\right]. \qquad (2.32)$$

which satisfies the boundary condition $I_\nu = I_\nu(0)$ at $l = 0$. So we are considering a situation in which some radiation of strength $I_\nu(0)$ is incident from the left on a slab of material which extends from $l = 0$ to $l = R$, say. The slab of material is characterised by an emissivity $j_\nu$ and opacity $\kappa_\nu = \alpha_\nu/\rho$. If the path length of the light through the source is $R$, then the emerging intensity at $l = R$ has two limiting forms:

$$I_\nu(R) = I_\nu(0)e^{-\alpha_\nu R} + \frac{j_\nu}{\alpha_\nu}\left[1 - e^{-\alpha_\nu R}\right] \cong \begin{cases} I_\nu(0) + j_\nu R & (\text{if } \tau \equiv \alpha_\nu R \ll 1) \\ (j_\nu/\alpha_\nu) & (\text{if } \tau \equiv \alpha_\nu R \gg 1) \end{cases}.$$
$$(2.33)$$

Equation (2.33) shows several interesting features. Let us first consider the case of $I_\nu(0) = 0$ so that there is no incident radiation and all the emission is from the slab itself. Since $\alpha_\nu^{-1}$ is the mean free path for the process (governed by the cross section $\sigma_\nu$) the condition $\tau_\nu = \alpha_\nu R \ll 1$ implies that the mean free path is much larger than the source size and the absorption is not very effective. (The source is then said to be *optically thin*.) In that case, the limiting form in Eq. (2.33) with $I_\nu(0) = 0$ gives $I_\nu = j_\nu R$ and we directly observe the source emissivity $j_\nu$ unhindered by absorption processes. In the other extreme limit, when $\alpha_\nu R \gg 1$, (called *optically thick*), the observed intensity is the ratio between the emissivity and the absorption coefficient of the source. If the emission arises from matter in thermal equilibrium, then this ratio is fixed and must be equal to the intensity of the black body spectrum, $B_\nu$, completely independent of the details of the process. (An optically thick medium in thermal equilibrium must always emit the blackbody spectrum irrespective of the specific process involved in the emissivity $j_\nu$.) We thus find the *universal* relation

$$j_\nu = B_\nu \alpha_\nu \tag{2.34}$$

between the absorption and emission characteristic at a given frequency $\nu$ for material in thermal equilibrium. For example, if we know the emission rate at a given frequency, we can determine the absorption rate — a feature which we will have occasion to use repeatedly. In thermal equilibrium, Eq. (2.33) with $I_\nu(0) = 0$ gives $I_\nu = B_\nu[1 - \exp(-\tau_\nu)]$ so that $I_\nu$ is always less than $B_\nu$. No material system *in thermal equilibrium* can emit at a given frequency $\nu$ more intensity than a blackbody can emit at the same frequency. Further, if $B_\nu$ can be approximated by the Rayleigh-Jeans Limit — as often happens in radio astronomy — then the same relation holds for the brightness temperature: $T = T_{source}[1 - \exp(-\tau_\nu)]$.

The situation becomes more interesting when we study Eq. (2.33) with $I_\nu(0) \neq 0$. Writing the solution as $I = I_\nu(0)e^{-\tau_\nu} + S_\nu(1 - e^{-\tau_\nu})$ where $\tau_\nu \equiv \alpha_\nu R$ is the optical depth through the medium, $S_\nu \equiv j_\nu/\alpha_\nu$ we find that the resulting intensity depends on relative values of $S_\nu$ and $I_\nu(0)$. Consider first a situation with $S_\nu > I_\nu(0)$ which could, for example, correspond to cool incident radiation passing through a hot gas with a source function $S_\nu$. Then Eq. (2.33) can be rewritten as $I_\nu - I_\nu(0) = (S_\nu - I_\nu(0))(1 - e^{-\tau_\nu})$. This reduces to the limiting form of $S_\nu - I_\nu(0)$ at large optical depths ($\tau_\nu \gg 1$) and $(S_\nu - I_\nu(0))\tau_\nu$ at small optical depths ($\tau_\nu \ll 1$). When the optical depth is small, one sees the continuum background with bright emission lines at those frequencies for which $S_\nu\tau_\nu$ is large. These intensities will increase with $\tau_\nu$ and eventually saturate at $S_\nu$ at large optical depths. In this case, we always have $I_\nu < S_\nu$.

The other limit of $S_\nu < I_\nu(0)$ corresponds to hot radiation propagating through cool gas which, for example, occurs in the case of Sun. Now $I_\nu - S_\nu = (I_\nu(0) - S_\nu)e^{-\tau_\nu}$ showing $S_\nu < I_\nu < I_\nu(0)$. In this case, we get absorption lines which dips below the continuum background of $I_\nu(0)$ at small optical depths; as the optical depth increases, the absorption becomes

more prominent and in the saturated case, we detect the intensity $S_\nu$ due
to the foreground.

The opacity of matter to the flow of radiation also measures the amount
of radiation that is absorbed by matter. When this happens, the radiation
will exert a pressure on matter depending on the opacity. To obtain an
expression for the radiative pressure, we note that if the flux of radiation
energy is $\mathcal{F}$ the corresponding momentum flux is $\mathcal{F}/c$. Since the momentum
absorbed by the matter per second is the force, which should also be equal
to the pressure gradient of the radiation, we get

$$\left(\frac{\mathcal{F}}{c}\right)n\sigma = \mathcal{F}\left(\frac{\rho\kappa}{c}\right) = -\nabla P_R = -\frac{4}{3}aT^3\nabla T \qquad (2.35)$$

where the first relation comes from Eq. (2.30) and the last relation comes
from $P_R = (1/3)aT^4$ for the radiation. (The sign shows that the flux is
from the high temperature region to the low temperature region.) This
allows one to relate the temperature gradient in matter to the radiation
flux

$$\nabla T = -\frac{3}{4}\frac{\rho\kappa}{ac}\frac{1}{T^3}\mathcal{F} \qquad (2.36)$$

which will be of use in the study of stellar structure (see chapter 4).

As a simple application of the radiation pressure arising from the opacity, consider the force exerted by the radiation emitted by a body of luminosity $L$ on matter around it. Since $\mathcal{F} = (L/4\pi r^2)$ at a distance $r$ from the
body, matter will be pushed out if the radiative force exceeds the gravitational attraction. The condition for this *not* to happen (so that matter *can*
accrete onto the central mass) is

$$\frac{L}{4\pi r^2}\left(\frac{\kappa\rho}{c}\right) < \frac{GM}{r^2}\rho \qquad (2.37)$$

which translates to the condition on luminosity

$$L < L_E = \left(\frac{4\pi Gc}{\kappa}\right)M. \qquad (2.38)$$

This quantity, $L_E$, is called *Eddington luminosity* and is the maximum possible luminosity that can be achieved in simple accretion process. (Strictly
speaking, astronomers define $L_E$ with $\kappa$ arising due to scattering by free
electrons in the medium; but, of course, the above argument is valid for
any $\kappa$.) It plays a crucial role in the study of quasars and active galactic
nuclei.

## 2.4 Astrophysical radiative processes

We shall now apply these considerations to some specific astrophysical processes in which radiation is emitted due to different acceleration processes. Among the classical (or nearly classical) radiation processes, the most important ones are thermal bremsstrahlung, synchrotron radiation and inverse Compton scattering with which we will begin our discussion.

### 2.4.1 *Thermal bremsstrahlung*

As a first example, let us study the radiation emitted by a fully ionised hydrogen plasma at temperature $T$, when the electrons are constantly scattered by the protons. In a scattering between an electron of mass $m_e$ and a proton, with an impact parameter $b$ and relative velocity $v$ the acceleration of the electron is $a \approx (q^2/m_e b^2)$ and lasts for a time $(b/v)$. Using Eq. (2.7) we can estimate the total energy emitted in such an encounter to be:

$$\mathcal{E} \approx (q^2 a^2/c^3)(b/v) \approx (q^6/c^3 m_e^2 b^3 v) \approx (q^6 n_i/c^3 m_e^2 v) \qquad (2.39)$$

since $b \approx n_i^{-1/3}$ on the average. (This ignores distant encounters and a more precise calculation will require integrating over all impact factors leading to a logarithmic correction, as in Eq. (1.64). This is, however, adequate for our purpose.) The factor $m_e^2$ in the denominator explains why we can ignore the radiation emitted by the ions in the scattering, since it is sub-dominant by a factor $(m_e/m_p)^2$. We have also ignored the scattering of electrons by electrons and ions by ions. The reason for concentrating on electron-ion collisions is the following: For a system of *non relativistic* charges with *same* charge-to-mass ratio, $(q/m)$, the second derivative of the dipole moment

$$\ddot{\mathbf{d}} = \sum q_i \ddot{\mathbf{x}}_i = \sum \left( \frac{q_i}{m_i} \right) m_i \ddot{\mathbf{x}}_i = \frac{q}{m} \sum \dot{\mathbf{p}}_i, \qquad (2.40)$$

vanishes due to momentum conservation. Hence the dipole radiation from electron-electron collision and ion-ion collision will be sub dominant to the effects from electron-ion collisions as long as the system is non relativistic. (You can't write $\dot{\mathbf{p}}_i = m_i \ddot{\mathbf{x}}_i$ for relativistic particle; we will not, however, discuss relativistic plasmas.)

The total energy radiated per unit volume will be $n_e \mathcal{E}$. Since each collision lasts for a time $(b/v)$, there will be very little radiation at frequencies greater than $(v/b)$. For $\omega < (v/b)$, we may take the energy emitted per unit frequency interval to be nearly constant. Further, in the case of plasma in thermal equilibrium, $v \cong (k_B T/m_e)^{1/2}$. Putting all these together, we get

$$j_\omega \equiv \left(\frac{d\mathcal{E}}{d\omega dt dV}\right) \simeq (m_e c^2) r_0^3 \left(\frac{m_e c^2}{k_B T}\right)^{1/2} n_e n_i = K n^2 T^{-1/2}; \quad (\hbar\omega \lesssim k_B T) \tag{2.41}$$

where $r_0 \equiv q^2/m_e c^2$ is called the *classical electron radius* and $K = (m_e c^2) r_0^3 (m_e c^2/k_B)^{1/2}$ is just a constant. This process is called *thermal bremsstrahlung*. The bremsstrahlung spectrum is nearly flat for $0 < \omega \lesssim (k_B T/\hbar)$ and will fall rapidly for $\omega \gtrsim (k_B T/\hbar)$, where the upper limit comes from the fact that an electron with a typical energy of $(k_B T)$ cannot emit photons with energy higher than $(k_B T/\hbar)$. A more precise expression is given by

$$j_\nu = K g(\nu, T, Z) Z^2 n_e n_i T^{-1/2} \exp\left(-\frac{h\nu}{k_B T}\right) \tag{2.42}$$

where $g$ is called the *Gaunt factor* and $Z^2$ takes care of the ionic charge being $Zq$ in case the gas is not hydrogen.

The total energy radiated, over all frequencies, from such a plasma can be found by integrating Eq. (2.41) over $\omega$ in the range $(0, k_B T/\hbar)$. This gives

$$\left(\frac{d\mathcal{E}}{dt dV}\right) = \int_0^{(k_B T/\hbar)} d\omega \left(\frac{d\mathcal{E}}{d\omega dt dV}\right) \simeq (m_e c^2) r_0^3 \frac{m_e c^2}{\hbar} \left(\frac{k_B T}{m_e c^2}\right)^{1/2} n_e n_i$$

$$= 1.4 \times 10^{-27} n_e^2 T^{1/2} \text{ erg s}^{-1}\text{cm}^{-3}. \tag{2.43}$$

The bremsstrahlung radiation emitted by the plasma takes away the energy of the system and thus can be an effective cooling process at high temperatures. The characteristic time scale for cooling of a thermal plasma is given by the the ratio between the thermal energy density $n_e k_B T$ and the energy lost per second per unit volume of the plasma obtained above. This gives

$$t_{\text{cool}} = \frac{n_e k_B T}{(d\mathcal{E}/dt dV)} \simeq \left(\frac{1}{n_e r_0^2 c}\right) \alpha^{-1} \left(\frac{k_B T}{m_e c^2}\right)^{1/2} \propto n_e^{-1} T^{1/2} \tag{2.44}$$

where $\alpha \equiv (q^2/\hbar c)$ is the *fine-structure constant*. Putting in numbers, we get

$$t_{\text{cool}} \sim \frac{n_e k_B T}{1.4 \times 10^{-27} T^{1/2} n_e^2} \simeq 3 \left(\frac{T}{10^8 \, \text{K}}\right)^{1/2} \left(\frac{n_e}{0.01 \, \text{cm}^{-3}}\right)^{-1} \text{ Gyr.} \tag{2.45}$$

Thermal bremsstrahlung occurs in several astrophysical systems leading to radiation in completely different wave-bands. One application of our result is in the case of regions around hot stars which are completely ionized by the UV photons from the star. Such a local island of plasma in the interstellar medium (ISM) is called a *HII region*.[1] They exist in the spiral arms of our galaxy and emit thermal bremsstrahlung radiation with a relatively flat spectrum in the radio band. For example, the HII regions in Orion have $T \simeq 10^4$ K, $n_i \simeq 2 \times 10^3$ cm$^{-3}$ and an effective line of sight thickness of $6 \times 10^{-4}$ pc. From Eq. (2.41) we can estimate the flux to be about $3 \times 10^{-21}$ erg m$^{-2}$ s$^{-1}$ Hz$^{-1}$ = 300 Jy where 1 Jy = $10^{-26}$ W m$^{-2}$ Hz$^{-1}$. This is typical of emission from HII regions.

Another prominent source of bremsstrahlung radiation is the hot gas in clusters of galaxies. Taking the cluster gas to be fully ionized hydrogen with a mass of about $M_{\text{gas}} = 10^{48}$ gm spread over a sphere of radius $R \simeq 3$ Mpc, one can estimate the mean number density of ions and electrons as $n_e \approx (M_{\text{gas}}/m_p V) \approx 10^{-4}$ cm$^{-3}$ where $V$ is the volume of the cluster. The virial temperature of the gas is $k_B T \simeq (GMm_p/R)$ giving $T \simeq 10^8$ K. Using Eq. (2.43) and the estimated values of $n_e$ and $T$, we get the total luminosity from the whole cluster to be $\mathcal{L} = LV \approx 5 \times 10^{44}$ erg s$^{-1}$. This radiation will be in the wave-band corresponding to the temperature of $10^8$ K which is in x-rays. Several such x-ray emitting clusters have been observed. In fact, the gas mass within about 500 kpc of the center of the cluster is fairly tightly correlated with the x-ray luminosity: $M_{\text{gas}}(< 500 \text{ kpc})/10^{14} M_\odot \approx (0.3 L_x/10^{45} \text{ erg s}^{-1})$.

Using Eq. (2.45) we can also estimate the cooling time for the X-ray emitting gas in such a cluster. While the mean number density of electrons computed above is about $10^{-4}$ cm$^{-3}$, the central densities can be higher by a factor of few hundreds giving $n_e \simeq 0.01$ cm$^{-3}$ which is used in Eq. (2.45). This will make central regions of the cluster to cool rapidly (within a lifetime of a few Gyr) thereby reducing the central pressure support. This, in turn, could cause flow of gas from the outer regions towards the center. Investigation related to such *cooling flows* (including the question of whether they really exist!) has been a major area of research in the study of clusters.

---

[1]This is a good place to let you know of one of the many crazy notations used by astronomers. The symbol HI stands for neutral hydrogen while HII stands for singly ionized hydrogen — that is, proton. Similarly CIV, for example, will stand for triply ionized carbon and CI will represent neutral carbon atom — which, mercifully, is not often used. In general, if you see (element symbol) (Roman numeral for n) you should take it to mean the element with (n-1) electrons removed.

We can use our results to study another important process called the *free-free absorption* which is a key source of opacity in plasma. This is the 'time reversed' process of bremsstrahlung, in which a photon is absorbed by an electron while in the proximity of an ion. In equilibrium, the rate for this free-free absorption should match that of thermal bremsstrahlung. We write the free-free absorption rate as $n\sigma_{ff}B(\nu)$ where $\sigma_{ff}$ is the free-free absorption cross section and $B(\nu) \propto \nu^3(e^{\hbar\nu/k_BT} - 1)^{-1} \propto \nu^2 T$ (when $\hbar\nu \lesssim k_BT$ which is usually the relevant range in the astrophysical contexts) is the intensity of thermal radiation. Equating this to the bremsstrahlung emissivity found in Eq. (2.41), $j_\nu \propto n^2 T^{-1/2}$, we get $\sigma_{ff} \propto (n/\nu^2 T^{3/2})$. With all the numerical factors added, this result corresponds to $\alpha_\nu = n\sigma_{ff}$ in the Rayleigh-Jeans regime given by:

$$\alpha_\nu \approx \frac{r_0^3}{2}\left(\frac{m_e c^2}{k_B T}\right)^{3/2}\frac{c^2}{\nu^2}n^2 \equiv Cn^2 T^{-3/2}\nu^{-2}. \qquad (2.46)$$

This result shows that bremsstrahlung photons emitted by a source can be self absorbed at small frequencies at which $\alpha \propto \nu^{-2}$ can be large. More specifically, consider a source of size $R$ at a distance $d$ emitting bremsstrahlung radiation. Using the result in Eq. (2.32), we find the flux to be

$$F_\nu = 2k_B T\left(\frac{\nu^2}{c^2}\right)\left(\frac{R}{d}\right)^2\left(1 - \exp\left[-CT^{-3/2}n^2\nu^{-2}R\right]\right). \qquad (2.47)$$

At low frequencies, the exponential term vanishes and $F_\nu \propto \nu^2$; at high frequencies, the exponential can be expanded in a Taylor series and the leading term will cancel the $\nu^2$ in the front leading to a constant flux independent of $\nu$. The turn over frequency is $\nu_c^2 = CT^{-3/2}n^2 R$ which can be determined entirely in terms of source parameters. Such a turnover in the spectrum with $\nu^2$ dependence changing over to constant flux has been observed in several astrophysical sources.

In most cases of interest, we can take the typical frequency of the photon involved in the process to be proportional to $T$, so that $\sigma_{ff} \propto nT^{-3.5}$. Adding the numerical constants, the corresponding opacity can then be written in the form

$$\kappa_{ff} \propto \rho T^{-3.5} \simeq 10^{23}\rho T^{-3.5} \text{ cm}^2 \text{ gm}^{-1}. \qquad (2.48)$$

This is usually called *Kramer's opacity* and plays a key role in stellar modeling.

## 2.4.2 Synchrotron radiation

Another major source of acceleration for charged particles is the magnetic fields hosted by plasmas. The radiation emitted by this process, called *synchrotron radiation*, can be estimated as follows: Consider an electron moving with velocity **v** in a magnetic field **B**. To use Eq. (2.7) we need to estimate the acceleration *in the instantaneous* rest frame of the particle. In that frame, the magnetic force $(q/c)[\mathbf{v} \times \mathbf{B}]$ is zero; but the magnetic field in the lab frame will lead to an electric field of magnitude $E' = (v/c)\gamma B$, in the instantaneous rest frame of the charge, inducing an acceleration $a' = (qE'/m_e)$. Accelerated by this field, the charged particle will radiate energy at the rate (which is the same in the rest frame and lab frame)

$$\frac{d\mathcal{E}}{dt} = \left(\frac{d\mathcal{E}'}{dt'}\right) = \frac{2}{3}\frac{q^2}{c^3}(a')^2 = \frac{2}{3}\frac{q^2}{c^3}\cdot\frac{q^2}{m_e^2}\frac{v^2}{c^2}\gamma^2 B^2. \tag{2.49}$$

Thus the power radiated by an electron of energy $\epsilon = \gamma m_e c^2$ scales with the energy as $(d\mathcal{E}/dt) \propto v^2\epsilon^2 B^2$. For slow speeds $(v \ll c)$, the energy $\epsilon \simeq m_e c^2$ is a constant and the power is proportional to the kinetic energy of the particle; at relativistic speeds, $v/c \approx 1$ and the power is proportional to the *square* of the particle energy. Hereafter, we shall consider the extreme relativistic case with $v/c \approx 1$ and $\gamma \gg 1$.

Since the energy density in the magnetic field is $U_B = (B^2/8\pi)$ we can also write the above result in a more suggestive form as:

$$\left(\frac{d\mathcal{E}}{dt}\right) = \frac{16\pi}{3}\left(\frac{q^2}{m_e c^2}\right)^2 \gamma^2 c U_B \simeq (\sigma_T c U_B)\gamma^2 \tag{2.50}$$

where $\sigma_T \equiv (8\pi/3)(q^2/m_e c^2)^2$ is called the *Thomson scattering cross section* [which we will discuss in the next section; see Eq. (2.61)]. The time-scale for energy loss by this process can be found by dividing the energy of the electron $(\gamma m c^2)$ by $(d\mathcal{E}/dt)$; this gives

$$t_{\text{sync}} \equiv \frac{\gamma m c^2}{(d\mathcal{E}/dt)} \simeq 5 \times 10^8 \text{s} \left[\gamma^{-1} B_{\text{gauss}}^{-2}\right] \cong 10^{10} \text{ yr} \left(\frac{B}{10^{-6}\text{G}}\right)^{-2}\left(\frac{\epsilon}{\text{GeV}}\right)^{-1}. \tag{2.51}$$

for the extreme relativistic case. (More formally, Eq. (2.50) for a single electron gives $(d\epsilon/dt) \propto \epsilon^2$ which integrates to give $\epsilon(t) = \epsilon(0)(1 + t/\tau)^{-1}$ where $\tau = 3m_e^2 c^3/4\sigma_T\epsilon(0)U_B$. This is essentially the same time scale as in Eq. (2.51)).

Let us next consider the *spectrum* of synchrotron radiation from a highly relativistic charged particle. For a non relativistic particle spiraling in a magnetic field, we have $(mv^2/r_L) = qB(v/c)$ where $r_L$, called the *Larmor radius*, is the the radius of the orbit. The characteristic angular frequency is $\omega_L = (r_L/v) = (qB/m_e c)$. For relativistic motion with constant $v^2$, the effective mass is $m_e(1 - v^2/c^2)^{-1/2} = m_e\gamma$ so that the corresponding angular frequency becomes $\omega_L = (qB/mc\gamma) = (qcB/\epsilon)$ for a particle of energy $\epsilon$ in a magnetic field field $B$. Based on this consideration, one would have expected the synchrotron radiation to peak at this frequency $\omega_L$ but actually the synchrotron radiation from an extreme relativistic particle peaks at the frequency

$$\omega_c \approx \omega_L \gamma^3 \propto B\gamma^2 \propto B\epsilon^2 \qquad (2.52)$$

where the extra factor $\gamma^3$ arises from special relativistic effects. One factor of $\gamma$ arises purely from time dilation factor while the factor $\gamma^2$ arises from the fact that radiation is beamed mostly in the forward direction. This latter factor can be understood as follows: From the discussion in Sec. 2.3, we know that most of the radiation from a relativistically moving particle is emitted in the forward direction into a cone with opening angle

$$\Delta\theta \simeq \left(1 - \frac{v^2}{c^2}\right)^{1/2} \simeq \gamma^{-1} = \frac{mc^2}{\epsilon}. \qquad (2.53)$$

An observer will receive the radiation only when the direction of observation is within a cone of angular width $\Delta\theta$. To relate the time of arrival of radiation, $t$, to the angle $\theta$, let us consider the radiation *emitted* at two instances $t_1 = 0$ and $t_2 = t$ when the charged particle was at positions $\theta_1 = 0$ and $\theta_2 = \theta$. Clearly $r_B\theta = v(t_2 - t_1) = vt$ giving $t = (r_B\theta/v)$. The arrival time of the pulses at the observer will be less by the amount of time $(r_B\theta/c)$ taken by the radiation to travel the extra distance $(r_B\theta)$. So $t_{obs} \simeq t - r_B\theta/c = (r_B\theta/v)(1 - v/c)$. Hence

$$\theta(t_{obs}) \simeq \frac{vt_{obs}}{r_B}\frac{1}{(1 - v/c)} \simeq \frac{2vt_{obs}}{r_B}\gamma^2 \propto \gamma^2. \qquad (2.54)$$

This gives an extra factor of $\gamma^2$ in Eq. (2.52) which essentially arises from the relativistic beaming. The exact spectrum for synchrotron, $F(\omega/\omega_c)$, behaves as

$$F(\omega/\omega_c) = \begin{cases} \exp(-\omega/\omega_c) & (\text{for } \omega \gg \omega_c) \\ (\omega/\omega_c)^{1/3} & (\text{for } \omega \ll \omega_c.) \end{cases} \qquad (2.55)$$

In fact, most of the energy will be emitted at frequencies close to $\omega_c \propto B\epsilon^2$ where $\epsilon$ is the energy of the charged particle. It follows that $\epsilon \propto \omega_c^{1/2}B^{-1/2}$. Numerically,

$$\epsilon \approx 4\nu_c^{1/2}B^{-1/2}\,\text{erg} \qquad (2.56)$$

when all quantities are in c.g.s. units and $\nu = \omega/2\pi$. Using this relation, Eq. (2.51) can be expressed as

$$t_{\text{sync}} \cong 5 \times 10^8 \text{ yr} \left(\frac{B}{10^{-6}\text{G}}\right)^{-3/2} \left(\frac{\nu_c}{1\text{ GHz}}\right)^{-1/2}. \qquad (2.57)$$

This is lifetime (of a synchrotron source emitting with the peak frequency $\nu_c$), over which the charged particles would have lost most of their energy, unless other physical processes replenish it.

Let us now move on from the radiation emitted by a single electron to that emitted by a bunch of electrons with a distribution in the energies. If $n(\epsilon)d\epsilon$ denotes the number of electrons with energies in the range $(\epsilon, \epsilon + d\epsilon)$ then the total radiation emitted by this bunch of electrons will be $j_\omega \propto (B^2\epsilon^2)(n(\epsilon))(d\epsilon/d\omega)$ where the first factor $B^2\epsilon^2$ is the energy emitted by a single electron, the second factor is the number of electrons with energy $\epsilon$ and the last factor is the Jacobian $(d\epsilon/d\omega) \propto \epsilon^{-1} \propto \omega^{-1/2}$ from $\epsilon$ to $\omega$. If the spectrum of electrons is a power law $n(\epsilon) = C\epsilon^{-p}$, then the radiation spectrum will be

$$j_\nu \approx \frac{e^3}{m_ec^2}\left(\frac{3e}{4\pi m_e^3 c^5}\right)^{(p-1)/2} C\,B^{(p+1)/2}\,\nu^{-(p-1)/2}, \qquad (2.58)$$

where we have reintroduced all the constants. (A more precise calculation multiplies the expression by a $p$ dependent factor which is about 0.1.) This leads to a power law spectrum $j_\nu \propto \nu^{-\alpha}$ with $\alpha = (p-1)/2$ allowing us to relate the observed spectral index $\alpha$ to the spectral index $p$ of the particles producing the radiation. This is a standard procedure used to explain sources which are observed with a power law spectrum.

Several discrete sources (supernova remnants, radio galaxies, quasars, ...) emit radio waves essentially through synchrotron process. As a specific example, consider the diffuse radio background in our galaxy due to synchrotron radiation from electrons in the interstellar medium. Observations indicate that the spectrum of the electrons is given by $(dN/dE) \simeq 3 \times 10^{-11}E^{-3.3}$ particles cm$^{-3}$ if $E$ is measured in GeV. If the magnetic field is about $6 \times 10^{-6}$ Gauss, Eq. (2.58) will predict a volume emissivity of

$$j_\nu \simeq 3 \times 10^{-38} \left( \frac{\nu}{10 \text{ MHz}} \right)^{-1.1} \text{erg s}^{-1} \text{ cm}^{-3} \text{ Hz}^{-1}. \qquad (2.59)$$

This is close to the observed background emission. Over a line-of-sight of 1 kpc, this will give a flux of $10^{-20}$ W m$^{-2}$ rad$^{-2}$ Hz$^{-1}$. Similar power law radio spectra are seen in the case of radio galaxies, quasars *etc.* and are thought to be due to synchrotron radiation.

An interesting application of synchrotron cooling is to the Crab nebula which is a supernova remnant that exhibits power law spectrum $F_\nu \propto \nu^{-\alpha}$ in radio frequencies. The spectrum has a slope $\alpha \approx 0.3$ for $\nu < 10^3$ GHz and a steeper slope of $\alpha \approx 0.8$ for $\nu > 10^3$ GHz. The flux is about $10^3$ Jansky at $\nu = 1$ GHz. This observed "break" in the spectral index (i.e. the change from $\alpha = 0.3$ to $\alpha = 0.8$) can be nicely understood in terms of a model in which a power law spectrum of electrons with the distribution $n(\epsilon) \propto \epsilon^{-p}$ are injected at $t = 0$ and are continuously losing energy due to synchrotron radiation till today $(t = T)$. If the synchrotron loss time is $t_{\text{sync}}$, then for $t_{\text{sync}} \gg T$, we expect the spectrum to retain the original shape with $n(\epsilon) \propto T\epsilon^{-p}$. On the other hand, at high energies at which $t_{\text{sync}} \ll T$, the equilibrium between injected energy and radiation loss will require $n(\epsilon) \propto t_{\text{sync}}\epsilon^{-p}$. But since $t_{\text{sync}} \propto \epsilon^{-1}$ (see Eq. (2.51)), we find that $n(\epsilon) \propto \epsilon^{-(p+1)}$ at high $\epsilon$, i.e., the electron spectrum steepens by unity. Since the spectral index of synchrotron emission is $\alpha = (p - 1/2)$ (see Eq. (2.58)), it steepens by 0.5 which matches with the observations. (While this idea is reasonable, it is correct only within an order of magnitude. This model predicts that the break frequency $\nu_b$ should occur at a point where $t_{\text{sync}} = T$. Setting $\nu_b$ to the characteristic frequency of synchrotron radiation, one can use this relation to estimate the age of the Crab nebula and we get $T \approx 2000$ years, which is somewhat on the higher side.)

These calculations assume that the radiation emitted by the source reaches us without any modification. In general, some of the radiation will be absorbed — especially at low frequencies — thereby modifying the spectrum. For thermal bremsstrahlung we found that this effect leads to a $\nu^2$ dependence at low frequencies (see Eq. (2.47)). In the case of synchrotron radiation, absorption makes the low frequency spectrum acquire a universal form, with $j_\nu \propto \nu^{5/2}$. This scaling has a simple physical interpretation. You will recall that the low frequency thermal radiation has the form $B_\nu \approx (2\nu^2/c^2)k_B T$ in which the first factor $(2\nu^2/c^2)$ comes from phase space and the second factor $(2 \times (1/2)k_B T)$ comes from the mean energy of the photons. In the case of a synchrotron source, the phase space factor, of course, remains the same but $k_B T$ should be replaced by the mean energy $\epsilon$ of a synchrotron electron emitting radiation at the frequency $\nu$. From Eq. (2.56) $\epsilon \propto \nu^{1/2} B^{-1/2}$ giving

$$I_\nu \propto B_0^{-1/2} \nu^{5/2} \approx \left(\frac{2\nu^2}{c^2}\right)\left(\frac{\nu}{\nu_L}\right)^{1/2} m_e c^2 \qquad (2.60)$$

which explains the $(5/2)$ power. In the last equality we have reintroduced the constants of proportionality and expressed the result in terms of the Larmor frequency $\nu_L$.

Synchrotron radiation will be linearly polarized with a high degree of polarization. To see this, consider the radiation emitted by a single electron circling in a magnetic field. When the electron is viewed in the orbital plane, the radiation is hundred percent linearly polarized with the electric vector oscillating perpendicular to the magnetic field. When viewed along the direction of the magnetic field, the radiation is circularly polarized. (If it is right handed circular polarization when viewed from the top, it will be left handed circular polarization when seen from the bottom). Consider now the relativistic motion of the electron which beams the radiation within a narrow cone in the direction of motion. The radiation is now mostly confined to the plane of the orbit and, in any realistic situation, there will be a bunch of charged particles with a distribution in the pitch angle. In that case, the two components of the circular polarization will effectively cancel while the the linear polarization will survive to a significant extent. Hence the radiation will be linearly polarized to a high degree. For example, the radiation at $\nu \simeq 10^{-3}\nu_c$ will be about 50 per cent polarized while the rare photons with $\nu \approx 10\nu_c$ will be almost completely polarized. If the emission arises from a power law spectrum of charged particles with index $p$, then the degree of polarization can be computed to be $(p+1)/(p+7/3)$.

### 2.4.3 *Inverse Compton scattering*

Another important process by which astrophysical systems generate high energy photons is by scattering of photons by highly relativistic electrons. In such a process, significant amount of energy can be transfered from the electrons to low energy photons thereby converting them into high energy photons. We will now see how this comes about.

The simplest form of scattering of photons by charged particles occurs when a charged particle — originally at rest — is accelerated by an electromagnetic wave which is incident on it. Consider a charge $q$ placed on an electromagnetic wave of amplitude $E$. The wave will induce an acceleration $a \simeq (qE/m)$ causing the charge to radiate. The power radiated will be $P = \left(2q^2 a^2/3c^3\right) = \left(2q^4/3m^2 c^3\right) E^2$. Since the incident power in the electromagnetic wave was $S = \left(cE^2/4\pi\right)$, the scattering cross section (for electrons with $m = m_e$) is

$$\sigma_T \equiv \frac{P}{S} = \frac{8\pi}{3} \left(\frac{q^2}{m_e c^2}\right)^2 \approx 6.7 \times 10^{-25} \text{cm}^2 \qquad (2.61)$$

which is called the *Thomson scattering cross section*. This cross section governs the basic scattering phenomena between charged particles and radiation. The corresponding mean free path for photons through a plasma will be $l_T = (n_e \sigma_T)^{-1}$ and the *Thomson scattering opacity*, defined to be $\kappa_T \equiv (n_e \sigma_T/\rho)$, is

$$\kappa_T = \left(\frac{n_e}{n_p}\right) \left(\frac{\sigma_T}{m_p}\right) = 0.4 \text{ cm}^2 \text{ gm}^{-1} \qquad (2.62)$$

for ionized hydrogen with $n_e = n_p$.

Incidentally, one key feature of any scattered electromagnetic radiation is that it will be polarized even if the original radiation is not. This feature can be understood fairly easily: Consider the electromagnetic radiation propagating along the positive $z-$axis and hitting an electron near the origin. If the incident radiation is unpolarized, its electric field vector could be anywhere in the $x-y$ plane. Under its influence, the charged particle will oscillate in the direction of the electric field vector in the $x-y$ plane. When viewed from a direction making angle $\theta$ with the $z-$axis, only the motion of the charge perpendicular to the line of sight will lead to the radiation field making the scattered radiation in the direction $\theta$ to be partially polarized. In fact, the radiation in the direction $\theta = \pi/2$, say along the $x-$axis, will be totally polarized. This is because, when viewed along the $x-$axis, only the component of the motion of the charged particle in the $y$ direction will contribute to radiation. (If you look at the sky at a direction 90° away from the Sun through a polaroid, you can easily verify that the scattered sunlight is totally polarized; it is the same effect.) In star forming regions, one can often map the polarization of light scattered by the interstellar cloud and verify this effect.

Let us proceed further and study the case in which a bunch of *highly relativistic* electrons are interacting with a radiation field of photons. We can actually use the Thomson scattering to study this case as well by the following trick: We first make a Lorentz transformation to the instantaneous rest frame of the electron, study the scattering using the result obtained above and then transform back to the lab frame.

In the lab frame, the radiation field may be thought of as a random superposition of electromagnetic waves with $\langle E^2 \rangle = \langle B^2 \rangle = 8\pi U$ where $U$ is the energy density. If a charged particle moves through this field with a velocity $\mathbf{v}$, then, in the rest frame of the charge, the electric field will have a magnitude $E' \simeq \gamma E$. Accelerated by this field, the charged particle will radiate energy at the rate

$$\frac{d\mathcal{E}}{dt} = \frac{d\mathcal{E}'}{dt'} \simeq \frac{q^2 a'^2}{c^3} \simeq \left(\frac{q^2}{mc^2}\right)^2 c\gamma^2 E^2 \simeq \sigma_T c U \gamma^2. \qquad (2.63)$$

The net effect is to convert a photon at frequency $\nu_0$ to a frequency $\gamma^2 \nu_0$ and is quite dramatic if $\gamma \gg 1$. Since $\gamma \approx 10^3$ for a 0.5 GeV electron, this process can convert a radio ($10^9$ Hz) photon to a UV ($10^{15}$ Hz) photon or a far infrared ($3 \times 10^{12}$ Hz) photon into an x-ray ($3 \times 10^{18}$ Hz) photon. You can see why *inverse Compton scattering* — as it is called — is very important in high energy astrophysics.

The lifetime for inverse Compton loss to significantly reduce the energy of the charged particle will be

$$\tau = \frac{E}{d\mathcal{E}/dt} \cong \frac{E}{\sigma_T c \gamma^2 U}. \qquad (2.64)$$

As an example, consider a 100 GeV electron moving through the cosmic background radiation, which has a temperature $T = 2.73$ K and the energy density $U_{CBR} = aT^4 \simeq 4 \times 10^{-13}$ erg cm$^{-3}$. Equation (2.64) shows that its life time is $\tau \lesssim 10^7$ yr. If you catch such an electron, it should have originated from within a region of size $c\tau \simeq 3$ Mpc.

Equation (2.63) is similar in form to Eq. (2.50) and the correspondence between the two processes runs deep. Just as in the case of synchrotron radiation, here also the spectral form is $\nu^{-(p-1)/2}$ if the electrons have a power law distribution in energy with index $p$. There are several astrophysical sources in which relativistic electrons of a given spectrum will be responsible for *both* synchrotron radiation and inverse Compton scattering at different frequencies. The electrons in the presence of a magnetic field will produce synchrotron photons (which could be, for example, in radio band) which will undergo inverse Compton scattering with the electrons thereby producing high energy radiation in another band, say x-rays. In general, when a radiation field and magnetic field coexist, the ratio between the power radiated by inverse Compton process and synchrotron process will be in the ratio $f = (U_{\text{sync}}/U_B)$ where $U_{\text{synch}}$ is the energy density of primary synchrotron photons.

In fact, one can envisage a situation in which the first generation photons produced by inverse Compton process undergoes a further inverse Compton scattering to produce second generation of high energy photons *etc.* Obviously this will increase the photon density as $(1 + f + f^2 + ...)$ which will converge only for $f < 1$. This leads to an interesting bound on the brightness temperature of a region in which the process takes place. To make an order of magnitude estimate of the maximum brightness temperature possible, let us first note that the flux of synchrotron radiation from a source of size $R$ at a distance $d$ can be written in two different forms as:

$$cU_{\text{sync}}\frac{R^2}{D^2} = \nu I_\nu \frac{\pi R^2}{D^2} \approx \nu_c \frac{2k_B T_b}{\lambda_c^2}\frac{\pi R^2}{D^2} \qquad (2.65)$$

where we have evaluated the synchrotron flux as $\nu I_\nu$ at the critical frequency at which the flux is maximum and has also introduced the brightness temperature through the relation $I = 2k_B T/\lambda^2$. This gives us the energy density of synchrotron photons in terms of the brightness temperature to be $U_{\text{sync}} \approx (2\pi\nu_c^3/c^3)k_B T_b$. On the other hand, the energy density of magnetic field is $U_B = B^2/8\pi$. We can first eliminate $B$ in terms of $\nu_c$ using Eq. (2.52), $\nu_c = (qB/2\pi m_e c)\gamma^2$ and express $\gamma$ in terms of the brightness temperature by $\gamma m_e c^2 \approx 3k_B T_b$. This gives the energy density of the magnetic field in terms of the brightness temperature to be $U_B \approx (\pi m_e^6 c^{10}/162q^2 k_B^4)(\nu_c^2/T_b^4)$. Thus we find that

$$f = \frac{U_{\text{sync}}}{U_B} = \left(\frac{324q^2 k_B^5}{c^{13}m_e^6}\right)\nu_c T_b^5. \qquad (2.66)$$

The condition for instability $f > 1$ translates numerically to

$$\left[\frac{T_b}{10^{12}\text{ K}}\right]^5 \left(\frac{\nu_c}{10^{8.5}\text{ Hz}}\right) > 1. \qquad (2.67)$$

It follows that if the brightness temperature greatly exceeds $10^{12}$ K at radio frequencies, the inverse Compton losses will be very high leading to a bound of $T_b \lesssim 10^{12}$ K for any radio source. The same result can be expressed as

$$k_B T_b > \left(\frac{\lambda_c}{r_0}\right)^{1/5} m_e c^2 \qquad (2.68)$$

where $\lambda_c$ is the wavelength at maximum frequency and $r_0$ is the classical electron radius.

The inverse Compton scattering obtained above, of course, is only a part of the story. When a radiation field is interacting with charged particles, there will be energy transfer from charged particles to photons as well as from photons to charged particles. The net effect, of course, will be to drive the system to thermal equilibrium with equipartition of energies. So far we only looked at the transfer of energy *from* charged particles *to* photons. To obtain the full effect we need to consider both processes which can be handled by the following argument.

Let the energy transferred from the photons to the electrons be $\Delta E$ on the average. (We shall omit the symbol $\langle \ \rangle$ for simplicity of notation, even though we are dealing with average values.) When $E \ll mc^2$ (where $E$ is the typical energy of the photon) and $k_B T \ll mc^2$, we can expand $(\Delta E/mc^2)$ in a double Taylor series in $(E/mc^2)$ and $(k_B T/mc^2)$, retaining up to quadratic order:

$$\frac{\Delta E}{mc^2} = c_1 + c_2 \left(\frac{E}{mc^2}\right) + c_3 \left(\frac{k_B T}{mc^2}\right) + c_4 \left(\frac{E}{mc^2}\right)^2$$
$$+ c_5 \left(\frac{E}{mc^2}\right) \left(\frac{k_B T}{mc^2}\right) + c_6 \left(\frac{k_B T}{mc^2}\right)^2 + \cdots . \qquad (2.69)$$

Interestingly enough, the coefficients $(c_1, \cdots, c_6)$ can be fixed by simple arguments: (i) Since $\Delta E = 0$ for $T = E = 0$, we must have $c_1 = 0$. (ii) Consider next the scattering of a photon with electrons at rest. This will correspond to $T = 0$ and $E \neq 0$. If the scattering angle is $\theta$, the wavelength of the photon changes by

$$\Delta \lambda = \left(\frac{h}{mc}\right) (1 - \cos\theta). \qquad (2.70)$$

(This is just the textbook case of Compton scattering.) Such a scattering of a photon, by an electron at rest, is symmetric in the forward - backward directions and the mean fractional change in the frequency is $(\Delta\omega/\omega) = -(\Delta\lambda/\lambda) = -(\hbar\omega/mc^2)$; so the average energy transfer to the electrons is $\Delta E = E^2/mc^2$. This implies that $c_2 = 0$ and $c_4 = 1$. (iii) If $E = 0$ and $T \neq 0$ photon has zero energy and nothing should happen; hence $c_3 = c_6 = 0$. So our expression reduces to

$$\frac{\Delta E}{mc^2} = \left(\frac{E}{mc^2}\right)^2 + c_5 \left(\frac{E}{mc^2}\right) \left(\frac{k_B T}{mc^2}\right)^2 . \qquad (2.71)$$

(iv) To fix the last constant $c_5$, we can consider the following thought experiment. Suppose there is a very *dilute* gas of photons at the *same* temperature as the electrons. Then the number density $n(E)$ of photons is given by the Boltzmann limit of the Planck distribution:

$$n(E)\, dE \propto E^2 \left(e^{\beta E} - 1\right)^{-1} dE \propto E^2 \exp\left(-E/k_B T\right) dE. \qquad (2.72)$$

In this case, since the temperatures are the same, we expect the net energy transfer between the electrons and photons to vanish. (You need to use the Wien limit rather than the exact Planck spectrum to ensure that photons behave as particles.) That is, we demand

$$0 = \int\limits_0^\infty dE\, n(E)\, \Delta E \qquad (2.73)$$

where $n(E) \propto E^2 \exp(-E/k_B T)$. Substituting for $\Delta E$ and $n(E)$ from Eq. (2.71) and Eq. (2.72), one can easily show that $(4k_B T + c_5 k_B T) = 0$ or $c_5 = -4$. Hence we get the final result

$$\frac{\Delta E}{E} = \frac{(E - 4k_B T)}{mc^2}. \qquad (2.74)$$

One can say that, in a typical collision between an electron and photon, the electron energy changes by $(E^2/mc^2)$ and photon energy changes by $(4k_B T/mc^2)E$. (To order $v/c$, the probability of photon getting redshifted or blueshifted is equal and $\Delta E/E = \pm v/c$ where $v$ is the velocity of the electron. On the average, this gives zero energy transfer to $\mathcal{O}(v/c)$. But to order $\mathcal{O}(v^2/c^2)$ the probability for blueshift is larger than the probability for redshift which is why you get $\Delta E/E \sim v^2/c^2$.)

Repeated scattering of photons by hot electrons can lead to a process called Comptonisation, which is of importance in some astrophysical systems. The mean-free-path of the photon due to Thomson scattering is $\lambda_\gamma = (n_e \sigma_T)^{-1}$. If the size of the region $l$ is such that $(l/\lambda_\gamma) \gg 1$ then the photon will undergo several collisions in this region; but if $(l/\lambda_\gamma) \lesssim 1$ then there will be few collisions. In terms of the optical depth $\tau_e \equiv (l/\lambda_\gamma) = (n_e \sigma_T l)$, the $\tau_e \gg 1$ limit implies strong scattering. If $\tau_e \gg 1$, then the photon goes through $N_s (\gg 1)$ collisions in traveling a distance $l$. From standard random walk arguments, we have $N_s^{1/2} \lambda_\gamma \simeq l$ so that $N_s = (l/\lambda_\gamma)^2 = \tau_e^2$. On the other hand, if $\tau_e \lesssim 1$, then $N_s \simeq \tau_e$; therefore we can estimate the number of scatterings as $N_s \simeq \max(\tau_e, \tau_e^2)$. The average fractional change in the photon energy, per collision, is given by the second term in Eq. (2.74), viz., $4(k_B T_e/m_e c^2)$. Hence the condition for significant change of energy is

$$N_s \left( \frac{4k_B T_e}{m_e c^2} \right) = 4 \left( \frac{k_B T_e}{m_e c^2} \right) \max \left( \tau_e, \tau_e^2 \right) \simeq 1. \qquad (2.75)$$

It is conventional to define a parameter $y$, (called *Compton-y parameter*) by

$$y = \left( \frac{k_B T_e}{m_e c^2} \right) \max \left( \tau_e, \tau_e^2 \right). \qquad (2.76)$$

Then we can write the condition for significant scattering as $y \simeq 1/4$.

Inverse Compton scattering leads to an interesting physical phenomena called *Sunyaev-Zeldovich effect* which is proving to be of great importance in the study of clusters of galaxies. The standard model of the universe, which we will study in detail in chapter 6, shows that we are permeated by a radiation background (CMBR) with a temperature of about 2.73 K. The large clusters of galaxies which contain hot electron gas, at temperature $T_e$, are embedded in this background radiation thereby leading to inverse Compton scattering of the microwave photons by the hot electrons. Each scattering will boost the photon energy by the fractional amount $(k_B T_e/m_e c^2)$ without, of course, creating new photons. Thus the photons in the low energy tail of the CMBR blackbody spectrum are moved to high energies thereby leading to a spectral distortion in the direction of the cluster, with less photons at low frequencies and more photons at high frequencies (compared to the Planck spectrum). At long wavelengths this will essentially appear as a uniform drop in the temperature given by

$$\frac{\Delta T}{T} = -2 \int \frac{k_B T_e}{m_e c^2} n_e \sigma_T \, dl \qquad (2.77)$$

where the integration is along the path of the photon through the hot gas. The factor 2 comes from a more detailed analysis but the rest of the factors are easy to understand as due to inverse Compton scattering. This effect has been measured for several large clusters and acts as a powerful diagnostic tool.

As a simple illustration, consider a cluster containing an intra cluster plasma which is distributed with a radial dependence given by

$$n(r) = \frac{n_c}{(1 + r^2/r_c^2)} \qquad (2.78)$$

where $r_c$ is a constant. The optical depth through the cluster which appears in Eq. (2.77) becomes

$$\tau_0 \equiv \sigma_T \int_{-\infty}^{\infty} n(r)dr = 6.4 \times 10^{-3} \left( \frac{n_c}{\text{cm}^{-3}} \right) \left( \frac{r_c}{\text{kpc}} \right). \qquad (2.79)$$

We can obtain some additional information about $n_c$ and $r_c$ from the x-ray emission of the plasma. Using the bremsstrahlung emissivity $\epsilon_{\text{ff}} \simeq 1.4 \times 10^{-27} T^{1/2} n^2$ erg s$^{-1}$ cm$^{-3}$ (see Eq. (2.43)) we find that the total x-ray luminosity is given by

$$L_X = 4\pi \int_0^{r_c} \epsilon_{\text{eff}}\, r^2 dr$$

$$= 2.5 \times 10^{41} f^{-1} \left( \frac{n_c}{\text{cm}^{-3}} \right)^2 \left( \frac{r_c}{\text{kpc}} \right)^3 \left( \frac{k_B T_e}{\text{keV}} \right)^{1/2} \text{ erg s}^{-1} \quad (2.80)$$

where $f$ is a "filling factor" which is a ratio of the x-ray emitting volume to the core volume of the cluster. Observations suggest that, for a typical cluster, $L_X \approx 10^{44} h^{-2}$ erg s$^{-1}$, $r_c \approx 200 h^{-1}$ kpc and $T_e \approx 4$ keV where $h$ is the dimensionless Hubble expansion rate of the universe. Using these in Eq. (2.80) we can estimate $n_c = 5 \times 10^{-3} h^{1/2} f^{-1/2}$ cm$^{-3}$. Substituting this result in Eq. (2.79) we get $\tau_0 = 6.4 \times 10^{-3} (hf)^{-1/2}$. Equation (2.77) now gives

$$\frac{\Delta T}{T} = -2\tau_0 \left( \frac{k_B T_e}{mc^2} \right) = -1.0 \times 10^{-4} (fh^{-1})^{1/2}. \qquad (2.81)$$

Though small, this is observable by current technology. Thus, the measurement of $(\Delta T/T)$ and modeling of $f$ will allow an estimate of $h$ which is extremely important in cosmology.

## 2.5 Radiative processes in quantum theory

We shall now take up the study of radiative processes in quantum theory which uses a somewhat different language. To work out the details of quantum radiative processes, we need to determine the relevant energy levels (which, in turn, will fix the frequency of photons which are emitted, through $\Delta E = \hbar\omega$) as well as the transition rate $Q$ between the two levels. Let us begin with a determination of the energy levels of simple atomic and molecular systems.

## 2.5.1    Energy levels

The Hamiltonian describing a nonrelativistic electron, moving in the Coulomb field of a nucleus of charge $Zq$ is given by $H_0 = (p^2/2m_e) - (Zq^2/r)$. If this electron is described a wave function $\psi(\mathbf{x}, L)$ where $L$ denotes the characteristic scale over which $\psi$ varies significantly, then the expectation value for the energy of the electron in this state is $E(L) = \langle\psi|H_0|\psi\rangle \approx (\hbar^2/2m_e L^2) - (Zq^2/L)$. The first term arises from the fact that $\langle\psi|p^2|\psi\rangle = -\hbar^2\langle\psi|\nabla^2|\psi\rangle \approx (\hbar^2/L^2)$, which is equivalent to the uncertainty principle stated in the form $p \cong \hbar/L$. This expression for $E(L)$ reaches a minimum value when $L$ is varied at $L_{\min} = (a_0/Z)$ with $E_{\min} = -Z^2\epsilon_a$ where

$$a_0 \equiv \frac{\hbar^2}{m_e q^2} \equiv \frac{\lambdabar_e}{\alpha} \approx 0.52 \times 10^{-8} \text{ cm}; \quad \epsilon_a \equiv \frac{m_e q^4}{2\hbar^2} = \frac{1}{2}\alpha^2 m_e c^2 \approx 13.6 \text{ eV}.$$

$$(2.82)$$

Here we have defined $\lambdabar_e \equiv (\hbar/m_e c)$ and $\alpha \equiv (q^2/\hbar c)$. The quantities $a_0$ and $\epsilon_a$ correspond to the size and ground state energy of a hydrogen atom with $Z = 1$. (These results are exact even though our analysis based on $p \simeq \hbar/L$ is approximate.) The wavelength $\lambda$ corresponding to the energy $\epsilon_a$ is $\lambda = (hc/\epsilon_a) = 2\alpha^{-2}\lambdabar_e \simeq 10^3$ Å and lies in the UV band. The *fine structure constant* $\alpha \approx 7.3 \times 10^{-3}$ plays an important role in the structure of matter and arises as the ratio between several interesting variables:

$$\alpha = \frac{v}{Zc} = \frac{2\mu_B}{qa_0} = \frac{r_0}{\lambdabar_e} = \frac{\lambdabar_e}{a_0} \qquad (2.83)$$

where $v$ is the speed of an electron in the atom, $\mu_B \equiv (q\hbar/2m_e c)$ is the *Bohr magneton* representing the magnetic moment of the electron and $r_0 \equiv (q^2/m_e c^2)$ is the classical electron radius. The actual energy levels of the hydrogen (like) atom are $E_n = -\epsilon_a/n^2$ with $n = 1, 2, \ldots$.

The transitions between these energy levels in an atom will lead to photon energies upwards of few electron volts and the corresponding wavelength will be in the optical and UV bands in most of the cases. Note that, while the characteristic size of the atom is $a_0$, the typical wavelength of radiation emitted from the atom is $\lambda \approx (2/\alpha)a_0$ which is about 300 times longer, due to the smallness of $\alpha$. The fact that the ground state energy scales as $E_{\min} \propto Z^2$ is also of importance while considering atoms with large atomic numbers (See Ex.2.15).

The actual Hamiltonian for an electron in the hydrogen atom is a bit more complicated than what we have used above and a more precise form

leads to some interesting effects. To begin with, the correct expression for
kinetic energy is not $(p^2/2m_e)$ but $E = (p^2c^2 + m_e^2c^4)^{1/2} - m_ec^2$. The
expression for $E(L)$ which we need to minimize is then

$$E(L) \cong \left[\frac{\hbar^2c^2}{L^2} + m_e^2c^4\right]^{1/2} - m_ec^2 - \frac{Zq^2}{L}. \qquad (2.84)$$

The minimum value now occurs at

$$L_{\min} = \frac{\hbar}{m_ec}\left[\frac{1}{(Z\alpha)^2} - 1\right]^{1/2}. \qquad (2.85)$$

As long as $Z\alpha \ll 1$, this will reproduce the previous results.[2] To see the
effect of relativistic correction we can expand the Hamiltonian in Eq. (2.84)
in a Taylor series in $(p/m_ec)$. The lowest order correction will be

$$E_1 \simeq \frac{p^4}{m_e^3c^2} \approx \frac{p^2}{m_e}\left(\frac{p}{m_ec}\right)^2 \approx \left(\frac{v}{c}\right)^2\frac{q^2}{a_0} \approx \alpha^2\epsilon_a \approx 10^{-3} \text{ eV}. \qquad (2.86)$$

The zeroth order term $H_0$ is of the order of $(q^2/a_0) \approx 10$ eV corresponding
to $\lambda \simeq 1200$ Å, while this correction gives $10^{-3}$ eV with the corresponding
photon wavelength of $\lambda \simeq 1$ mm. (It is worth remembering that 1 eV
corresponds to a photon wavelength of 12400 Å.)

There is another effect which leads to a correction of the *same* order
of magnitude $\alpha^2E_0$. This arises from a term in the Hamiltonian, $H_{\text{sp-or}}$,
due to the coupling between the spin magnetic moment of the electron
$\mu_e \simeq (q\hbar/2m_ec)$ and the magnetic field $B \cong (v/c)E \cong (v/c)(Zq/r^2)$ in
the instantaneous rest frame of the electron, obtained by transforming the
Coulomb field. This should have the magnitude

$$\bar{H}_{\text{sp-or}} = \mu B = \frac{Zq^2}{2r^2}\frac{\hbar v}{m_ec^2} = \frac{Zq^2}{m_e^2c^2r^3}(\mathbf{l}\cdot\mathbf{s}) \qquad (2.87)$$

where $\mathbf{l}$ and $\mathbf{s}$ are the orbital $(m_evr)$ and spin $(\hbar/2)$ angular momentum of
the electron. (The actual result is half of this value with the extra factor
arising due to an interesting phenomenon called *Thomas precession* which

[2] As an aside, let us note two curious aspects of Eq. (2.85). First, you see that $L_{\min}$
will become imaginary for $Z > \alpha^{-1} \approx 137$. This is a *real* relativistic instability which
arises in heavy ion collisions. Second, if we did the same analysis for a gravitating system
made of $N$ nucleons, then $Z\alpha = Zq^2/\hbar c$ will be replaced by $(GMm_p/\hbar c) = (Gm_p^2/\hbar c)N$.
Clearly we expect some instability when $N \sim (\hbar c/Gm_p^2)$. We will see this effect in chapter
5 in the study of stellar remnants, though a precise analysis leads to the instability when
$N \sim (\hbar c/Gm_p^2)^{3/2}$.

we will not discuss). This correction, with $l \simeq s \simeq \hbar$, is

$$\frac{q^2 \hbar^2}{m_e^2 c^2 a_0^3} \approx \frac{q^2}{a_0} \left( \frac{\hbar}{m_e c a_0} \right)^2 \approx \alpha^2 \epsilon_a \qquad (2.88)$$

which is of the same order as the correction found in Eq. (2.86). These two corrections in Eq. (2.86) and Eq. (2.88) together are called *fine structure corrections* (since, historically, this is related to the fine structure observed in the spectral lines).

The angular momentum conservation requires that a fine structure transition will require $\Delta l = 1$. In the case of hydrogen, one rarely finds an electron in the $n = 2, l = 1$ state allowing a fine structure transition to the ground state. But the situation is different for heavier elements like, for example, singly ionized carbon $C^+$. Since $C^+$ has 5 electrons, the fifth one will end up at $n = 2, l = 1$ state; further, the fine structure energy difference for carbon is about $10^{-2}$ eV because the expression in Eq. (2.86) scales as $Z^4$. This corresponds to a temperature of about 50 K which exists in diffuse molecular clouds in the interstellar medium. Collisions between atoms can excite transitions between these levels. De-excitation will then occur through emission of radiation which escapes from the cloud. The net effect is the loss of kinetic energy of the atoms. Hence, fine structure transitions of heavier elements like carbon can be a main source of cooling in the interstellar medium.

Finally, one also needs to take into account the direct interaction between the magnetic moments of the electron and proton. This is represented by an extra term in the Hamiltonian

$$H_{\text{sp-sp}} = \vec{\mu}_e \cdot \left[ \frac{\vec{\mu}_N}{r^3} - \frac{3\mathbf{r} \cdot \vec{\mu}_N}{r^5} \mathbf{r} \right] \qquad (2.89)$$

which gives the coupling between the magnetic moments of the nucleus and the electron. These magnetic moments are given by

$$\vec{\mu}_e = - \left( \frac{q\hbar}{2m_e c} \right) \mathbf{s} \equiv -g_e \mu_B \mathbf{s}; \qquad \vec{\mu}_p \simeq 5.6 \left( \frac{q\hbar}{2m_p c} \right) \mathbf{S} \equiv g_N \mu_N \mathbf{S} \quad (2.90)$$

where $\mu_B$ is the Bohr magneton, $\mu_N$ is the corresponding quantity for the proton and $(\mathbf{s}, \mathbf{S})$ are the spin vectors of electron and proton. This correction is of the order of

$$\frac{\mu_B \mu_N}{a_0^3} \approx \frac{m_e}{m_p} \frac{\mu_B^2}{a_0^3} \approx \frac{m_e}{m_p} \left( \frac{\hbar}{m_e c a_0} \right)^2 \frac{q^2}{a_0} \approx 10^{-3} \alpha^2 E_0 \approx 10^{-6} \text{ eV} \quad (2.91)$$

(corresponding to $\lambda \simeq 10^2$ cm) which is smaller than the fine structure correction discussed above and is called *hyperfine correction*. A more precise calculation gives the wavelength of radiation emitted in the hyperfine transition of hydrogen to be about 21 cm. This radiation, which is in the radio band, is used extensively in astronomy as a diagnostic of atomic (neutral) hydrogen.

As a further example, consider the corresponding hyperfine transition in $^3\mathrm{He}^+$. Since the frequency scales as $\mu_{\mathrm{nucl}}/r^2$, we need to estimate how $\mu_{\mathrm{nucl}}$ and $r$ vary when we go from hydrogen to singly ionized helium. Since magnetic moment of the proton has $q/m_p$, we can take $\mu_{\mathrm{nucl}} \propto Z/A$; further $r$ scales as $1/Z$. Hence we expect the hyperfine energy to vary as $Z^4/A$. The wavelength of transition in helium will therefore be 21 cm $\times (A/Z^4) = (21 \times 3)/2^4 \approx 3.9$ cm which is pretty close to the observed value of 3.46 cm.

Let us next consider the radiation arising from the molecular energy levels. The simplest molecular structure consists of two atoms bound to each other in the form of a diatomic molecule. The effective potential energy of interaction $U(r)$ between the atoms in such a molecule arises from a residual electrostatic coupling between the electrons and the nuclei. The resulting potential energy has a minimum at a separation $r \simeq a_0$ (which is typically of the order of the size of the atom) and the depth of the potential well at the minimum is comparable to the electronic energy level, $\epsilon_a$, of the atom. The atoms of such a molecule can vibrate at some characteristic frequency $\omega_{\mathrm{vib}}$ about the mean position along the line connecting them. Treated as a harmonic oscillator, this will lead to the vibrational energy levels separated by $E_{\mathrm{vib}} \approx \hbar\omega_{\mathrm{vib}}$. If the displacement $x$ of the oscillator is $x \simeq a_0$ from the minimum, we expect the vibrational energy $E_{\mathrm{vib}}$ to be about $\epsilon_a$. Writing $\epsilon_a \approx (1/2)\mu\omega_{\mathrm{vib}}^2 a_0^2 \cong (\hbar^2/m_e a_0^2)$, where $\mu$ is the reduced mass of the two atoms making the molecule, we can determine $\omega_{\mathrm{vib}}$. We get

$$E_{\mathrm{vib}} = \hbar\omega_{\mathrm{vib}} \approx \frac{\hbar^2}{(\mu m_e)^{1/2} a_0^2} \approx \left(\frac{m_e}{\mu}\right)^{1/2} \epsilon_a \simeq 0.25 \text{ eV} \qquad (2.92)$$

if $\mu \simeq m_p$. This is the characteristic energy of photons emitted in vibrational transitions.

Such a diatomic molecule can also rotate about an axis perpendicular to the line joining them. If the rotational angular momentum is $J$, this will contribute an energy of about

$$E_{\text{rot}} \approx \left( \frac{J^2}{\mu a_0^2} \right) \approx \left( \frac{\hbar^2}{\mu a_0^2} \right) \approx \left( \frac{m_e}{\mu} \right) \epsilon_a \approx 10^{-3} \epsilon_a \simeq 10^{-2} \text{ eV} \qquad (2.93)$$

if $J \simeq \hbar$ and $\mu \simeq m_p$. It follows from Eq. (2.92) and Eq. (2.93) that $E_{\text{rot}} : E_{\text{vib}} : E_0 \approx (m_e/\mu) : (m_e/\mu)^{1/2} : 1$ and $E_{\text{vib}} \approx \sqrt{\epsilon_a E_{\text{rot}}}$. Since $(m_e/\mu) \approx 10^{-3}$, the wavelength of radiation from vibrational transitions are about 40 times larger than those of electronic transitions; similarly, the rotational transitions lead to radiation with wavelengths about 1000 times larger than electronic transitions.

Taking $\mu \approx m_p$, of course, is only true if you are thinking of $H_2$ molecule, which unfortunately has no permanent dipole moment because of its symmetry. Considering the fact that, after hydrogen and helium, some of the most abundant elements in the universe are carbon, oxygen, silicon ... *etc.*, one might try CO molecule as a good choice. For CO, $\mu \approx 30 m_p$ which gives the wavelength of transition to be $\lambda_{\text{co}} \approx 2.6$ mm which is in the infrared band.

The energy level structure of common atoms and molecules is of considerable importance for life on Earth in which the atmosphere shields most of the photons from reaching the ground. To begin with, very high energy photons ($\epsilon \gtrsim 2 m_e c^2 \approx 1$ MeV) lose their energy due to production of $e^+ e^-$ pairs when they pass through matter. (You can easily show that this process cannot take place in the vacuum since you cannot conserve both energy and momentum for the process $2\gamma \to e^+ e^-$ in the vacuum; however, in the presence of a nucleus to absorb the momentum, this process can convert radiation into electron-positron pairs.) For photons in the energy range 50 keV $\lesssim \epsilon \lesssim 1$ MeV, the most dominant energy loss process is Compton scattering with the electron. Since the atomic binding energies are usually small compared to this range of energies, the photon essentially sees the electron as "free" particle. Below about 50 keV, photo ionisation starts dominating over Compton scattering as the energy loss mechanism. This continues till the energy drops to about 3 eV. It turns out, miraculously enough, that there are no viable transitions in matter in the energy range of about 1.3 eV $\lesssim \epsilon \lesssim 3$ eV. This band of photons, therefore, can propagate through matter in the atmosphere fairly freely and constitutes the "visible" band of photons. Below 1.3 eV down to about $10^{-4}$ eV, some amount of absorption takes place because of molecular energy levels. Further down, we get another transparent window between $10^{-4}$ eV and $10^{-8}$ eV (corresponding to the wavelength range of 1 cm to 100 m; the radio band). Below $10^{-8}$

eV, radiation cannot propagate through the ionospheric plasma (you will see why in chapter 3, Sec.3.4.2) thereby closing the window permanently. Thus ground based astronomy is mostly confined to optical and radio bands essentially because of the nature of atomic and molecular energy levels.

### 2.5.2 Transition rates and cross sections

The energy levels of atomic systems determine the *frequency* of radiation. But the *intensity* of radiation, related to the number of photons emitted per second, is determined by the transition rate $Q$ (see Eq. (2.9)), which we shall now estimate for simple cases.

The basic rate, governed by $Q$, can be estimated using the correspondence with classical theory. If the rate of transition between two energy levels is $Q$ and the energy of the photon emitted during the transition is $\hbar\omega$, then the rate of energy emission is $Q\hbar\omega$. Classically, the same system will emit energy at the rate $(d^2\omega^4/c^3)$ where $d$ is the electric dipole moment of the atom. (See Eq. (2.8)). This assumes that the radiation is predominantly due to direct coupling between the electric field and the dipole moment of the atom; if not, you have to use the relevant moment — like the electric quadrapole moment or magnetic dipole moment, instead of $d$ — in this equation. Writing $Q\hbar\omega \approx (d^2\omega^4/c^3)$, we find that

$$Q \cong \frac{\omega^3 d^2}{\hbar c^3} \simeq \frac{\alpha^5}{8}\left(\frac{\lambda}{c}\right)^{-1} \simeq \frac{\alpha^5}{8}\left(\frac{m_e c^2}{\hbar}\right) \approx 10^9 \text{ s}^{-1} \qquad (2.94)$$

where we have used $d = qa_0 \simeq (q\lambda_e/\alpha)$ and $\hbar\omega \simeq (1/2)\alpha^2 m_e c^2$. The rate $Q$ can also be expressed in different forms as

$$Q \approx \frac{q^2}{\hbar c}\frac{a_0^2}{c^2}\omega^3 = \frac{q^2}{m_e c^3}\omega^2 = \omega^2\frac{r_0}{c} \qquad (2.95)$$

where $r_0$ is the classical electron radius. A more precise quantum mechanical calculation corrects this by contributing an extra numerical factor $f$, called *oscillator strength* which is usually of order unity for strong lines.

The transition rate for other processes, like e.g. the 21 cm radiation, can be estimated in the same manner using Eq. (2.8). In the relation $Q\hbar\omega_{21} \approx (\omega_{21}^4 d^2/c^3)$, we now have to use the magnetic dipole moment of electron $d \approx (q\hbar/m_e c)$. This gives $Q \approx \alpha(\lambda_e/c)^2\omega_{21}^3$. Estimating $\hbar\omega_{21}$ from the expression (2.91) derived above for the hyperfine structure energy level, this transition rate can be evaluated to be

$$Q_{21\text{cm}} \simeq \left(\frac{r_0\omega_{21}}{c}\right)\left(\frac{\hbar\omega_{21}}{m_ec^2}\right)\omega_{21} \approx 10^{-15} \text{ s}^{-1}. \tag{2.96}$$

The ratio between this rate and the transition rate between the primary energy levels of hydrogen atom, computed earlier in Eq. (2.94), is

$$\frac{Q_{21\text{cm}}}{Q} \simeq \left(\frac{\omega_{21}}{\omega}\right)^3\left(\frac{\lambda_e}{a_0}\right)^2 \approx 10^{-24}. \tag{2.97}$$

Clearly, hyperfine transitions are very slow processes. It would not have been useful astrophysically but for the fact that there are so many neutral hydrogen atoms out there.

Let us verify these ideas with two more examples. First, consider the corresponding hyperfine transition in $^3$He$^+$ for which we estimated the wavelength of radiation to be 3.9 cm earlier (see discussion after Eq. (2.91)). Since $Q$ varies as $\omega^3 \propto \lambda^{-3}$, we will expect the rate of hyperfine transition in $^3$He$^+$ to be $(21/3.9)^3$ times larger than the $Q_{21}$ cm. This gives about $6 \times 10^{-13}$ s$^{-1}$. This is off from the true value only by a factor three. As a second example, consider the molecular transitions in CO which is given by an expression similar to Eq. (2.94) with $d$ representing the permanent dipole moment of the molecule and $\omega = 2\pi c/\lambda_{\text{co}}$ with $\lambda_{\text{co}} \approx 2.6$ mm as estimated earlier (see discussion after Eq. (2.93)). It turns out that the dipole moment of CO is 0.1 debye where 1 debye $= 10^{-18}$ in the c.g.s units for measuring dipole moment. This is a convenient unit because $qa_0 = 2.5$ debye. (The lower dipole moment of CO is a peculiarity arising from the strong double bond connecting C and O — most other molecules like $H_2O$, SiO, Cs .... all have dipole moments which are of the order of 1 debye which is what one would have naively expected.) Using these values we get a transition rate of $Q_{\text{co}} \approx 8.7 \times 10^{-8}$ per second which is pretty close to the observed value of $6 \times 10^{-8}$ per second.

One can also estimate the width of the spectral lines from the decay rate. Since every excited state has a probability to decay spontaneously to the ground state with a decay rate $Q$, the lifetime $\Delta t$ of the excited state is about

$$\Delta t \approx Q^{-1} \approx \left(\frac{q^2}{\hbar c}\right)^{-1}\left(\frac{a_0}{\lambda}\right)^{-2}\omega^{-1} \approx \left(\frac{10^8}{\omega}\right) \approx 10^{-9} \text{ s} \tag{2.98}$$

where the numerical estimate is for the optical radiation. (It is correspondingly larger, and is about $\Delta t = Q_{21\text{ cm}}^{-1} \simeq 10^{15}$ s for the 21 cm radiation.)

From the uncertainty principle between the energy and time, it follows that
the energy level of the excited state will be uncertain by an amount $\Delta E$
of the order of $(\hbar/\Delta t) \approx \hbar Q$. In the absence of such an uncertainty, the
transition between the two energy levels could result in an infinitely sharp
spectral lines at a specific frequency $\omega$. The width of the excited states
leads to a corresponding width to the spectral line called the *natural width*
$\Delta\omega$ where $(\Delta\omega/\omega) = (Q/\omega) \approx 10^{-8}$ for the main hydrogen lines. In terms
of wavelength, the natural width $\Delta\lambda = 2\pi c(\Delta\omega/\omega^2)$ is of the order of the
classical electron radius: $\Delta\lambda \approx 2\pi r_0$ providing another interpretation for
$r_0$.

The same rate $Q$ also governs the absorption of radiation at the transition frequency by the system. To take into account the finite line width,
it is convenient to introduce a (frequency dependent) *bound-bound* cross
section $\sigma_{bb}(\omega)$ for the absorption of radiation by the system. The function
$\sigma_{bb}(\omega)$ is expected to be sharply peaked at $\omega = \omega_0$ with a width of $\Delta\omega$.
If the phase space density of photons is $dN = n[d^3x d^3p/(2\pi\hbar)^3]$, then the
number density of photons per unit volume participating in the absorption
is $\mathcal{N} = nd^3p/(2\pi\hbar)^3 = (n\omega_0^2/\pi c^3)(\Delta\omega/2\pi)$ if we assume that the photons
in the band $\Delta\omega$ around $\omega_0$ are absorbed. The corresponding flux is $\mathcal{N}c$ and
the rate of absorption is

$$\mathcal{N}\sigma_{bb}c = \frac{n\omega_0^2}{\pi c^2}\,\sigma_{bb}\,\frac{\Delta\omega}{2\pi} \qquad (2.99)$$

which we know should be equal to

$$Qn = \omega_0^2\left(\frac{r_0}{c}\right)n \qquad (2.100)$$

giving $\sigma_{bb}(\Delta\omega/2\pi) = \pi r_0 c$. Since we expect $\sigma_{bb}$ to be a sharply peaked
function at $\omega_0$ with a width $\Delta\omega$, this result can be stated more formally as,

$$\int_0^\infty \sigma_{bb}(\omega)\,\frac{d\omega}{2\pi} = \pi r_0 c = \frac{\pi q^2}{m_e c} = (\pi r_0^2)\left(\frac{c}{r_0}\right). \qquad (2.101)$$

This — in turn — suggests expressing the cross section as

$$\sigma_{bb}(\omega) \equiv \frac{\pi q^2}{m_e c}\phi_\omega \qquad (2.102)$$

where $\phi_\omega$ is called the *line profile function*; the integral of $\phi_\omega$ over $(d\omega/2\pi)$
is unity to satisfy Eq. (2.101). When the line widths are ignored, $\phi_\omega$ is
proportional to the Dirac delta function; when the finite width of the energy
levels is taken into account $\phi_\omega$ will become a function which is peaked

at $\omega_0$ with a narrow width $\Delta\omega$, such that, at the transition frequency, $\phi(\omega_0) \cong (\Delta\omega)^{-1}(2\pi)$. The absorption cross section at the center of the line $\omega = \omega_0$ is given by $\sigma \simeq 2\pi^2(rc/\Delta\omega) \simeq (\lambda_0^2/2)$ where $\lambda_0$ is the wavelength of the photon. A more rigorous quantum mechanical theory of transition rates is needed to determine the explicit form for $\phi(\omega)$ and overall multiplicative numerical factor $f$ called the *oscillator strength*. It can be shown that the functional form of $\phi(\omega)$ is a Lorentzian so that the final cross section in Eq. (2.102) is usually written in the form:

$$\sigma_{bb} = \frac{\pi q^2}{mc}f\phi_\omega \simeq \frac{\pi(q^2/mc)\Gamma f}{(\omega - \omega_0)^2 + (\Gamma/2)^2}; \quad \Gamma \approx \Delta\omega. \tag{2.103}$$

The line width described above is usually called *intrinsic* or *natural* line width. (The Lorentzian profile can be most easily understood by treating the dynamics of the electron in an atom as that of a damped harmonic oscillator with a natural frequency $\omega_0$ and damping constant $\Gamma$; such a classical model captures the essence of quantum mechanical calculations with surprising accuracy.)

### 2.5.3 *Ionisation and recombination*

So far we discussed the transitions between bound states with discrete energy levels. Another important class of radiative process involve transitions between a bound state of an atom and the continuum. We shall now discuss this process, and the related opacities.

To study these processes which essentially corresponds to a photon ionizing an atom, we will begin by relating the photo-ionisation rate to the rate for the time reversed process, which is *recombination* rate in which an ion and electron gets bound together, releasing the excess energy as radiation. The recombination rate per unit volume of a plasma will be proportional to (a) number density of electrons, $n_e$, (b) number density of ions, $n_i$, (c) the relative velocity of encounter, $v$ and, finally, (d) the cross section for the process, which will be about $\pi l^2$, where $l$ gives the effective range of interaction between electron and proton. There are two possible choices for $l$. From Coulomb interaction, we get a characteristic length scale for charged particles approaching each other with relative velocity $v$, which is about $l_1 \cong (2Zq^2/m_e v^2)$; this is the scale at which electrostatic interaction energy becomes comparable to the kinetic energy. On the other hand, an electron of speed $v$ has a de Broglie wavelength of $l_2 \cong (\hbar/m_e v)$ and will make its presence felt within a region of this size. Clearly should choose $l$

to be the larger of $l_1$ or $l_2$ depending on the context.

Let us first consider the case with $l = l_2$, which corresponds to $(v/c) \gtrsim (q^2/\hbar c)$. Since each recombination releases an energy of about $(1/2)m_e v^2$ we can estimate the rate of energy release per unit volume to be:

$$\left(\frac{dE_{\rm rec}}{dV\,dt}\right) \propto \left(\frac{m_e v^2}{2}\right)(n_e n_i)\left(\frac{h}{m_e v}\right)^2 v \propto n_e n_i T^{1/2} \propto \rho^2 T^{1/2} \quad (2.104)$$

where we have assumed that $m_e v^2 \approx k_B T, n_e \propto \rho, n_i \propto \rho$. In equilibrium, the photo-ionisation rate (which removes energy from the radiation field) should match the recombination rate. The amount of energy removed by photo-ionisation is proportional to $dE_{\rm ion} \propto n_{\rm atom}\sigma E_{\rm rad}$ where $\sigma$ is the photo-ionisation cross section which needs to be determined. Equating $dE_{\rm ion}$ to $dE_{\rm rec}$ and using $E_{\rm rad} \propto T^4$, we get

$$n_{\rm atom}\,\sigma\,T^4 \propto n_e n_i T^{1/2} \quad (2.105)$$

so that $\sigma \propto \rho T^{-3.5}$. Introducing the *bound-free opacity* $\kappa_{\rm bf}$ by the definition $\kappa_{\rm bf} = (n_{\rm atom}\sigma/\rho)$ and taking $n_e \propto \rho, n_i \propto \rho$ we find that

$$\kappa_{\rm bf} \propto \rho T^{-3.5} \quad (2.106)$$

which scales just like the free-free opacity in Eq. (2.48).

In a radiation bath with temperature $T$, the typical energy of photons is $h\nu \simeq k_B T$. The result Eq. (2.106) suggests that, in a more general context, the frequency dependence of this cross section will be of the form $\sigma(\nu) = \sigma_{\rm bf}(\nu/\nu_I)^{-s}$ for $\nu > \nu_I$ (and zero otherwise) with $s \approx 3$ and $\nu_I = (\epsilon_a/h)$ is the frequency corresponding to the ionisation energy $\epsilon_a$ of the atom. The prefactor $\sigma_{\rm bf}$ can be estimated using the fact that any absorption cross section $\sigma(\nu)$ satisfies the constraint Eq. (2.101):

$$\int_0^\infty \sigma(\nu)\,d\nu = \pi r_0 c f = \left(\frac{\pi e^2}{m_e c}\right) f \quad (2.107)$$

where the term within the bracket comes from classical theory and the oscillator strength $f$ needs to be supplied by the full quantum theory. Using this, we find:

$$\sigma_{\rm bf} = \pi(s-1)f\,r_0\lambda_I \simeq 2\pi f r_0\lambda_I \quad (2.108)$$

with $\lambda_I = (c/\nu_I), s \simeq 3$. Therefore, the photo-ionisation cross section is essentially the product of the classical electron radius and the wavelength at ionisation threshold. Using $\lambda_I = 912$ Å, we get $\sigma_{\rm bf} \approx 10^{-17}$ cm$^2$ for hydrogen. (Incidentally, this shows that a column density $N_c$ of hydrogen which leads to unit optical depth, $N_c \sigma_{\rm bf} = 1$, is given by $N_c \simeq \sigma_{\rm bf}^{-1} \simeq 10^{17}$ cm$^{-2}$. This is useful in studying quasar absorption systems; see Chapter 9.) For heavier elements $\sigma$ will scale as

$$\sigma(\nu) = \sigma_{\rm bf} \left(\frac{\nu_I}{\nu}\right)^3 \propto fr_0 \left(\frac{c}{\nu_I}\right) \left(\frac{\nu_I}{\nu}\right)^3 \propto \left(\frac{\nu_I^2}{\nu^3}\right) \propto Z^4 \nu^{-3} \qquad (2.109)$$

if we take $s = 3$.

Given the form of $\sigma_{\rm bf}$, we can determine the cross section for recombination, $\sigma_{\rm rec}$. Recombination can occur with an electron having any momentum while the ionisation can be caused by photon of any energy. Equilibrium therefore requires $4\sigma_{\rm rec} d^3 p_e = 8\sigma_{\rm bf} d^3 p_\gamma$ where $4 = 2 \times 2$ is due to electron and proton spins, $8 = 2 \times 2 \times 2$ is due to electron, proton and photon spins and $p_e, p_\gamma$ refer to the momenta of the electron and photon. If the momentum of the photon arises from that of the electron in the recombination, we have $dp_e \simeq dp_\gamma$. This gives $\sigma_{\rm rec} = 2\sigma_{\rm bf}(p_\gamma/p_e)^2$; if the plasma is at some temperature $T$, then

$$\sigma_{\rm rec} = \frac{2(\hbar\omega)^2}{c^2(m_e v)^2}\sigma_{\rm bf} \cong 2\left(\frac{\epsilon_a}{m_e c^2}\right)\left(\frac{\epsilon_a}{k_B T}\right)\sigma_{\rm bf} \simeq 10^{-22}\left(\frac{k_B T}{10 \text{ eV}}\right)^{-1} \text{cm}^2.$$
$$(2.110)$$

The rate of recombination per unit volume per second is $n^2 \sigma_{\rm rec} v \equiv n^2 \alpha_R$ where $\alpha_R$ (called *recombination coefficient*) is given by

$$\alpha_R = \sigma_{\rm rec} v \simeq 2 \times 10^{-14}(k_B T/10 \text{ eV})^{-1/2} \text{ cm}^3 \text{ s}^{-1}. \qquad (2.111)$$

This result is extensively used in the study of astrophysical plasmas.

In the analysis so far we have assumed that the recombination proceeds directly to the ground state. Such a process, however, will release a photon with energy $\hbar\omega \gtrsim \epsilon_a$ which can immediately ionize another atom. The net recombination usually proceeds through electron and proton combining to form an excited state of the atom which decays to ground state later on. All the photons emitted in the process will then have $\hbar\omega < \epsilon_a$ and cannot ionize other neutral atoms in the ground state. In this case, the speeds are usually low ($v/c \lesssim q^2/\hbar c$) and we can write a relation similar to Eq. (2.104) but with $l \neq l_1 = (2Zq^2/m_e v^2)$. Then the energy loss due to recombination is:

$$\left(\frac{dE_{\rm rec}}{dV dt}\right) \propto \left(\frac{1}{2}m_e v^2\right)(n_e n_i)\left(\frac{Zq^2}{(1/2)m_e v^2}\right)^2 v \propto n_e n_i v^{-1} \propto n^2 T^{-1/2}.$$
$$(2.112)$$

This can be a source of cooling for plasma at $T \gtrsim 10^4$ K, in addition to the bremsstrahlung cooling $(dE_{\text{bre}}/dV dt) \propto n^2 T^{1/2}$ which was obtained earlier in Eq. (2.43). Since both processes scale as $n^2$ the total rate of cooling for a plasma can be expressed in the form $(dE/dV dt) = n^2 \Lambda(T)$ with $\Lambda(T) = aT^{1/2} + bT^{-1/2}$ where $a$, $b$ are numerical constants; the first term due to Bremsstrahlung dominates for $T \gtrsim 10^6$ K. The function $\Lambda(T)$ is called the *cooling curve* of the plasma and it incorporates the effects of both line cooling as well as continuum radiation in the form of Bremsstrahlung. Incidentally, at very low temperatures, neither process is effective and the gas simply cannot cool since collisions cannot excite the atomic energy levels. To capture the behaviour at low temperatures, one needs to take into account the fact that the gas will be mostly neutral. A reasonable fitting formula for the hydrogen plasma, taking all these effects into account is given by

$$\Lambda(T) = \frac{10^{-24}}{[1 + aT_5^{-7/6}\exp(\epsilon_a/k_BT)]^2} \left[2.1T_5^{-7/6} + 0.44T_5^{1/2}\right] \qquad (2.113)$$

where $\epsilon_a = 13.6$ eV, $a = 2.8 \times 10^6$ and all quantities are in cgs units with $T_5 = (T/10^5 \text{ K})$.

As a specific example involving photo-ionisation and recombination, let us apply these ideas in the case of a gas of hydrogen atoms, ionized by a flux of radiation $F$ from some source (like the $HII$ regions around stars). Let $n_H, n_i, n_e$ be the number densities of hydrogen atoms, protons and electrons respectively. The steady state will be reached when the rate of ionisation, $\sigma_{\text{bf}} n_H F$, is balanced by the rate of recombination, $n_e n_i \alpha_R$. Writing $n_e n_i \alpha_R \simeq \sigma_{\text{bf}} n_H F$ where $F$ is the flux of ionizing photons with frequency higher than the ionisation threshold (with $\nu > \nu_I$) and taking $n_e = n_i = xn_0, n_H = (1-x)n_0$, we get:

$$\frac{x^2}{(1-x)} \simeq \left(\frac{\sigma_{\text{bf}}c}{\alpha_R}\right)\left(\frac{F}{n_0 c}\right) \simeq 5 \times 10^4 \left(\frac{T}{10^4 \text{ K}}\right)^{1/2}\left(\frac{F}{n_0 c}\right). \qquad (2.114)$$

This relation determines the ionisation fraction $x$ in many astrophysical contexts. If a source of photons emitting $\dot{N}_\gamma$ photons per second (with $\nu > \nu_I$) ionizes a region of volume $V$ around it, then the same argument gives $n_e^2 \alpha_R V \simeq \dot{N}_\gamma$. Taking $n_e = xn_0 \simeq n_0$, we can determine the volume of the region which is ionized to be $V = (\dot{N}_\gamma/\alpha_R n_0^2)$.

To get the feel for the numbers involved, consider the region around a hot star with luminosity $L = 3.5 \times 10^{36}$ erg s$^{-1}$ and a surface temperature

$T_s \simeq 3 \times 10^4$ K. The number of ionizing photons, $\dot{N}_\gamma$, (with $\nu > \nu_I$) emitted by such a star can be estimated by integrating the Planck spectrum, from $\nu = \nu_I$ to $\nu = \infty$, and will be about $3 \times 10^{48}$ s$^{-1}$. Using the result $V = (\dot{N}_\gamma / \alpha n_0^2)$ obtained above, with $n_0 \simeq 10$ cm$^{-3}$, we find that matter will be fully ionized for a region of radius $R = (3V/4\pi)^{1/3} \simeq 10$ pc. At a distance of 5 pc from the star Eq. (2.114) gives $(1 - x) \simeq 10^{-3}$ indicating nearly total ionisation. The radiation from such HII regions is one of the probes of the interstellar medium.

Finally, we mention another important complication in the case of bound-free transitions of hydrogen atom which has astrophysical significance. It is possible for the *neutral* hydrogen atom to capture another electron and become a negative hydrogen ion. This $H^-$ ion is a a bound state made of two electrons and a proton, with a binding energy of about 0.75 eV. (The presence of an extra electron polarizes the hydrogen atom leading to a residual electric field in the vicinity which, in turn, binds the electron.) The bound-free absorption from this state gives rise to continuous opacity at wavelengths shorter than 16502 Å corresponding to the energy 0.75 eV. For conditions prevalent in sun-like stars, the opacity due to hydrogen ion can be approximated by the expression $\kappa_{H^-} \approx 2.5 \times 10^{-31}(Z/0.02)\,\rho^{1/2}\,T^9$ cm$^2$ g$^{-1}$ where $Z$ is the fractional abundance of elements heavier than H (which usually supplies the extra electron) and all the quantities are in c.g.s. units. You will see in Chapter 4 that this can be a very significant source of opacity in stars like the Sun.

### 2.5.4 *Spectral line profiles*

We saw earlier that the spectral lines have a natural line width which is described by a Lorentzian line profile [see Eq. (2.102)]. In astrophysical contexts, there are two other processes which contribute to the line widths. The first is due to the collisions between the atoms while the second is a Doppler shift arising from the random motion of atoms in thermal equilibrium.

The collisions will change the phase of the electric field of the radiation in a random manner, so that the electric field will now become $E(t) = E_0 e^{i\omega_0 t - \Gamma t/2} e^{i\psi(t)}$ where $\psi(t)$ is a random function of time such that it changes completely randomly at each collision and remains the same between collisions. The net effect of averaging $e^{i\psi}$ over such collisions is to introduce a term $\exp(-\nu_c t)$ into the decay rate of excited atoms where $\nu_c$ is the collision rate (see Ex. 2.16). Therefore collisions can be accounted for by the same line profile as in Eq.(2.103) with the replacement $\Gamma/2 \to \Gamma/2 + \nu_c = \Gamma'/2$.

The thermal broadening of the lines, due to the Doppler effect, leads to a different line profile: In thermal equilibrium, the probability distribution for the line-of-sight velocity (say, $v_x$) is a Gaussian proportional to $\left[\exp - \left(v_x^2/v_T^2\right)\right]$ with $v_T^2 = (2k_B T/m)$ where $m$ is the mass of the atom emitting the radiation. When an atom moving with velocity $v_x$ emits light, there will be a Doppler shift $(\Delta\omega/\omega) \simeq (v_x/c)$. Hence the line profile will be

$$ I(\omega) \propto \exp - \left[ \frac{(\omega - \omega_0)^2}{\omega_0^2} \left( \frac{c}{v_T} \right)^2 \right]. \qquad (2.115) $$

The fractional line width $(\Delta\omega/\omega)$ is $(v_T/c)$. Since $v_T \simeq 0.1 \, \text{km s}^{-1} \, (T/1\,\text{K})^{1/2}$ for hydrogen and $(\Delta\omega/\omega) \simeq 10^{-7}$ for the natural line width, it is clear that Doppler width will be larger than the natural line width for $v_T > 10^{-7}c \simeq 0.003 \, \text{km s}^{-1}$. So one might think that this will completely dominate the natural and collisonal line widths. However, the Gaussian width due to Doppler effect falls much more rapidly than the Lorentzian line profile. As you can easily verify, a Lorentzian $\Delta^2(x^2 + \Delta^2)^{-1}$ will dominate over a Gaussian $\exp(-x^2/2\sigma^2)$ for $(x/\sigma) \gtrsim 5$ even if $\Delta \simeq 10^{-3}\sigma$. Hence the natural width will dominate over the Doppler width away from the center of the line.

When all the three sources of line broadening are present, the net line profile can be determined by taking the Lorentzian profile due to natural and collisional broadening and convolving it with the probability distribution for the velocities. Since the probability for the velocity to be $v_z$ is given by a Maxwellian distribution at some temperature $T$, the net line profile can be easily shown to be

$$ \phi_{\text{tot}}(\omega) = \frac{2}{\sqrt{\pi}} \frac{a}{(\Delta\omega)_D} \int\limits_{-\infty}^{\infty} dq \frac{e^{-q^2}}{(q - u)^2 + a^2} \qquad (2.116) $$

where

$$ u = \left( \frac{\omega - \omega_0}{\omega_0} \right) \left( \frac{c}{v_T} \right); \quad a = \left( \frac{\Gamma'}{4\omega_0} \right) \left( \frac{c}{v_T} \right); \quad (\Delta\omega)_D = \omega_0 \left( \frac{v_T}{c} \right). \qquad (2.117) $$

The function in the right hand side of Eq. (2.116) is called the *Voigt function* which is Gaussian near the origin and Lorentzian away from the origin. A good approximation to the Voigt profile with correct asymptotic limits is

$$\phi = \left(\frac{2\sqrt{\pi}}{\Delta\omega_D}\right)\left[e^{-u^2} + \frac{a}{\sqrt{\pi}u^2}\right]. \tag{2.118}$$

This profile is used extensively in the study of spectral lines from astrophysical systems.

## Exercises

### Exercise 2.1 *Numerical estimates for light:*

Estimate (i) the total intensity $I_\nu d\nu$ (in Watts cm$^{-2}$), (ii) amplitude of electric field (in Volts cm$^{-1}$) and (iii) number density of photons in logarithmic band of frequency, $n_\nu(d\nu/\nu)$, for: (a) A mercury lamp, (b) Laser.

### Exercise 2.2 *Aspects of intensity:*

A spherical source of radiation (with radius $R$) has a uniform intensity $I_\nu$. (a) Show that the total flux of radiation from the source at a distance $r$ from the center of the source will be

$$S_\nu \equiv \int_\Omega I_\nu \cos\theta d\Omega = \pi I_\nu \left(\frac{R}{r}\right)^2. \tag{2.119}$$

(b) The flux radiated from the surface (at $r = R$) is $\pi I_\nu$ while the total solid angle is $4\pi$. Where did the factor 4 vanish?

### Exercise 2.3 *How dark is it?*

Compare (i) the energy incident on your retina per second when your eye is closed due to photons inside your eye (by treating the cavity of the eye as a black body at body temperature) with (ii) the energy flux of photons in your eye when you are viewing a 100 W bulb at a distance of 1 m. Explain the result. [Hint: You will find that (i) is larger than (ii) by a factor of about 6500. The reason it is dark when you close your eye has to do with the frequency band in which the two radiations peak and the sensitivity of the eye rather than to the incident energy densities. If you compute the energy density between (400 – 800) nm in the two cases, you will find that the light bulb dominates.]

### Exercise 2.4 *Power in/of radio astronomy:*

(a) Consider a radio telescope of collecting area $A$ (about $\pi 150^2$ m$^2$ for Arecibo) that is used to observe Jupiter (brightness temperature $T = 122$ K) at the frequency $\nu = 30$ GHz. Compute the total power in Watts received from Jupiter which has a radius $r_J \approx 7.1 \times 10^4$ km and is at a distance of $d \approx 4.2$ AU. [Hint:

Jupiter subtends a solid angle $\Omega \propto (r_J/d)^2$. In the Rayleigh-Jeans approximation, the specific intensity is $B_\nu = 2k_B T/\lambda^2$ Jy ster$^{-1}$; the observed flux density is $B_\nu \Omega$ and the received flux is $\nu B_\nu \Omega$. Hence the total power received is $\nu B_\nu \Omega A$.]

(b) Why should stimulated emission be important in the radio but not in ultra violet?

### Exercise 2.5    *By Jove, a planet:*

Jupiter can be approximated as a blackbody with a temperature of 122 K. If it has a radius of $7.1 \times 10^4$ km, calculate the bolometric luminosity of Jupiter. How does it compare with that of Sun? Suppose you have a satellite based telescope, operating at the wavelength at which Jupiter's emission is peaked, and is sensitive up to apparent magnitude of 30 (this is typically the sensitivity Hubble Space Telescope has but, of course, Hubble is sensitive at visible band). Will a planet like Jupiter be bright enough for detection at a distance of 10 pc? If the angular resolution of the telescope is 0.1 arcsec, what is the maximum distance from which it can resolve Jupiter and Sun (they are separated by 5.2 AU)?

### Exercise 2.6    *Doppler effect on temperature:*

An observer is moving with speed $v$ through a region containing isotropic blackbody radiation of temperature $T$. Show that the intensity of radiation measured by this observer, in a direction making angle $\theta$ with respect to the direction of motion, will be Planckian in form but with a Doppler shifted temperature

$$T' = T \left( \frac{\sqrt{1 - (v^2/c^2)}}{1 - (v/c)\cos\theta} \right). \tag{2.120}$$

(Incidentally, since $T'$ depends on $\theta$, this result shows that there is no such thing as 'Lorentz transformation law for temperature'.) (a) Expand this result in a Taylor series in $v/c$ and show that the lowest order term is a dipole. The motion of our local group through the Cosmic Microwave Background Radiation (CMBR) will lead to such a dipolar anisotropy of the CMBR temperature. Assuming that this anisotropy has an amplitude of about $(\Delta T/T) = 10^{-3}$, estimate the speed with which the local group is moving through CMBR. (b) Show that, to quadratic order in $(v/c)$, you get a quadrupole moment for temperature distribution. Determine the direction at which the quadrupole produces maximum temperature and its amplitude.

### Exercise 2.7    *Going forward:*

Consider a source of radiation moving at $v/c = 0.999$. Show that: (i) radiation originally emitted at 45 degrees to the direction of motion is seen within 1 degree of the direction of motion. (ii) Let us assume that there was no radiation at 90 degrees to the direction of motion. (This is what will happen, for example, if the charge is moving in a circular path with the acceleration being perpendicular to the velocity.) Show that this "null direction" is at 2.5 degrees with respect to the direction of motion.

## Exercise 2.8    *Clouds and sunsets:*

In an one dimensional geometry with genuine absorption of radiation,the intensity decays exponentially with the optical depth as $e^{-\tau}$. But if the medium is only scattering the photons rather than actually absorbing them, then the situation is different. Since the photons are just performing one dimensional random walk in this case, it will satisfy the diffusion equation. The steady state solution to an one dimensional diffusion equation $d^2n/dl^2 = 0$ varies linearly with $l$. Use this and appropriate boundary condition to show that the transmission coefficient of a rectangular slab of material decreases only as $\tau^{-1}$ (But for this, cloudy days would have been much darker.) Do you think similar results will hold in 3-dimensional geometry?

## Exercise 2.9    *Pushed away lightly:*

Consider an optically thin cloud of material that is ejected by the radiation pressure from a nearby luminous object with a mass-to-luminosity ratio $M/L$. Show that the terminal velocity acquired by the ejected blob is given by

$$v^2 = \frac{2GM}{R} \left[ \frac{\kappa}{4\pi Gc} \frac{L}{M} - 1 \right]. \qquad (2.121)$$

Hence obtain the bound on $M/L$ for the ejection to occur.

## Exercise 2.10    *Temperature is not everything?*

The HII regions have temperatures very similar to effective temperatures of stars. Then how come they don't shine in the visible band as brightly as stars?

## Exercise 2.11    *Dissecting Orion nebula:*

Orion nebula has a radio flux which is $F \propto \nu^2$ for $\nu < \nu_c \approx 1$ GHz and a nearly flat flux for $\nu > \nu_c$. At $\nu = \nu_c$, the flux is about $F_c \approx 300$ Jy. The nebula subtends 4 arcminute at a distance of 500 pc. You are expected to model this as bremsstrahlung arising from a region of ionized hydrogen. (Hint: Use Eq. (2.47).) Given the data on Orion nebula, estimate the temperature and the electron density.

## Exercise 2.12    *Deep trouble:*

An electron moves in a circular orbit under the influence of a magnetic field. Because this is a accelerated motion, the electron dutifully emits synchrotron radiation. But, wait a second. Where does this energy come from, since magnetic field cannot do any work on charged particles?

**Exercise 2.13** *Cosmic rays:*

Electrons in the cosmic rays have been detected with energies up to $3 \times 10^{20}$ eV. The diffuse magnetic field in the Local Group of galaxies (made of Milky Way, Andromeda and a few tens of other smaller galaxies), over a region of size of 1 Mpc, is about $10^{-7}$ Gauss. What is the synchrotron energy loss time compared with the time to cross the Local Group, for the cosmic ray electron traveling perpendicular to the magnetic field. For comparison, compute the same for the galactic disk with a size of 1 kpc and magnetic field of $3 \times 10^{-6}$ Gauss. What is the moral of the story?

**Exercise 2.14** *Angular size of synchrotron source:*

Argue that if Compton losses should not be too large, then the angular size of a synchrotron source of total flux $F_\nu$ is bounded by $\theta > 10^{-3}$ arc sec $(F_\nu/\mathrm{Jy})^{1/2} (\nu_c/\mathrm{GHz})^{-9/10}$.

**Exercise 2.15** *Ironing it out:*

Compute the energy difference between $n = 1$ and $n = 2$ energy levels of an iron atom which has been stripped off all but but one of its electrons. (This is the $K\alpha$ line used in x-ray astronomy.) Also find the ionisation energy of this system. (This is called the K-edge.)

**Exercise 2.16** *Collisions and the Lorentzian profile:*

Using the argument given in the text, prove that collisions also lead to a Lorentzian profile for the line width. [Hint: What is the Fourier transform of $\theta(t) \exp(-\Gamma t/2)$?]

**Exercise 2.17** *Paradox?*

We argued that surface brightness is independent of distance. Then how come we use an inverse square law to describe the decrease of stellar flux? Why isn't stellar surface brightness constant making the flux independent of distance?

**Exercise 2.18** *Actually, it is quite simple:*

When photons and electrons interact through Compton and inverse Compton scattering, the evolution of photon number density $n(\nu, t)$ in the frequency space is described by the equation

$$\frac{\partial n}{\partial t} = \frac{h n_e \sigma_T}{m_e c} \frac{1}{\nu^2} \frac{\partial}{\partial \nu} \nu^4 \left[ \frac{k_B T}{n} \frac{\partial n}{\partial \nu} + n + n^2 \right] \qquad (2.122)$$

called Kompaneets equation. This, rather formidable looking, equation is quite easy to obtain if you go about it the right way. (a) Show that Compton processes conserve the number of photons requiring the evolution equation to have the general form

$$\frac{\partial n}{\partial t} = -\frac{1}{\nu^2}\frac{\partial}{\partial \nu}\nu^2 j(\nu) \tag{2.123}$$

where $j(\nu)$ is the "radial" current in the moment space. (b) Using the results obtained in the text, you can now show that there are two pieces to this current. The Compton scattering, leading to $\langle \delta\nu/\nu \rangle \approx -h\nu/m_e c^2$, will give a drift velocity $d\nu/dt = -n_e \sigma_T h\nu^2/m_e c$. Show that this gives a current $nd\nu/dt$. (b) Argue that, in addition, there is also a diffusion in the frequency space giving $j = -Ddn/d\nu$ with the diffusion coefficient $D = (1/3)n_e \sigma_T c\langle (\Delta\nu)^2\rangle$. Finally show that $\Delta\nu = \nu(v/c)$ and combining the two terms, derive the Kompaneets equation.

## Notes and References

There are several text books which deal with radiative processes in great detail. For conceptual clarity, the best one is Landau and Lifshitz (1975a). Astrophysical applications are stressed in Shu (1991) and Rybicki and Lightman (1979). Detailed derivations of many of the results that are quoted without proof in this section can be found in these references.

1. *Section 2.1.1:* This derivation of radiation field is based on Thomson (1907). A more accessible reference in which this derivation is repeated is Berkeley Volume 3, Crawford (1968).

# Chapter 3

# Matter

## 3.1 Equations of state

In the study of different astrophysical systems, you have to deal with matter differing widely in its physical properties. Take, for example, the density: the intergalactic medium has $10^{-28}$g cm$^{-3}$, sun-like star has 1g cm$^{-3}$ and a neutron star has $10^{14}$g cm$^{-3}$ — with a dynamic range of 42 orders of magnitude! Similar variation occurs in many other parameters and — as a consequence — astrophysical systems come with matter existing in all the four different states. How do we tackle this enormous diversity ?

It turns out that, by and large, the description of matter can be simplified by the use of *equation of state*. (There are systems with large number of particles — like galaxies made of stars — for which one cannot write down an equation of state; such systems are, indeed, more difficult to model.) In general, equation of state allows you to determine the pressure in terms of other physical parameters, most importantly density and temperature. Since pressure determines the force balance, which determines the motion and dynamics, we can make considerable progress once we know the equation of state.

Let us begin with a system of $N$ particles ($N \gg 1$), each of mass $m$, occupying a region of volume $V$. The pressure exerted by a sufficiently large number of particles can be obtained as the momentum transfered per second normal to a (fictitious) surface of unit area (which we take to be in the yz plane). The contribution to rate of momentum transfer (per unit area) from particles of energy $\epsilon$ will be $n(\epsilon)p_x(\epsilon)v_x(\epsilon)$ where $n(\epsilon)$ denotes the number of particles per unit volume with momentum $\mathbf{p}(\epsilon)$ and velocity $\mathbf{v}(\epsilon)$. The net pressure is obtained by averaging this expression over the angular distribution of velocities and summing over all values of the energy.

Since the momentum and velocity are parallel to each other, the product $p_x v_x$ averages to $(1/3)pv$ (in 3-dimensions) giving

$$P = \frac{1}{3} \int_0^\infty n(\epsilon)\, p(\epsilon)\, v(\epsilon) d\epsilon \qquad (3.1)$$

where the integration is over all energies. This result is of very general validity since it is purely kinematical.

Of course, you can do something with it only if you know the functions $p(\epsilon)$ and $n(\epsilon)$. This is easiest for the systems called *ideal* in which the kinetic energy dominates over the interaction energy of the particles. In that case $\epsilon$ is essentially the kinetic energy of the particle. Further, since we may want to deal with both relativistic and nonrelativistic particles, it is good to use the exact expression for kinetic energy. Using the relations

$$p = \gamma m v; \quad \epsilon = (\gamma - 1)mc^2; \quad \gamma \equiv \left(1 - \frac{v^2}{c^2}\right)^{-1/2}, \qquad (3.2)$$

where $\epsilon$ is the *kinetic* energy of the particle, the pressure can be expressed in the form

$$P = \frac{1}{3} \int_0^\infty n\epsilon \left(1 + \frac{2mc^2}{\epsilon}\right) \left(1 + \frac{mc^2}{\epsilon}\right)^{-1} d\epsilon. \qquad (3.3)$$

This expression simplifies significantly in the following two limits: non-relativistic (NR) and extreme relativistic (ER). In the NR limit (with $mc^2 \gg \epsilon$), Eq. (3.3) gives $P_{\rm NR} \approx (2/3)\langle n\epsilon \rangle = (2/3)U_{\rm NR}$ where $U_{\rm NR}$ is the energy *density* (i.e., energy per unit volume) of the particles. In the ER case (with $\epsilon \gg mc^2$ or when the particles are massless), the corresponding expression is $P_{\rm ER} \approx (1/3)\langle n\epsilon \rangle = (1/3)U_{\rm ER}$. Hence, in general, $P \approx U$ up to a factor of order unity. In these two limits, the kinetic energy has the limiting forms:

$$\epsilon = \sqrt{p^2c^2 + m^2c^4} - mc^2 = \begin{cases} p^2/2m & (p \ll mc; NR) \\ pc & (p \gg mc; ER). \end{cases} \qquad (3.4)$$

To proceed further and determine the actual behaviour of the system we need to know the origin of the momentum distribution of the particles. The familiar situation is the one in which short-range interactions ('collisions') between the particles effectively exchange the energy so as to randomize the momentum distribution. This will happen if the effective mean free path of the system $l$ is small compared to the length scale $L$ at which physical parameters change. When such a system is in steady state, one can assume that local thermodynamic equilibrium, characterized by a local temperature $T$, exists in the system. Then, the probability for occupying a state with energy $E$ is given by the usual Boltzmann distribution, $P(E) \propto \exp[-(E/k_BT)]$, and the mean energy will be of the order of $k_BT$. Using Eq. (3.4) with $\epsilon \simeq k_BT$ we get:

$$p \cong \begin{cases} (2mk_BT)^{1/2} & (k_BT \ll mc^2; \text{ NR}) \\ (k_BT/c) & (k_BT \gg mc^2; \text{ ER}). \end{cases} \tag{3.5}$$

In this case, the momentum and kinetic energy of the particles vanish when $T \rightarrow 0$.

This limit contains, in particular, the important special case of a non relativistic ideal gas with the equation of state $PV = Nk_BT$ which corresponds to $\epsilon = (3/2)k_BT$ and $U_{NR} = N\epsilon$. Considering its importance, we will a spend a moment, recalling its key properties. Depending on the context, astronomers write the equation of state for ideal gas in several equivalent forms:

$$P = \frac{N}{V}k_BT = \left(\frac{Nm_H}{V\rho}\right)\left(\frac{k_B}{m_H}\right)\rho T = \frac{\mathcal{R}}{\mu}(\rho T) \tag{3.6}$$

with

$$\mu \equiv \frac{V\rho}{Nm_H}; \qquad \mathcal{R} \equiv \frac{k_B}{m_H} \equiv N_A k_B \tag{3.7}$$

where we have defined the gas constant $\mathcal{R} \equiv (k_B/m_H) \equiv N_A k_B$, the Avogadro number $N_A = m_H^{-1}$ and the mean molecular weight $\mu$. The inverse of the molecular weight gives the effective number of particles per hydrogen atom of the gas. Clearly the value of $\mu$ depends on the number of *particles* contributed by each atom which, in turn, will depend on whether the atom is ionized or not. (Note that both free electrons and ions contribute to pressure while the mass is mostly contributed by the ions.) It is therefore easy to see that

$$\frac{1}{\mu m_H} = \frac{1}{m_H}\sum_j \left(\frac{1+Z_j}{A_j}\right)X_j, \tag{3.8}$$

where $A_j$ is the atomic weight of the $j$th species of the particles, $X_j$ is the fraction of the total mass contributed by the $j$th species and $Z_j$ is the effective number of electrons supplied by each atom of the $j$th species. If we are interested only in the pressure contributed by electrons (or ions), we can use the equation Eq. (3.6) with $\mu$ replaced by $\mu_{\text{ele}}$ (or $\mu_{\text{ion}}$) respectively.

For example, the number density of electrons in a hydrogen plasma with the ionisation fraction $x_H$ will be

$$n_e = \left(\frac{\rho}{m_H}\right)\mu_e^{-1} \cong \left(\frac{\rho}{2m_H}\right)(1 + X_H) \tag{3.9}$$

which defines $\mu_e$ and is a useful relation in estimating processes related to electrons. Finally, we recall for future reference that the entropy of an ideal gas can be expressed in the form:

$$S = C_V \ln[TV^{(\gamma-1)}] \propto \ln[T^{1/(\gamma-1)}V] \propto \ln[PV^\gamma] \propto \ln[PT^{\gamma/(\gamma-1)}] \tag{3.10}$$

where $\gamma = C_P/C_V$ is the ratio of the specific heats. For an ideal mono-atomic gas, $\gamma = 5/3$. It follows that for adiabatic processes, $PV^\gamma$ as well as the other combinations inside the square brackets in Eq. (3.10) are constant. The equivalent relation, $P \propto \rho^\gamma$, for adiabatic processes, will be used extensively in the later chapters.

All this is fine for classical particles for which momentum and energy vanish at zero temperature. Quantum mechanical fermions, like electrons, behave differently. The mean energy of a system of electrons will *not* vanish even at zero temperature, because electrons obey *Pauli exclusion principle* which requires that the maximum number of electrons which can occupy any quantum state is 2, one with spin up and another with spin down. Since uncertainty principle requires $\Delta x \Delta p_x \gtrsim 2\pi\hbar$, we can associate $(d^3x d^3p)/(2\pi\hbar)^3$ micro-states with a phase space volume $d^3x\, d^3p$. Therefore, the number of quantum states with momentum *less* than $p$ is $[V(4\pi p^3/3)/(2\pi\hbar)^3]$ where $V$ is the spatial volume available for the system. The lowest energy state for a system of $N$ electrons will be the one in which the electrons fill all the levels up to some momentum $p_F$ (called *Fermi momentum*). This gives,

$$n = \left(\frac{N}{V}\right) = 2\frac{(4\pi p_F^3/3)}{(2\pi\hbar)^3} = \frac{1}{3\pi^2}\left(\frac{p_F}{\hbar}\right)^3 \tag{3.11}$$

so that $p_F \equiv \hbar(3\pi^2 n)^{1/3}$. Hence, at $T = 0$, all the states with $p < p_F = \hbar(3\pi^2 n)^{1/3}$ will be occupied by electrons; that is, even at $T = 0$ electrons will have nonzero momentum (and energy). Further if $p_F$ is of the order of or greater than $mc$ (which can happen if $n$ is large enough), then a fraction of electrons will be relativistic, even at $T = 0$ !

The above analysis, of course, is strictly valid only at $T = 0$. At finite temperatures, electrons will also have a thermal energy $k_B T$ which should be compared with the energy $\epsilon_F$ (called *Fermi energy*) corresponding to

$p_F$ to decide whether the system is classical or quantum mechanical. The quantum mechanical effects will be dominant if $\epsilon_F \gtrsim k_B T$ (and the system is called *degenerate*) while the classical theory will be valid for $\epsilon_F \ll k_B T$ (*non degenerate*). This energy $\epsilon_F$ will, in turn, depend on whether the system is NR or ER:

$$\epsilon_F = \begin{cases} \frac{p_F^2}{2m} = \left(\frac{\hbar^2}{2m}\right)\left(3\pi^2 n\right)^{2/3} & \text{(NR)} \\ p_F c = (\hbar c)\left(3\pi^2 n\right)^{1/3} & \text{(ER).} \end{cases} \tag{3.12}$$

(Note that the result $p_F = \hbar(3\pi^2 n)^{1/3}$ is independent of whether the particles are relativistic or not. But the relation between the energy and the momentum has different limiting forms in these two cases and hence the dependence of $\epsilon_F$ on $n$ is different.) The ratio $(\epsilon_F/k_B T)$, which determines the degree of degeneracy (that is, how quantum mechanical is the system under consideration) is

$$\left(\frac{\epsilon_F}{k_B T}\right) \simeq \begin{cases} \frac{1}{2}\left(3\pi^2\right)^{2/3}\left(\frac{\hbar^2}{m}\frac{n^{2/3}}{k_B T}\right) & \text{(NR)} \\ \left(3\pi^2\right)^{1/3}\left(\frac{\hbar c}{k_B T}n^{1/3}\right) & \text{(ER)} \end{cases} \tag{3.13}$$

with the two limiting forms being valid for $n \ll (\hbar/mc)^{-3}$ (NR) and $n \gg (\hbar/mc)^{-3}$ (ER) respectively. In the first case with

$$n \ll \left(\frac{\hbar}{m_e c}\right)^{-3} \simeq 10^{31} \text{ cm}^{-3}; \quad \rho = m_p n \ll 10^7 \text{ gm cm}^{-3}, \tag{3.14}$$

the system is non relativistic; it will also be degenerate if $(\epsilon_F/k_B T) \gg 1$ and classical otherwise. The transition occurs at $\epsilon_F \simeq k_B T$ which corresponds to

$$nT^{-3/2} = \frac{(mk_B)^{3/2}}{\hbar^3} = 3.6 \times 10^{16} \tag{3.15}$$

in c.g.s. units. To see what these numbers mean, let us estimate the ratio $\mathcal{R} \equiv (k_B T/\epsilon_F) \propto (mk_B T/\hbar^2 n^{2/3})$ for some astrophysical systems. The solar wind ($n \simeq 10$ cm$^{-3}$, $T \simeq 10^6$ K) has $\mathcal{R} \simeq 10^{20}$ which makes it totally classical; at the center of the Sun ($n \simeq 10^{26}$ cm$^{-3}$, $T \simeq 10^7$ K) we get $\mathcal{R} \simeq 5$. This is uncomfortably close to unity and one should realize that quantum effects are not completely ignorable at the center of the Sun. At the center of a typical white dwarf ($n \simeq 10^{30}$ cm$^{-3}$, $T \simeq 10^6$ K) we have $\mathcal{R} \simeq 10^{-3}$ making it totally quantum mechanical.

In the second case in Eq. (3.12) (with $n \gg (\hbar/m_e c)^{-3} \simeq 10^{31}$ cm$^{-3}$; $\rho \equiv m_p n \gg 10^7$ gm cm$^{-3}$), electrons have $p_F \gg m_e c$ and are relativistic *irrespective of the temperature*. The quantum effects will dominate thermal effects if $k_B T \ll (\hbar c) n^{1/3}$ and we will have a relativistic, degenerate gas. Figure 3.1 summarizes the different regions in the $n - T$ plane.

In general, the kinetic energy of the particle will have contributions from both the temperature as well as Fermi energy and one needs to use the full Fermi-Dirac distribution. But if we are only interested in the asymptotic limits, one can approximate the total kinetic energy per particle to be $\epsilon \approx \epsilon_F(n) + k_B T$. Note that such a system has a *minimum* energy $N\epsilon_F(n)$ even at $T = 0$.

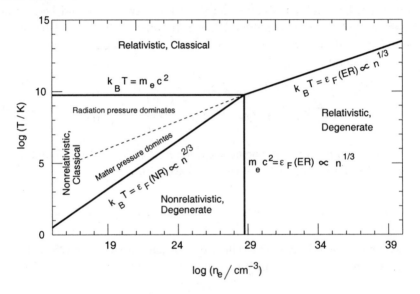

Fig. 3.1    Regions with different equations of state in the $n - T$ plane.

Let us now get back to the pressure exerted by such a gas of electrons. Using our general result $P \simeq n\epsilon$ in Eq. (3.3), we can obtain the equation of state for the different cases discussed above. First, if the system is classical with $k_B T \gg \epsilon_F$, so that $\epsilon \cong k_B T$ then $P \simeq nk_B T$ in both NR and ER limits. This is just the ideal classical gas of particles or, say, photons. Second, for a quantum mechanical gas of fermionic particles with $k_B T \ll \epsilon_F$, and $\epsilon \approx \epsilon_F$, it follows from Eq. (3.12) that $P \simeq n\epsilon_F$. This pressure varies as the (5/3) rd power of density in the non-relativistic case and as $(4/3)$ rd power of density in the relativistic case. [Remember that whether the system is relativistic or not is decided by the ratio $(p_F/mc)$ or — equivalently — the ratio $(\epsilon_F/mc^2)$ with the transition occurring at $n \approx (\hbar/mc)^{-3}$.] Including all the proportionality factors, we get, in the non relativistic limit:

$$P_{\text{NR}} = \frac{(3\pi^2)^{2/3}}{5} \frac{\hbar^2}{m_e} \left(\frac{\rho}{m_p}\right)^{5/3} \equiv \lambda_{\text{NR}} \rho^{5/3} \qquad (3.16)$$

and in the relativistic limit:

$$P_R = \frac{1}{4}(3\pi^2)^{1/3} \hbar c \left(\frac{\rho}{m_p}\right)^{4/3} \equiv \lambda_R \rho^{4/3}. \qquad (3.17)$$

In general, if one writes the pressure in terms of density as $P \propto \rho^\gamma$, then $\gamma$ is $(5/3)$ in the non relativistic limit ($p_F \ll mc$) all the way close to about $p_F \approx 0.6mc$; it decreases rapidly to $\gamma = 4/3$ in the transition range $(0.6 - 10)mc$ and stays close to $4/3$ for $p_F \gtrsim 10mc$.

## 3.2 Self-gravitating barotropic fluids

Let us now consider an important application of the above results. One of the common situations in astrophysics corresponds to the case in which the matter remains in static equilibrium with the pressure gradient balancing the gravitational force. In this case, the conditions of equilibrium become

$$\nabla P = -\rho \nabla \phi; \qquad \nabla^2 \phi = 4\pi G \rho \qquad (3.18)$$

where the second equation assumes that the fluid is self-gravitating; ie., the gravitational force is only due to the mass density of the fluid. Combining these equations we get $\nabla \cdot (\nabla P/\rho) = -4\pi G \rho$. For a non rotating body in thermal equilibrium, $P$ and $\rho$ will be spherically symmetric and this equation reduces to:

$$\frac{1}{r^2} \frac{d}{dr} \left(\frac{r^2}{\rho} \frac{dP}{dr}\right) = -4\pi G \rho. \qquad (3.19)$$

We cannot, of course, integrate this equation until another relation between $P$ and $\rho$ is given. Usually, such a relation will involve the temperature thereby requiring further equations to close the systems. But there is one special situation of considerable importance in astrophysics in which pressure can be expressed purely as a function of density like in the two cases in Eq. (3.16) or Eq. (3.17). (Such fluids are called *barotropic*.) If the system is barotropic, we will have a relation $P = P(\rho)$ and the above equation can be immediately integrated.

One simple choice, which has applications in several astrophysical contexts, is given by the *polytropic equation of state* for which

$$P = K\rho^{(n+1)/n} = K\rho^{\gamma} \qquad (3.20)$$

where $K$ and $\gamma = (n+1)/n$ are constants. In this case, it is possible to use scaling arguments to obtain several important conclusions regarding the equilibrium configuration.

To do this, let us transform the variables in Eq. (3.19) to an independent variable $\xi$ and a dependent variable $\theta$ by the relations $\rho = \rho_c \theta^n, r = \alpha \xi$ where $\rho_c = \rho(0)$ is the central density and the constant $\alpha$ is chosen to be

$$\alpha = \left[ \frac{n+1}{4\pi G} K \rho_c^{(1-n)/n} \right]^{1/2}. \qquad (3.21)$$

Straightforward algebra now converts Eq. (3.19) to a differential equation for $\theta(\xi)$:

$$\frac{1}{\xi^2} \frac{d}{d\xi} \left( \xi^2 \frac{d\theta}{d\xi} \right) = -\theta^n \qquad (3.22)$$

called the *Lane-Emden* equation. We need two boundary conditions for integrating this equation which are easy to determine: From the definition, it is clear that we need $\theta = 1$ at $\xi = 0$ in order to produce the correct central density. Further, since the pressure gradient $(dP/dr)$ has to vanish at the origin, it is necessary that $(d\theta/d\xi) = 0$ at $\xi = 0$. Given these two boundary conditions at the origin, Lane-Emden equation can be integrated to find $\theta(\xi)$. The integration proceeds from the origin till a point $\xi = \xi_1$ at which $\theta$ vanishes for the first time. At this point, the density and pressure vanishes showing that the surface of the physical system has been reached. All other physical variables can be expressed in terms of the behavior of $\theta(\xi)$ near $\xi = \xi_1$.

Analytic solutions to Lane-Emden equations are known for the cases with $n = 0, 1, 5$ corresponding to $\gamma = \infty, 2, 6/5$. Denoting the solution for a particular value of $n$ by $\theta_n(\xi)$, you can easily verify that

$$\theta_0 = 1 - \xi^2/6; \qquad \theta_1 = \frac{\sin \xi}{\xi}; \qquad \theta_5 = \left( 1 + \xi^2/3 \right)^{-1/2}. \qquad (3.23)$$

The solutions for $n = 0$ and $n = 1$ vanish at $\xi_1 = \sqrt{6}$ and $\xi_1 = \pi$ respectively. The case $n = 0$ corresponds to a sphere of constant density and the case $n = 1$ (corresponding to $\gamma = 2$) is of some relevance in modeling planetary interiors and brown dwarfs (see Exercise 3.2). On the other hand, for $n = 5$, $\theta_5(\xi)$ does not vanish for any finite value of $\xi$ and the matter distribution extends to infinity. Such a system is not of much relevance in describing gaseous spheres like stars but is used — in a different context — to model galaxies.

Using the transformation equation between the original variables and $\theta$ and $\xi$, we can express all other physical parameters in parametric form. The radius and mass of the system are given by

$$R = \alpha \xi_1 = \left[ \frac{(n+1)K}{4\pi G} \right]^{1/2} \rho_c^{(1-n)/2n} \xi_1, \qquad (3.24)$$

$$M = 4\pi \alpha^3 \rho_c \left[ -\xi^2 \frac{d\theta}{d\xi} \right]_{\xi=\xi_1} = 4\pi \left[ \frac{(n+1)K}{4\pi G} \right]^{3/2} \rho_c^{(3-n)/2n} \left[ -\xi^2 \frac{d\theta}{d\xi} \right]_{\xi=\xi_1}. \qquad (3.25)$$

The mean density $\bar{\rho} = (3M/4\pi R^3)$ is related to the central density by

$$\bar{\rho} = \rho_c \left[ -\frac{3}{\xi} \frac{d\theta}{d\xi} \right]_{\xi=\xi_i}. \qquad (3.26)$$

The central pressure and the gravitational binding energy of this sphere are

$$P_c = \frac{GM^2}{R^4} \left[ 4\pi(n+1) \left( \frac{d\theta}{d\xi} \right)^2_{\xi=\xi_1} \right]^{-1}; \qquad (3.27)$$

$$U_g \equiv \frac{1}{2} \int \rho\phi \, d^3\mathbf{x} = -\frac{3}{5-n} \frac{GM^2}{R}. \qquad (3.28)$$

Eliminating $\rho_c$ between Eq. (3.24) and Eq. (3.25) we can find the mass-radius relationship for these systems:

$$M \propto R^{(3-n)/(1-n)}. \qquad (3.29)$$

Relation like Eq. (3.29) can also be obtained more easily from scaling arguments. In this particular case, for example, equating $\nabla P \simeq (P/R)$ to $(GM\rho/R^2)$ and using $P \propto \rho^{[1+(1/n)]}$, we get $\rho^{1/n} \propto (M/R)$. Further, using $\rho \propto (M/R^3)$, one obtains the scaling $M \propto R^{(3-n)/(1-n)}$. This relation is independent of the detailed nature of the solution and reflects the fact that Lane-Emden equation allows self similar solutions. Note that, for $n < 1$, the

$M(R)$ is an increasing function while for $1 < n < 3$, the mass $M$ *decreases* with increasing $R$.

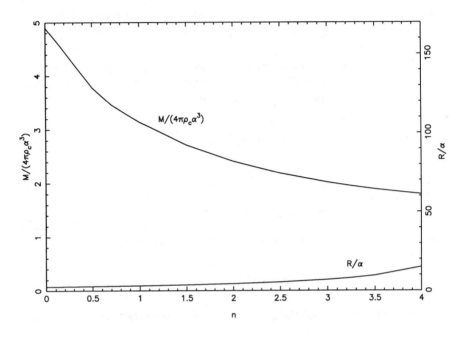

Fig. 3.2   The variation of the radius and mass of Lane-Emden spheres of different indices parameterized by the central density $\rho_c$.

There are two cases of astrophysical significance in which the pressure is provided by degenerate fermions. This situation arises in white dwarfs and neutron stars at the end of stellar evolution. The equation of state is then given by Eq. (3.16) or Eq. (3.17) depending on whether the system is non relativistic or relativistic. In the first case, $\gamma = 5/3$ and $n = (\gamma-1)^{-1} = 3/2$. Then Eq. (3.29) gives $M \propto R^{-3}$ with more massive systems having smaller radius. On the other hand, for $\gamma = 4/3$, we have $n = 3$ and the scaling in Eq. (3.29) shows that $M$ is independent of $R$; this value represents a transition point. We will see in chapter 5 that the limiting value of $M$ plays an important role in the study of neutron stars.

When the Lane-Emden index $n$ tends to infinity, the polytropic index $\gamma$ goes to unity and the equation of state becomes $P = K\rho$. Such a system is called an *isothermal sphere* (since $P = \rho k_B T$ with a constant $T$ leads to this relation). Since polytropic solutions become difficult to handle as

$n \rightarrow \infty$, it is convenient to approach the isothermal sphere in a different manner.

Consider a self gravitating ideal gas with constant temperature $T$, density $\rho(r)$, pressure $P(r) = \rho(r)k_B T$ and a gravitational potential $\phi(r)$ such that the density at any location is given by

$$\rho(r) = \rho_c \exp(-\beta[\phi(r) - \phi(0)]) \qquad (3.30)$$

with $\rho_c$ denoting the central density. This density distribution corresponds to a system in Boltzmann distribution at constant temperature. Combining Poisson equation with Eq. (3.30), we get a differential equation for the gravitational potential:

$$\nabla^2 \phi = 4\pi G \rho_c e^{-\beta[\phi(\mathbf{x}) - \phi(0)]}. \qquad (3.31)$$

In practical applications, isothermal sphere arises when the source (gas of molecules or 'gas' of stars, ....) is described by a distribution function

$$f(\epsilon) = \frac{\rho_0}{(2\pi\sigma^2)^{3/2}} \exp\left(-\frac{\epsilon}{\sigma^2}\right), \qquad (3.32)$$

where $\epsilon = (1/2)v^2 + \phi$. This distribution is parameterized by two constants $\rho_0$ and $\sigma$. One can easily verify that the mean square velocity $< v^2 >$ is $3\sigma^2$ and that the density distribution is $\rho(r) = \rho_0 \exp(-\phi/\sigma^2)$ with the central density $\rho_c = \rho_0 \exp(-\phi(0)/\sigma^2)$. It is conventional to define a *core radius* $r_0$ and a set of dimensionless variables by

$$r_0 = \left(\frac{9\sigma^2}{4\pi G\rho_c}\right)^{1/2}; \quad l = \frac{r}{r_0}; \quad \xi = \frac{\rho}{\rho_c}. \qquad (3.33)$$

Then the Poisson equation $\nabla^2 \phi = 4\pi G\rho$ can be rewritten in the form

$$\frac{1}{l^2}\frac{d}{dl}\left(\frac{l^2}{\xi}\frac{d\xi}{dl}\right) = -9\xi. \qquad (3.34)$$

This equation has to be integrated numerically with the boundary condition $\xi(0) = 1$ and $\xi'(0) = 0$, to determine the density profile. Given the solution to this equation, all other quantities can be determined.

Let us consider the nature of solutions to this equation. Remarkably enough, there exists a simple exact solution to this equation. By direct substitution, you can check that

$$\xi_{sing}(l) = \frac{2}{9l^2} \qquad (3.35)$$

satisfies this equation. This solution, however, is singular at the origin
and hence is not physically admissible. (It clearly violates the boundary
conditions at the origin.) The importance of this solution lies in the fact
that other (physically admissible) solutions tend to this solution for large
values of $r$. Thus all the exact solutions have the behaviour $\rho \propto r^{-2}$,
$M(r) \propto r$ at large $r$; the model has to be cut off at some radius to provide
a finite mass. The numerical solution to Eq. (3.34) starts at unity near
$l = 0$ and behaves like the solution in Eq. (3.35) for large $l$. For $l \lesssim 2$, the
numerical solution is well approximated by the function $\xi(l) = (1 + l^2)^{-\frac{3}{2}}$.

Very often in astronomy, what is measured is the 2-dimensional density
distribution obtained by projecting the 3-dimensional density distribution
on to the plane of the sky. The projected, two dimensional, surface density
corresponding to this $\xi(l)$ is $S(R) = 2(1 + R^2)^{-1}$, where $R$ is the projected
dimensionless radial distance corresponding to $l$. It follows that the (true)
central surface density $2\rho_c r_0$ where $\rho_c$ is the central volume density and $r_0$
is the core radius.

## 3.3  Flows of matter

The entire subject of fluid dynamics revolves around understanding the
motion of fluids in different circumstances which is essentially governed by
a continuity equation and Euler equation. You will find that, curiously
enough, the simple, tractable, forms of fluid motion are rarely of interest
in astrophysics! Either the motion is supersonic or it is turbulent or there
is no motion at all (like in the cases discussed in the last section, in which
the gravitational force is balanced by the pressure gradient). We shall
now describe some of non trivial examples of fluid flow which occur in
astrophysics, beginning with the important example of accretion.

### 3.3.1  *Spherical accretion*

Accretion of matter plays an important role in several astrophysical systems. It also provides a simple example in which a transition from subsonic
to supersonic flow occurs under appropriate conditions and illustrates some
peculiar features of supersonic flow compared to subsonic ones.

In the normal subsonic situations, a fluid flowing through a nozzle of
variable area will speed up in narrow constrictions and slow down when the
cross-sectional area is large. In supersonic flow, a fluid will actually *speed*

*up* when the area of cross section increases! To see this, let us write down the equation governing the one dimensional fluid flow

$$\frac{dv}{dt} = v\frac{dv}{dx} = -\frac{1}{\rho}\frac{dP}{dx} = \frac{c_s^2}{\rho}\frac{d\rho}{dx} \qquad (3.36)$$

where $c_s = (dP/d\rho)^{1/2}$ is the speed of sound and we have written $P = c_s^2\rho$ for the equation of state. This gives, $d\ln\rho = -(v^2/c_s^2)d\ln v$. On the other hand, from the conservation of mass, we have $\rho v A = $ constant where $A$ is the cross-sectional area. Taking the logarithmic derivatives of this equation and using the equation for $d\ln\rho$, we get $d\ln v = -(1 - v^2/c_s^2)^{-1}d\ln A$. This shows that if $v^2 > c_s^2$, then decrease in the cross-sectional area leads to decrease in flow velocity and vice-versa. You can see that our usual intuition is built on subsonic flows and needs to be corrected for supersonic phenomena.

A simple example of supersonic flow occurs in the case matter of accreting to a central object. Consider a spherical object of mass $M$ and radius $R$ embedded inside a gas cloud. The gas flows towards the central mass due to gravitational attraction. In a highly idealized situation, we may consider the flow to be radial at some steady rate $\dot{M}$. Let us further assume that the gas is ideal and polytropic with an index $\gamma$ where $1 < \gamma < 5/3$. For a steady, spherically symmetric flow, mass conservation requires $\rho v r^2$ be a constant, where $v$ is the radial component of the velocity. We will set $\dot{M} \equiv -4\pi r^2 \rho v$ to be the constant accretion rate. Logarithmic derivatives of this equation give:

$$\frac{1}{\rho}\frac{\partial\rho}{\partial r} = -\frac{1}{vr^2}\frac{\partial}{\partial r}\left(vr^2\right). \qquad (3.37)$$

In the Euler equation, $\rho(dv/dt) = -\nabla_r P - \nabla_r\phi$ we can write $(dv/dt) = (\partial v/\partial t) + v(\partial v/\partial r) = v(\partial v/\partial r)$ for a steady, time independent flow. Further, if the force arises from the gravitational field of the central mass then $-\nabla_r\phi = -GM/r^2$. Therefore, the Euler equation becomes:

$$v\frac{\partial v}{\partial r} + \frac{1}{\rho}\frac{\partial P}{\partial r} + \frac{GM}{r^2} = 0. \qquad (3.38)$$

Writing $(\partial P/\partial r) = (\partial P/\partial\rho)(\partial\rho/\partial r) = c_s^2(\partial\rho/\partial r)$, where $c_s = (\partial P/\partial\rho)^{1/2}$ is the speed of sound and using Eq. (3.37), Eq. (3.38) becomes:

$$v\frac{\partial v}{\partial r} - \frac{c_s^2}{vr^2}\frac{\partial}{\partial r}\left(vr^2\right) + \frac{GM}{r^2} = 0. \qquad (3.39)$$

This can be rearranged into a more transparent form:

$$\frac{1}{2}\left(1 - \frac{c_s^2}{v^2}\right)\frac{\partial v^2}{\partial r} = -\frac{GM}{r^2}\left(1 - \frac{2c_s^2 r}{GM}\right) \qquad (3.40)$$

that allows us to understand several features of accretion.

To begin with, note that $c_s^2$ will tend to some constant value at very large distances from the central object. (In general, of course, $c_s^2$ depends on $\rho$ and hence on $r$). Therefore, the factor $\left(1 - 2c_s^2 r/GM\right)$ will be negative definite at large $r$, making the right hand side of Eq. (3.40) positive at large $r$. In the left hand side, we want $\left(\partial v^2/\partial r\right)$ to be negative because we expect the gas to be at rest at large distances and flow with an acceleration as it approaches the central body. Hence $v^2 < c_s^2$ at large $r$; ie., the flow is sub-sonic at large distances. As the gas approaches the central object, the factor $\left(1 - 2c_s^2 r/GM\right)$ will go to zero at some critical radius $r_c$. For the polytropic equation of state $P \propto \rho^\gamma$ we have $c_s^2 = \gamma P/\rho = \gamma k_B T$; then, this critical radius is about

$$r_c = \frac{GM}{2c_s^2\left(r_c\right)} \simeq 7.5 \times 10^{13}\left(\frac{T}{10^4 K}\right)^{-1}\left(\frac{M}{M_\odot}\right) \text{cm}. \qquad (3.41)$$

This radius is much larger than the size of typical central masses encountered in astrophysics around which accretion takes place. Hence we must consider the flow at $r < r_c$ as well. Repeating the above analysis for $r < r_c$ we can conclude that $v^2 > c_s^2$. Hence the flow must make a transition from sub-sonic to supersonic speeds at $r = r_c$. Further, at $r = r_c$, we must have either $v^2 = c_s^2$ or $\left(\partial v^2/\partial r\right) = 0$

All the solutions to Eq. (3.40) can now be formally classified based on their behaviour near $r = r_c$ and are shown schematically in Fig. 3.3: (1) The first type of solutions has $v^2 = c_s^2$ at $r = r_c, v^2 \to 0$ as $r \to \infty$; further, $v^2 < c_s^2$ for $r > r_c$ and $v^2 > c_s^2$ for $r < r_c$. This is the relevant solution that describes accretion which starts out as sub-sonic at large distances and becomes supersonic near the compact object. (2) Consider next the solutions with $v^2 = c_s^2$ at $r = r_c, v^2 \to 0$ as $r \to 0; v^2 > c_s^2$ for $r > r_c$ and $v^2 < c_s^2$ for $r < r_c$. This represents a time reversed situation to spherical accretion describing a wind flowing from the compact object. (We shall discuss this case briefly in Sec. 4.4 as a model for the solar wind.) The flow is sub-sonic near the object and becomes supersonic at large distances. There are also solutions with either (3) $\left(\partial v^2/\partial r\right) = 0$ at $r = r_c; v^2 < c_s^2$ for all $r$ or with (4) $\left(\partial v^2/\partial r\right) = 0$ at $r = r_c; v^2 > c_s^2$ for all $r$. These solutions are either sub-sonic or supersonic in the entire range and hence cannot describe

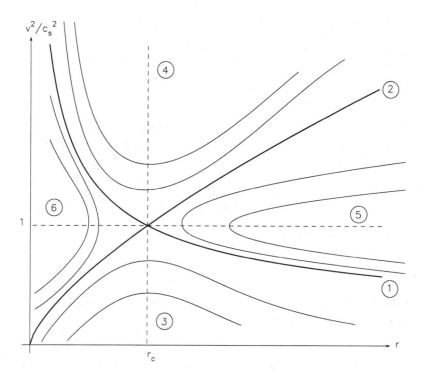

Fig. 3.3 Classification of possible fluid flow patterns in the case of spherical accretion.

accretion onto a compact object of sufficiently small radius. Finally one can
have solutions with either (5) $\left(\partial v^2/\partial r\right) = \infty$ when $v^2 = c_s^2; r > r_c$ always or
with (6) $\left(\partial v^2/\partial r\right) = \infty$ when $v^2 = c_s^2; r < r_c$ always. These two solutions
are unphysical because they give two possible values for $v^2$ at any given $r$.
Thus we conclude that only the type 1 solution is relevant to the problem
of spherical accretion on to a compact object.

To proceed further, we integrate Eq. (3.38) using $(\partial P/\partial \rho) = \gamma (P/\rho) = c_s^2$ to obtain:

$$\frac{v^2(r)}{2} + \frac{c_s^2(r)}{(\gamma - 1)} - \frac{GM}{r} = \text{constant.} \qquad (3.42)$$

Since we must have $v^2 \to 0$ as $r \to \infty$, the constant must be $c_s^2 (\infty) / (\gamma - 1)$.
On the other hand $v^2 (r_c) = c_s^2 (r_c)$ and $(GM/r_c) = 2c_s^2 (r_c)$. Using these

relations we can relate $c_s(r_c)$ to $c_s(\infty)$ getting

$$c_s(r_c) = c_s(\infty) \left[\frac{2}{(5-3\gamma)}\right]^{1/2} \tag{3.43}$$

provided $1 < \gamma < (5/3)$. Since $c_s^2 \propto \rho^{\gamma-1}$, we also get

$$\rho(r_c) = \rho(\infty) \left[\frac{c_s(r_c)}{c_s(\infty)}\right]^{2/(\gamma-1)} . \tag{3.44}$$

Finally, since $r^2\rho v$ is a constant , we can write $\dot{M} = 4\pi r_c^2 \rho(r_c) c_s(r_c)$, using which we can relate the accretion rate $\dot{M}$ to the conditions of the gas at infinity. Straight forward algebra gives

$$\dot{M} = \pi G^2 M^2 \frac{\rho(\infty)}{c_s^3(\infty)} \left[\frac{2}{5-3\gamma}\right]^{(5-3\gamma)/2(\gamma-1)} . \tag{3.45}$$

Numerically, this corresponds to the accretion rate

$$\dot{M} = 1.4 \times 10^{11} \text{g s}^{-1} \left(\frac{M}{M_\odot}\right)^2 \left(\frac{\rho(\infty)}{10^{-24}\text{g cm}^{-3}}\right) \left(\frac{c_s(\infty)}{10\,\text{km s}^{-1}}\right)^{-3} \tag{3.46}$$

if $\gamma = 1.4$, say. To complete the solution we find $v(r)$ is terms of $c_s(r)$ from the conservation of $\dot{M}$. This gives

$$-v = \frac{\dot{M}}{4\pi r^2 \rho(r)} = \frac{\dot{M}}{4\pi r^2 \rho(\infty)} \left(\frac{c_s(\infty)}{c_s(r)}\right)^{2/(\gamma-1)} . \tag{3.47}$$

Substituting this relation into Eq. (3.42) gives an algebraic equation for $c_s^2(r)$. Since this equation has fractional exponents it has to be solved numerically. Given $c_s^2(r)$, Eq. (3.47) determines $v(r)$ and the relation $\dot{M} = r^2\rho(r)v(r)$ fixes $\rho(r)$. This completes the solution to our problem. We see that there is a steady flow of matter to the central object which starts out subsonically but makes a transition to supersonic flow at a critical radius.

The spherically symmetric accretion is idealistic for several reasons; one reason being that we ignored the existence of magnetic fields in the central regions close to the compact objects. If the infalling matter is a plasma with free charged particles (as it is very likely to have) the magnetic fields will also exert a force on the accreting matter. If we take the field to be that of a dipole with $B(r) = B_s(R_s/r)^3$ where $B_s$ is the magnetic field at the surface of the central mass of radius $R_s$, then the magnetic pressure (which is equal to the energy density in the magnetic field) is $P_{mag}(r) = B^2/8\pi \propto r^{-6}$. The infalling matter exerts an inward ram pressure of the order of $\rho v^2$ which is balanced by the magnetic pressure at some radius $r = r_m$ (called *Alfven radius*) at which $\rho v^2 = B^2/8\pi$. Using $\dot{M} = 4\pi r^2 \rho v$ and $v^2 = 2GM/r$ we get

$$r_m = \left(\frac{1}{8G}\right)^{1/7} \left(\frac{B_s^4 R_s^{12}}{M \dot{M}^2}\right)^{1/7}. \tag{3.48}$$

For a $1 M_\odot$ neutron star with $L = L_E$, (see Eq. (2.38)) $B_s = 10^{12}$ G and $R_s = 10$ km, we find the Alfven radius to be about $r_m = 1200$ km $= 120$ $R_s$. Obviously, one cannot ignore magnetic fields close ($r \lesssim r_m$) to the compact object.

### 3.3.2 *Accretion disks*

In several astrophysical contexts, the infalling matter will have non-zero angular momentum which implies that the accretion cannot be spherically symmetric. The opposite extreme — which is often closer to reality — is *disk accretion* in which the matter forms a, planar, disk like structure around the central object and undergoes accretion. Obviously, there must be some processes which transfer angular momentum outward from the infalling matter, if this has to occur. We shall now describe some basic features of such a *thin* accretion disk in which the particles at any given radius move in a nearly Keplerian orbit to a high degree of accuracy.

Let us consider the radial structure of such a disk *in steady state*, assuming that some mechanism for angular momentum loss exists in the disc. Because the disk is assumed to be thin, the velocity of gas has only the radial component $v_r$ and azimuthal component $v_\phi$ with $v_r \ll v_\phi$. The amount of mass $\dot{M}$ crossing any radius $R$ per unit time is given by $-2\pi R \Sigma v_r(R) = \dot{M}$, where $\Sigma(R)$ is the *surface* density of the disk (obtained by integrating the mass density along the direction normal to the disc, say the $z$−direction). In a Keplerian disc, in which the particles move in nearly circular Keplerian orbits at any radius, the angular velocity of rotation varies as $\Omega(R) = (GM/R^3)^{1/2}$, so that the specific angular momentum varies as $R^2 \Omega \propto R^{1/2}$ which decreases with decreasing $R$. Therefore, accretion through successive Keplerian orbits towards the central mass is possible only if the gas can constantly lose angular momentum due to some viscous torque. If we assume that the viscous force *per unit length* around the circumference at any $R$ is given by $\mathcal{F} = \nu\Sigma(Rd\Omega/dR)$ where $\nu$ is some unspecified coefficient of kinematic viscosity, then the viscous torque around the whole circumference will be $G(R) = (2\pi R\mathcal{F})R = \nu\,\Sigma\,2\pi R^3(d\Omega/dR)$.

The direction of the torque is such that the fluid at radius less than $R$, which is rotating more rapidly, feels a backward torque and loses angular momentum while the fluid at radius larger than $R$ gains the angular momentum.

To determine the radial structure of the disk we have to equate this torque to the rate of loss of specific angular momentum. Consider an annular ring of material located between the radii $R$ and $R + dR$. In unit time, a mass $\dot{M}$ enters the ring at $R + dR$ with specific angular momentum $(R + dR)^2[\Omega(R + dR)]$ and leaves at $R$ with specific angular momentum $R^2[\Omega(R)]$. Thus the net angular momentum lost by the fluid per unit time in this ring is $\dot{M}[d(R^2\Omega)/dR]dR$. This angular momentum is lost because of the torque acting at both $R$ and $R + dR$ whose net effect is $(dG/dR)dR$. This leads to the equation

$$\dot{M}\frac{d(R^2\Omega)}{dR} = -\frac{d}{dR}\left[\nu\Sigma\,2\pi R^3\cdot\frac{d\Omega}{dR}\right] \qquad (3.49)$$

Using $\Omega(R) = (GM/R^3)^{1/2}$ and integrating, we get

$$\nu\,\Sigma\,R^{1/2} = \frac{\dot{M}R^{1/2}}{3\pi} + \text{constant.} \qquad (3.50)$$

The constant of integration needs to be fixed using an appropriate boundary condition. One possible way is to assume the existence of an inner radius $R_*$ at which the shear vanishes. So the right hand side of the above equation should vanish at $R_*$ which determines the integration constant. We get

$$\nu\Sigma = \frac{\dot{M}}{3\pi}\left[1 - \left(\frac{R_*}{R}\right)^{1/2}\right]. \qquad (3.51)$$

The actual value of $R_*$ will depend on the specific scenario. For example, if the compact object is a black hole, it is natural to take $R_*$ to be the radius of the smallest stable circular orbit, $R_* = 6(GM/c^2)$, (as we shall see in chapter 5). In the case of accretion on to an unmagnetized white dwarf or neutron star, $R_*$ can be taken to be the radius of the compact object. In the case of a magnetized compact object, the situation is much more complicated and $R_*$ is determined by the balance between the pressure of the magnetic field and ram pressure and will be approximately equal to the Alfven radius.

Given the radial structure of the disc, one can immediately compute the viscous dissipation rate per unit area, given by

$$D(R) = \nu \Sigma \left( R \frac{d\Omega}{dR} \right)^2 = \frac{3GM\dot{M}}{4\pi R^3} \left[ 1 - \left( \frac{R_*}{R} \right)^{1/2} \right]. \qquad (3.52)$$

The energy released between $R$ and $R + dR$ is about $(3GM\dot{M}/2R^2)dR$ for $R \gg R_*$ which is 3 times larger than the change in the orbital energy $(GM\dot{M}/2R^2)dR$. The excess actually comes from the energy released at smaller radii and transported to $R$ by viscous forces. The corresponding total luminosity, obtained by integrating $D(R)$, is

$$L = \int_{R_*}^{\infty} D(R)\, 2\pi R\, dR = \frac{GM\dot{M}}{2R_*}. \qquad (3.53)$$

Note that $L$ is half of the total gravitational potential energy lost in the drop from infinite distance to $R_*$. The other half of the potential energy is actually present in the form of kinetic energy of the fluid at $R_*$ and is not available for dissipation as heat. In the case of accretion to a black hole, this energy is lost inside the hole while for accretion onto a compact object with a surface, this energy is released in a boundary layer or in an accretion column. In general, such a dissipation is fairly complex and depends on the details.

A thin disk in steady state radiates energy through its top and bottom surfaces. If one assumes that the spectrum is a blackbody with surface temperature $T_s(R)$, then we must have $2\sigma T_s^4(R) = D(R)$, (the factor 2 arising from the existence of *two* sides to the disc), implying

$$T_s = \left\{ \frac{3GM\dot{M}}{8\pi R^3 \sigma} \left[ 1 - \left( \frac{R_*}{R} \right)^{1/2} \right] \right\}^{1/4}$$

$$= 6.8 \times 10^5 \eta^{-1/4} \left( \frac{L}{L_{\rm Edd}} \right)^{1/2} L_{46}^{-1/4} \mathcal{R}^{1/4}(x) x^{-3/4} \text{ K} \qquad (3.54)$$

where $\eta \equiv (L_E/\dot{M}_E c^2)$ is the radiative efficiency in the rest mass units (expected to be about 0.1), $x = (c^2 R/2GM)$, $\mathcal{R} = [1 - (R_*/R)^{1/2}]$ and $L_{46} = (L/10^{46} \text{ erg s}^{-1})$. Far from the inner edge, $\mathcal{R} \approx 1$, $T_s \propto x^{-3/4} \propto R^{-3/4}$ which is the characteristic dependence of temperature on the radius for the thin accretion discs. The overall temperature scale varies as $(\dot{M}/M^2)^{1/4} \propto (L/L_{\rm Edd})^{1/2} L^{-1/4}$.

This analysis applies to a thin accretion disk in which we are ignoring the vertical thickness of the gaseous disk. To see what this approximation means, note that the vertical $z-$component of the acceleration at the height $z$ above the plane of the disk at a distance $R$ from the center, is given by $g_z = (GM/R^2)(z/R)$. In pressure equilibrium, we must have $dP/dz = -\rho g_z$. Taking an isothermal equation of state $P = \rho c_s^2$, this equation can be easily integrated to give $\rho(z) = \rho(0)\exp[-(\Omega^2 z^2/2c_s^2)]$ where $\Omega$ is the angular velocity in the disk. The vertical scale height $h = \sqrt{2}\,c_s/\Omega = \sqrt{2}\,c_s R/v_\phi$ determines whether the disk is sufficiently thin; the condition being $h/R \approx c_s/v_\phi$ should be small compared to unity.

### 3.3.3   *Shock waves and explosions*

Another kind of flow, that occurs frequently in astrophysical systems, is the propagation of shock waves. Shocks arise in several contexts in which there is a supersonic flow of gas relative to a solid obstacle. In the laboratory frame, this could arise either because of an object moving through a fluid with supersonic velocities or because gas — flowing with supersonic velocities — suddenly encounters an obstacle. To obey the boundary conditions near the solid body, the flow has to change from supersonic to sub-sonic somewhere near the object. But this information cannot reach upstream where the flow is supersonic. Hence the pattern of flow must change very rapidly over a short region resulting in a shock wave. In the frame in which the gas is at rest, we may consider the solid body to be moving with supersonic speed and pushing the gas. This necessarily leads to steepening of density profile at some place ahead of the body. We shall now consider some of the properties of such shock fronts.

Let us idealize the surface of discontinuity as a plane with the gas flow occurring perpendicular to it with the flow being supersonic on the left side of the shock, say. It is convenient to study the situation in a frame in which the shock front is at rest. Let $P_1, \rho_1, u_1$ denote the state of the gas to the left of the shock front and let $P_2, \rho_2, u_2$ be the corresponding values on the right side. The physical quantities on the two sides of the shock front are related by the conservation of mass, energy and momentum. The conservation of mass gives $\rho_1 u_1 = \rho_2 u_2 \equiv j$. The momentum density of the fluid changes from $(\rho_1 u_1)u_1 = \rho_1 u_1^2$ to $(\rho_1 u_1)u_2$ across the shock; since this has to be balanced by the discontinuity in the force per unit area (i.e., discontinuity in the pressure) acting on the fluid element, we get the second conservation law $P_1 + \rho_1 u_1^2 = P_2 + \rho_2 u_2^2$. Finally let us consider the conservation of energy. The energy flux on either side is given by

$$E_n = \rho_n u_n \left( \frac{1}{2} u_n^2 + \frac{1}{\gamma - 1} \frac{P_n}{\rho_n} \right); \quad n = 1, 2 \tag{3.55}$$

where the first term in the bracket is the kinetic energy per unit mass and the second term is the internal energy per unit mass (see Eq. (1.53); $\epsilon$ was internal energy per unit *volume* so that $\epsilon/\rho$ is internal energy per unit *mass*). When $\gamma = 5/3$ we get the standard result of $(3/2)(P/\rho) = (3/2)k_B T$. The discontinuity in the energy across the shock must be equal to the net rate of work done by the pressure difference, requiring $E_2 - E_1 = P_1 u_1 - P_2 u_2$. Simple algebra shows that this condition can be rewritten as a conservation law:

$$\frac{1}{2} u^2 + \frac{\gamma}{\gamma - 1} \frac{P}{\rho} = \text{constant.} \tag{3.56}$$

These three conservation laws — $\rho u = $ constant, $P + \rho u^2 = $ constant and Eq. (3.56) — are good enough to determine the three quantities $(P_2/P_1), (\rho_2/\rho_1)$ and $(T_2/T_1)$ in terms of $\gamma$ and the Mach number $\mathcal{M}_1 = u_1/c_s$ on one side where $c_s^2 = \gamma P/\rho$ is the speed of sound in the medium. The result of solving the conservation equations, by straightforward (but a bit lengthy; try it!) algebra, is

$$\frac{\rho_2}{\rho_1} = \frac{(\gamma + 1) \mathcal{M}_1^2}{(\gamma - 1) \mathcal{M}_1^2 + 2} = \frac{u_1}{u_2}, \tag{3.57}$$

$$\frac{P_2}{P_1} = \frac{1 + \gamma \left( 2\mathcal{M}_1^2 - 1 \right)}{(\gamma + 1)}, \tag{3.58}$$

$$\frac{T_2}{T_1} = \frac{[(\gamma - 1) \mathcal{M}_1^2 + 2][1 + \gamma \left( 2\mathcal{M}_1^2 - 1 \right)]}{(\gamma + 1)^2 \mathcal{M}_1^2}. \tag{3.59}$$

Since we have assumed the flow to be supersonic on side 1 we have $\mathcal{M}_1 > 1$; it is clear from the above equations that $P_2 > P_1, \rho_2 > \rho_1, u_2 < u_1$ and $T_2 > T_1$. The strongest shock which one can have corresponds to the limit $\mathcal{M}_1 \to \infty$. In this case $P_2 \gg P_1$ and $\rho_2/\rho_1 \to (\gamma + 1)/(\gamma - 1)$. If we transform to a frame in which the shock is moving with a speed $v_2 = u_2 - u_1$ then the above conditions give

$$v_2 = \frac{2}{(\gamma + 1)} u_1; \quad \rho_2 = \rho_1 \frac{(\gamma + 1)}{(\gamma - 1)}; \quad P_2 = \frac{2}{(\gamma + 1)} \rho_1 u_1^2. \tag{3.60}$$

A simple example of such a shock wave occurs in the case of an "explosion". An explosion releases large amount of energy $E$ into an ambient gas in a time-scale which is small. This will result in the propagation of a spherical shock wave centered at the location of the explosion. The resulting flow pattern can be determined everywhere if we make the following assumptions: (i) The explosion is idealized as one which releases energy $E$ *instantaneously* at the origin. The shock wave is taken to be so strong that the pressure $P_2$ behind the shock is far greater than the pressure $P_1$ of the undisturbed gas. (ii) The original energy of the ambient gas is negligible compared to the energy $E$ which it acquires as a result of the explosion. (iii) The gas flow is governed by the equations appropriate for an adiabatic polytropic gas with index $\gamma$.

For such a strong shock, $P_2 \gg P_1$ and we can ignore $P_1$. In the same limit, $(\rho_2/\rho_1) \simeq (\gamma + 1)/(\gamma - 1)$ and hence $\rho_2$ is completely specified by $\rho_1$. Thus the strong explosion is entirely characterized by the total energy $E$ and the density $\rho_1$ of the unperturbed gas. Let the radius of the shock front at time $t$ be $R(t)$. Using $E, \rho_1$ and $t$ we can find only one quantity with the dimension of length, viz. $(Et^2/\rho_1)^{1/5}$. Hence, we *must* have

$$R(t) = R_0 \left(\frac{Et^2}{\rho_1}\right)^{1/5}. \qquad (3.61)$$

where $R_0$ is a constant to be determined by solving the equations of motion. All other physical quantities can now be determined in terms of $R(t)$. The speed of shock wave with respect to the undisturbed gas is

$$u_1 = \frac{dR}{dt} = \frac{2R}{5t} = \frac{2}{5} R_0 E^{1/5} \rho_1^{-1/5} t^{-3/5}. \qquad (3.62)$$

The pressure $P_2$, density $\rho_2$ and the speed of shock propagation, $v_2 = u_2 - u_1$, (relative to a fixed coordinate system at the back of the shock) can be determined using our junction conditions in Eq. (3.60) across the shock front. The density remains constant while $v_2$ and $P_2$ decrease as $t^{-3/5}$ and $t^{-6/5}$ respectively. We will find these results (usually called the *Sedov solution*) are useful in modeling supernova remnants (see chapter 5).

These scaling laws obtained above can be given a qualitative interpretation along the following lines. During the time interval $t$ after the explosion, the mass swept up by the fluid will be about $\rho_1 R(t)^3$ and the fluid velocity will be about $R(t)/t$. Therefore, the kinetic energy of the swept up gas will be $\rho_1 R^5 t^2$. On the other hand, the internal energy density will be of the order of $\rho_1 \dot{R}^2$ giving the total internal energy inside the expanding shock to be $\rho_1 \dot{R}^2 R^3$. Equating either of these energies to the total energy $E$ will lead to the scaling relation in Eq. (3.61).

### 3.3.4 *Turbulence*

So far we discussed the flow of fluids ignoring their viscosity. In real life, viscosity can introduce some non trivial effects.A dimensionless parameter of considerable importance in the study of flow of viscous fluids is the Reynolds number $R = (UL/\nu)$ where $U$ and $L$ are the characteristic velocity and length scales of the flow and $\nu \equiv (\mu/\rho)$ is the kinematic viscosity. The equations for viscous flow of a fluid can — in principle — be solved when the flow is steady, for any given boundary condition, for all values of $R$. Such formal solutions, however, may not actually be realized in nature because most of them are unstable. When the steady flow is unstable, any small perturbation will grow in time and destroy the original characteristics of the flow.

The mathematical investigation of this — rather generic — instability is extremely difficult. It is certain that steady flow is stable for sufficiently small Reynolds numbers. Experiments in fluid mechanics suggest that when the Reynolds number $R$ increases, we eventually reach a critical value $R_{\text{crit}}$ beyond which the flow becomes unstable; it is usual to say that *turbulence* sets in at $R = R_{\text{crit}}$. The value of $R_{\text{crit}}$ depends on the nature of the flow as well as the boundary condition and seems to be in the general range of $(10 - 100)$. This issue is of tremendous practical importance in, say, estimating the drag force which acts on a material body moving through the fluid (say, an airplane). For example, the drag force per unit length, $f_{\text{drag}}$ acting on a cylinder of diameter $d$, moving with speed $U$ through a fluid of density $\rho$ can be expressed in the form $f_{\text{drag}} = C_{\text{drag}}(R)(\rho U^2 d/2)$ where $C_{\text{drag}}(R)$, called *drag coefficient*, is only a function of the Reynolds number.

Though a highly turbulent flow cannot be analyzed quantitatively from the basic equations of fluid mechanics, it is possible to make some progress by introducing reasonable physical assumptions regarding the behaviour of energy in the turbulent flow. We assume that, in the mean rest frame of the fluid, the turbulent velocity is homogeneous and isotropic and can be described in terms of coexisting "eddies" of different sizes. Let $\lambda$ be the size of a generic eddy and let $v_\lambda$ and $\epsilon_\lambda$ denote the typical change in the velocity (across the eddy) and energy per unit mass contained in the eddy.

At very large scales, it seems reasonable to assume that viscosity does not play any crucial role and the energy which is injected into the system is merely transferred to the next level of smaller eddies. The rate at which energy can cascade from larger scales to smaller scales is given by

$$\dot{\epsilon} \simeq \left(\frac{1}{2}v_\lambda^2\right)\left(\frac{v_\lambda}{\lambda}\right) \simeq \frac{v_\lambda^3}{\lambda}. \tag{3.63}$$

The first factor gives the energy per unit mass in the eddies of scale $\lambda$ and the second factor $(\lambda/v_\lambda)^{-1}$ gives the inverse time-scale at which the transfer of energy occurs at this scale. (This transfer of energy arises due to the nonlinear term $(\mathbf{v} \cdot \nabla)\mathbf{v}$ in the Euler equation.) In steady state, the energy injected into the system at large scale $L$ has be transferred to some small scale $\lambda_s$, at which it can be dissipated as viscous heat. For the intermediate scales $\lambda_s \ll \lambda \ll L$, the energy cannot be accumulated and must be transferred at a constant rate $\dot{\epsilon}$ all the way down to the smallest scales; it then follows that $\left(v_\lambda^3/\lambda\right)$ is a constant in the intermediate scales allowing us to write

$$v_\lambda \approx U \left(\frac{\lambda}{L}\right)^{1/3} \tag{3.64}$$

where $U$ is the velocity at some macroscopically large scale $L$. Note that the largest eddies carry most of the velocity but the smallest ones carry most of the rotation $\mathbf{\Omega} = \nabla \times \mathbf{v}$ (called *vorticity*), which is about $\Omega \approx v_\lambda/\lambda \propto \lambda^{-2/3}$.

To estimate the smallest scale $\lambda_s$ at which this process stops we have to calculate the energy dissipation due to viscosity and equate it to $\dot{\epsilon}$. The above analysis shows that $v_\lambda = (\dot{\epsilon}\lambda)^{1/3}$. If the shear $\sigma$ of the velocity field at some scale $\lambda$ is of the order of $(v_\lambda/\lambda)$, viscous heating rate at scale $\lambda$ will be

$$h_\lambda \simeq \eta\sigma^2 \simeq \eta\left(\frac{v_\lambda}{\lambda}\right)^2 \simeq \eta\dot{\epsilon}^{2/3}\lambda^{-4/3} \tag{3.65}$$

where $\eta$ is the coefficient of viscosity. At large scales $h$ is negligible compared to $\dot{\epsilon}$ and hence the same amount of energy will be transmitted from scale to scale. As we reach smaller scales, viscous heating becomes important and when $h_\lambda \simeq \rho\dot{\epsilon}$, at some scale $\lambda = \lambda_s$, the cascading of energy stops. Equating the expressions for $h_\lambda$ and $\rho\dot{\epsilon} = \rho(U^3/L)$ we find

$$\lambda_s \simeq \left(\frac{\eta}{\rho\dot{\epsilon}^{1/3}}\right)^{3/4} \simeq L\left(\frac{\eta}{\rho UL}\right)^{3/4} \simeq \frac{L}{R^{3/4}} \tag{3.66}$$

where $R$ is the Reynolds number evaluated at the largest scale. As long as the Reynolds number is large, we have $\lambda_s \ll L$ the range of scales $\lambda_s \ll \lambda \ll L$, over which Eq. (3.64) is valid, is quite large.

More formally, one can describe the turbulent velocity field in terms of the power spectrum or the correlation function. If the velocity field is $\mathbf{v}(\mathbf{x})$, then the (scalar) *correlation function* of $\mathbf{v}$ can be defined as

$$\xi(\mathbf{x}) = \langle \mathbf{v}(\mathbf{x} + \mathbf{y}) \cdot \mathbf{v}(\mathbf{y}) \rangle \qquad (3.67)$$

and the *power spectrum* $S(\mathbf{k})$ is defined to be the Fourier transform of $\xi(\mathbf{x})$. From Eq. (3.64) we have $v(x) \simeq \dot{\epsilon}^{1/3} x^{1/3}$ so that we expect $\xi(x)$ to scale as $\xi(x) \propto \dot{\epsilon}^{2/3} x^{2/3}$. The Fourier transform of $\xi(x)$ will vary (with $k \simeq x^{-1}$) as

$$S(k) \propto x^3 \xi(x) \propto \dot{\epsilon}^{2/3} x^{11/3} \propto \dot{\epsilon}^{2/3} k^{-11/3}. \qquad (3.68)$$

This is known as *Kolmogorov's spectrum*. Usually, what is quoted in the literature is not the 3-dimensional Fourier transform $S(k)$ but the "one dimensional" power, defined as $k^2 S(k)$ which varies as $k^{-5/3}$. You will often hear that Kolmogorov spectrum has an index $5/3$, referring to this quantity.

One application of this spectrum is in the modeling of atmospheric turbulence. The atmosphere consists of large cells of gas at very different temperatures and turbulence mixes the hot and cold air down to very fine scales. The resulting temperature distribution has a Kolmogorov spectrum and two parcels of air separated by distance $l$ will have rms temperature difference of about $\langle (\delta T/T)^2 \rangle^{1/2} \approx 5 \times 10^{-5} (l/1 \text{ cm})^{1/3}$. (See Eq. (3.64)) which explains the $(1/3)$ scaling. This leads to fluctuations in the refractive index of the air. If $n$ is the refractive index of air, then $(n-1) \propto \rho \propto T^{-1}$ so that $\delta n = (n-1)\delta T/T$. Since $(n-1) = 3 \times 10^{-4}$ for air in visible band, $\delta n = 1.5 \times 10^{-8} (l/1 \text{ cm})^{1/3}$.

When a wavefront arriving from a distant source encounters a turbulent medium, with a randomly varying refractive index, the effective flux received by an observer will show spatial and temporal variations. The relative phase difference induced on the light ray crossing a blob of air of size $l$ will be $\delta\phi = 2\pi(l/\lambda)\delta n$. If the scale height of the atmosphere is $D \approx 10$ km, then the light will pass through about $N = D/l$ blobs, each of which will induce a random phase of about $\delta\phi$. The *net* phase difference is $\Delta\Phi = \sqrt{N}\delta\phi = 2\pi\sqrt{N}(l/\lambda)\delta n$. Using our result for $\delta n$, we get

$$\Delta\Phi = 2\pi(1.5 \times 10^{-8}) \, D^{1/2} \, l^{5/6} \, \lambda^{-1} \qquad (3.69)$$

when $l, D$ and $\lambda$ are expressed in cm. We will expect coherence to be lost when $\Delta\Phi \approx \pi$, which occurs at a critical length scale of about $l_c \approx 10(\lambda/5000 \text{ Å})^{6/5}$ cm. This is the characteristic scale of fluctuations which has significant effect on light transmission; note that this scale is strongly achromatic with a $\lambda^{6/5}$ dependence on the wavelength. The angular resolution of a ground based telescope, for example, will be $\Delta\theta \approx \lambda/R$ where $R$ is the aperture of the telescope, *only if* $R \lesssim l_c$. When $R > l_c$, you are not limited by the aperture but by the atmospheric fluctuations, so that $\Delta\theta \approx \lambda/l_c$. For $\lambda = 4000$ Å, $l_c = 10$ cm, we get $\Delta\theta \approx 1$" which is the intrinsic limitation of ground based telescopes. (This is referred to as *seeing* in astronomy.)

This random phase $\Delta\Phi$ will tilt the wave front thereby making a plane wave surface corrugated. The tilt angle is just $\Delta\theta = (\Delta\Phi/2\pi)(\lambda/l) \approx 1.5 \times 10^{-5} l^{-1/6}$ (for $D = 10$ km with $\lambda$ expressed in cm). The Kolmogorov spectrum continues down to about $l \simeq 1$ mm with most of the effects arising from the fluctuations in the range 1 mm to 5 cm. Because of the weak dependence on $l$, we can assume that the typical angular scale at which atmospheric turbulence starts smearing an image is $\Delta\theta \approx 10^{-5}$ rad $\approx 4$ arcsec. Incidentally, the angular size of a star is much less than $\Delta\theta$ so that it will twinkle due to the atmospheric turbulence; planets, on the other hand, subtend angles much larger than $\Delta\theta$ so that they do not twinkle. As you can see, the explanation for this well known fact is rather non trivial.

## 3.4   Basic plasma physics

In several astrophysical contexts, one needs to deal with matter in the state of complete or partial ionisation. When the gaseous system has significant amount of ions and free electrons, it is usual to call it a plasma and such a medium exhibits a wide variety of electromagnetic phenomena. We shall concentrate on some of them in the next few subsections.

### 3.4.1   *Ionisation equilibrium of plasma*

Neutral hydrogen atom, in its ground state, contains an electron and proton bound together with an energy of $E_0 = -13.6$ eV which is equivalent to a temperature of $T_I \simeq 1.6 \times 10^5$ K. At temperatures $T \simeq T_I$, fair fraction of the atoms will be ionized thereby leading to a composition made of protons, electrons and neutral hydrogen atoms. Our first task is to determine this ionisation fraction as a function of density and temperature.

Let $n_p, n_e$ and $n_H$ denote the number densities of protons, electrons and neutral hydrogen atoms (with $n_p = n_e$) and let the total number density be $n = n_p + n_H$. The ionisation fraction is defined to be the ratio $y \equiv [n_p/(n_p + n_H)] = (n_p/n) = (n_e/n)$. We are interested in determining $y$ as a function of the temperature $T$ and density $n$. For a bunch of non relativistic particles in thermal equilibrium, the number density $n$ as a function of energy $E$ can be expressed as

$$n \simeq \lambda^{-3} \exp\left(-\beta E\right); \qquad \lambda \equiv \left(\frac{2\pi\hbar^2}{mk_BT}\right)^{1/2}, \qquad \beta = (k_BT)^{-1} \qquad (3.70)$$

where $\lambda$ (called *thermal wavelength*) essentially decides the number of phase cells per unit volume with $\lambda^{-3} \simeq (\bar{p}/\hbar)^3$ where $\bar{p} \simeq (2mk_BT)^{1/2}$ is the mean momentum at temperature $T$. Applying Eq. (3.70) to the electrons, protons and neutral hydrogen atoms involved in the reaction $p + e \Longleftrightarrow H$, we get

$$\frac{n_e n_p}{n_H} = \left(\frac{\lambda_H}{\lambda_e \lambda_p}\right)^3 e^{-\beta E_I} \simeq \lambda_e^{-3} e^{-\beta E_I} \qquad (3.71)$$

where $E_I$ is the ionisation potential for the atom. In arriving at the last relation, we have set $m_p \approx m_H$ in the pre-factor. To be precise, the right hand side of Eq. (3.70) should be multiplied by the degeneracy factor, $g_i$, for each constituent. But $g_e = 2, g_p = 2$ and $g_H = 2 \times 2$ (due to spins) and this factor cancels out in Eq. (3.71). We will comment on this factor in detail later on (see Eq. (3.76) below.)

Using the definition of the ionisation fraction $y = (n_p/n) = (n_e/n)$, Eq. (3.71) can be written as

$$\frac{y^2}{1-y} = \frac{1}{n}\left(\frac{m_e k_B T}{2\pi\hbar^2}\right)^{3/2} e^{-\beta|E_o|} = \frac{1}{n\lambda_e^3} \exp(-\beta E_I). \qquad (3.72)$$

This relation is called *Saha's equation*. You see that the pre-factor $(n\lambda_e^3)^{-1}$ is the ratio between the number of phase cells available in a given volume to the number of particles in that volume. Our classical discussion is valid only if this factor is large compared to unity. For pure hydrogen, Eq. (3.72) becomes

$$\frac{y^2}{1-y} = \frac{4.01 \times 10^{-9}}{\rho} T^{3/2} \exp\left(-\frac{1.578 \times 10^5}{T}\right) \qquad (3.73)$$

where $\rho$ is in gm cm$^{-3}$ and $T$ is in Kelvin. To understand the temperature dependence of ionisation better, let us determine the temperature at which half the atoms are ionized [i.e., $y = (1/2)$] as a function of the density. Setting $y = 1/2$ gives the following curve in the $\rho - T$ plane,

$$\rho_{1/2}(T) = 0.51 \ \text{gm cm}^{-3} \left( \frac{T}{1.6 \times 10^5 \ \text{K}} \right)^{3/2} \exp \left( -\frac{1.578 \times 10^5 \ \text{K}}{T} \right) \tag{3.74}$$

which is shown in Fig. 3.4.

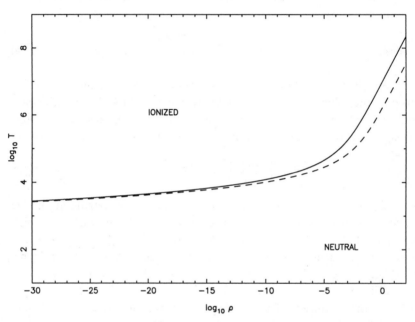

Fig. 3.4   The ionisation structure of pure hydrogen plasma in thermal equilibrium at different temperatures. The solid line corresponds to an ionisation fraction of $y = 0.9$ and the dashed line corresponds to $y = 0.5$.

You see that the temperature changes only by a factor 3 or so when the density changes by about 15 orders of magnitude! Thus, for a wide range of interesting densities, $y = 1/2$ corresponds to a temperature of about $10^4$ K. Also note that significant ionisation (50 per cent in the above case) occurs at a temperature which is well below the temperature corresponding to the binding energy $E_0$. This is a direct consequence of the fact that $(n\lambda_e^3)^{-1} \gg 1$ and can be seen from the condition $y = 0.5$ expressed in the form

$$k_B T_{0.5} = E_I \left[ \ln \frac{2}{n\lambda^3} \right]^{-1} \ll E_I. \tag{3.75}$$

The temperature at which half the atoms are ionized, $T_{1/2}$ is of the order of $k_B T_{1/2} \approx |E_0|/10$ which is a good thumb rule.

Similar analysis can be performed for more complicated ionisation scenarios. If $E_i$ is the energy needed to remove an electron from an atom changing its ionisation state from $i$ to $(i+1)$, then the corresponding result is

$$\frac{n_{i+1}}{n_i} = \left(\frac{g_e}{n_e}\right)\left(\frac{Z_{i+1}}{Z_i}\right)\left(\frac{m_e k_B T}{2\pi\hbar^2}\right)^{3/2} e^{-\beta E_i}$$

$$= \left(\frac{2k_B T}{P_e}\right)\left(\frac{Z_{i+1}}{Z_i}\right)\left(\frac{m_e k_B T}{2\pi\hbar^2}\right)^{3/2} e^{-\beta E_i} \qquad (3.76)$$

where we have introduced the effective degeneracy factors $Z_i, Z_{i+1}, g_e = 2$ and written $n_e = (P_e/k_B T)$. The $Z_i$ is the effective degeneracy factor computed by summing $g_n \exp(-\beta E_n)$ over all states of the atom.

In the discussion above we have assumed that there is a well-defined notion of temperature which is the same for electrons and ions in a plasma. There is some subtlety in this assumption because these particles have different inertia. To see this recall that thermal equilibrium is reached by energy transfer due to collisions. For a *neutral* gas of molecules, this is essentially determined by molecular collisions with the cross section $\sigma_0 \approx \pi a_0^2 \approx 8.5 \times 10^{-17}$ cm$^2$ and the corresponding mean free path $l = (n\sigma_0)^{-1}$. The time scale for establishment of Maxwellian distribution of velocities will be about $\tau_{\text{neu}} \simeq l/v \propto n^{-1} T^{-1/2}$. For an *ionized* classical gas, on the other hand, the cross section for scattering is decided by Coulomb interaction between charged particles. Since an ionized plasma is made of electrons and ions with vastly different inertia, the inter-particle collisions can take different time scales to produce thermal equilibrium (i) between electrons, (ii) between ions and (iii) between electrons and ions. Each of these need to be discussed separately. Consider first the time scale for electron-electron scattering. The typical impact parameter between two electrons is $b \approx (2q^2/m_e v^2)$ where $v$ is the typical velocity of an electron. The corresponding electron-electron scattering cross section is

$$\sigma_{\text{coul}} \approx \pi b^2 \approx \pi \left(\frac{q^2}{m_e}\right)^2 \frac{1}{v^4} \approx 10^{-20} \text{ cm}^2 \left(\frac{T}{10^5 \text{ K}}\right)^{-2} \qquad (3.77)$$

and the mean free path varies as $l = (n\sigma_{\text{coul}})^{-1} \propto (T^2/n)$. The mean free time between the electron-electron scattering will be $\tau_{ee} \approx (n\sigma v)^{-1}$ where $n$ is the number density of electrons and $\sigma \approx \pi b^2$. This gives

$\tau_{ee} \approx (m_e^2 v^3 / 2\pi q^4 n)$ which is the leading order dependence. A more precise analysis changes the numerical coefficient and introduces an extra logarithmic factor just as in the case of gravitational scattering studied in chapter 1; see Eq. (1.65). So a more precise value is $(\tau_{ee}/L)$ where $L \simeq 10$. Since $\tau \propto m^2 v^3 \propto T^{3/2} m^{1/2}$ at a given temperature $T \propto (1/2)mv^2$ the ion - ion collision time-scale $\tau_{pp}$ will be larger by the factor $(m_p/m_e)^{1/2} \simeq 43$ giving $\tau_{pp} = (m_p/m_e)^{1/2}\tau_{ee} \simeq 43\tau_{ee}$. Finally, consider the time-scale for significant transfer of energy between electrons and ions is still larger because of the following fact. When two particles of unequal mass scatter off each other, there is no energy exchange in the center of mass frame. In the case of ions and electrons, the center of mass frame differs from the lab frame only by a velocity $v_{CM} \simeq (m_e/m_p)^{1/2} v_p \ll v_p$. Since there is no energy exchange in the center of mass frame, the maximum energy transfer in the lab frame (which occurs for a head-on collision), is about $\Delta E = (1/2)m_p(2v_{cm})^2 = 2m_p v_{CM}^2 \simeq 2m_e v_p^2$ giving, $[\Delta E/(1/2)m_p v_p^2] \simeq (m_e/m_p) \ll 1$. Therefore it takes $(m_p/m_e)$ times more collisions to produce equilibrium between electrons and ions. That is, the time-scale for electron-ion collision is $\tau_{pe} = (m_p/m_e)\tau_{ee} \simeq 1836\tau_{ee}$. The plasma will completely relax to a Maxwellian distribution in this time-scale.

In a high temperature plasma discussed above, it is the collisions between the atoms in the high energy tail of the Maxwell distribution that cause the ionisation. The electrons can also be stripped off the atoms in different context. This occurs if the matter density is so high and the atoms are packed so close to each other that the electrons form a common pool. The Fermi energy of these electrons will be comparable to the ionisation energy of the atoms: $\epsilon_F \gtrsim E_I$. In this case, the electrons will be quantum mechanical and the relevant energy scale for them will be $\epsilon_F$. The temperature does not enter into the picture if $k_B T \ll \epsilon_F$ and we may call this a *zero temperature plasma*. (More conventionally, such systems are, of course, called *degenerate*.) For normal metals in the laboratory the Fermi energy is comparable to the atomic binding energy within an order of magnitude. If the temperature is below $10^4$ K, the properties of the system are essentially governed by Fermi energy.

You will recall that the derivation of equation of state based on the pressure given by Eq. (3.3) assumed that the *neutral* gas is ideal; i.e., the mutual interaction energy of the particles is small compared to the kinetic energy. To treat a *plasma* as ideal, it is necessary that Coulomb interaction energy of ions and electrons is negligible. The typical Coulomb potential energy between the ions and electrons in the plasma is given by

$\epsilon_{Coul} \approx Zq^2 n^{1/3}$ since the typical inter-particle separation is $n^{-1/3}$. If the *classical* high temperature plasma is to be treated as an ideal gas, this energy should be small compared to the energy scale of the particle $\epsilon \approx k_B T$, which requires the condition $nT^{-3} \ll (k_B/Zq^2)^3 \simeq 2.2 \times 10^8 Z^{-3}$ (in c.g.s. units) to be satisfied. On the other hand, to treat the high density quantum gas as ideal, we should require that the Coulomb energy $\epsilon_{Coul}$ is small compared to the non-relativistic *Fermi* energy $\epsilon_F \approx (\hbar^2/2m)n^{2/3}$. The condition now becomes $n \gg 8Z^3 a_0^{-3} \approx Z^3 \times 10^{26}$ cm$^{-3}$. You see that such a system becomes *more* ideal at *higher* densities; this — somewhat counter intuitive — result arises because the Fermi energy rises faster with density than the Coulomb interaction energy.

### 3.4.2 *Electromagnetic fields in plasma*

Since plasma has free charged particles, it has very complex electromagnetic properties, most of which play some role or the other in astrophysics. We shall discuss a few of these phenomena.

The simplest non trivial electromagnetic phenomenon that occurs in a plasma is called *shielding*. You know that, in the vacuum, the Coulomb field of a charged particle has long range and falls as $r^{-2}$. But if a charged particle is introduced into a plasma, its electric field decays exponentially with distance so that one could say the electric field is 'shielded' by the plasma. This arises because a positively charged particle, say, introduced into the plasma will polarize the medium gathering around it a cloud of negatively charged particles; the net effect of which is to reduce the effective charge as seen from a distance. The shielding would have been perfect at zero temperature but the finite temperature of the plasma will not allow the negatively charged particles to completely neutralize the positive charge. Hence, at non zero temperature we get an exponential decay of the electric field with a finite correlation length.

This result can be obtained fairly easily. The Poisson equation for the plasma, with an extra charge $Q$ added at the origin will be

$$\nabla^2 \phi = -4\pi q \left( n_+ - n_- \right) - 4\pi Q \delta \left( \mathbf{r} \right) \qquad (3.78)$$

where $n_\pm(\mathbf{x})$ is the number density of ions (electrons) in the plasma. In thermal equilibrium we can assume that $n_\pm \propto \exp(\mp q\phi/k_B T)$ giving

$$n_+ - n_- = \bar{n} \left[ \exp\left( -q\phi/k_B T \right) - \exp\left( q\phi/k_B T \right) \right] \qquad (3.79)$$

where $\bar{n}$ is the uniform density of particles in the absence of $\phi$. Far away from the charge $Q$, we will have $(q\phi/k_BT) \ll 1$ and $n_+ - n_- \simeq$ $-(2\bar{n}q/k_BT)\phi$. Hence Eq. (3.78) becomes

$$\nabla^2\phi = \frac{8\pi q^2\bar{n}}{k_BT}\phi - 4\pi Q\delta(\mathbf{r}) \tag{3.80}$$

which has the solution

$$\phi = \frac{Q}{r}\exp\left(-\frac{\sqrt{2}r}{\lambda_D}\right) \tag{3.81}$$

where $\lambda_D \equiv \left(k_BT/4\pi q^2\bar{n}\right)^{1/2}$ is called *Debye length*. Numerically,

$$\lambda_D = 6.9\text{cm}\left(\frac{T}{1\,\text{K}}\right)^{1/2}\left(\frac{n}{1\text{cm}^{-3}}\right)^{-1/2}. \tag{3.82}$$

The corresponding time-scale is given by $t_D = (\lambda_D/v)$ where $v$ is the typical velocity of the electron. This is the time it takes for the electrons to move over the distance $\lambda_D$ neutralising a random fluctuation. Using $v^2 \simeq (k_BT/m)$ we get

$$t_D^2 = \left(\frac{k_BT}{4\pi q^2n}\right)\left(\frac{m}{k_BT}\right) = \frac{m}{4\pi q^2n} \equiv \omega_p^{-2} \tag{3.83}$$

where $\omega_p$ is called the *plasma frequency*. You will see that it plays an important role in the propagation of electromagnetic waves in plasma.

For a plasma with number density $n$, the mean inter-particle distance is $n^{-1/3}$ and the typical electrostatic energy of nearest neighbours is $q^2n^{1/3}$. The ratio $\mathcal{R}$ between this electrostatic energy and the thermal energy $k_BT$ is

$$\mathcal{R} = \frac{q^2n^{1/3}}{k_BT} \simeq r_cn^{1/3} \simeq \left[\left(\frac{r_c^3}{\lambda_D^3}\right)N_D\right]^{1/3} \simeq N_D^{-2/3}. \tag{3.84}$$

Hence condition $N_D \gg 1$ implies that the typical electrostatic energy is small compared to $k_BT$ which will ensure that the plasma will behave collectively as an ideal gas. Numerically,

$$N_D \equiv \left(n\lambda_D^3\right) \simeq 360\left(\frac{T}{1\,\text{K}}\right)^{3/2}\left(\frac{n}{1\text{cm}^{-3}}\right)^{-1/2}. \tag{3.85}$$

We shall next consider two important electromagnetic wave phenomena that occur in a plasma. The first one is called *plasma dispersion* which introduces a frequency dependent velocity of propagation for electromagnetic radiation thereby making radiation of different frequencies propagate with different speeds through a plasma. The second effect is called *Faraday rotation* which involves a rotation of the plane of polarization of an electromagnetic wave that is propagating along the direction of a magnetic field embedded in the plasma. Both these effects have been used as useful diagnostics of astrophysical plasma and magnetic fields.

Let us begin with the concept of plasma dispersion. Consider an electromagnetic field which is propagating through a plasma and assume that all the relevant quantities vary in space and time as $\exp i(\mathbf{k} \cdot \mathbf{x} - \omega t)$. In free space, $k = |\mathbf{k}|$ and $\omega$ will be related by $\omega = kc$ and both the phase velocity $v_{\mathrm{ph}} = \omega/k$ and the group velocity $(d\omega/dk)$ will be equal to speed of light and a constant. The situation changes in the presence of a plasma and the new relation between $\omega$ and $k$ can be determined as follows. The Maxwell's equations in Fourier space has the form

$$i\mathbf{k} \cdot \mathbf{E} = 4\pi \rho\,, \qquad\qquad i\mathbf{k} \cdot \mathbf{B} = 0\,,$$
$$i\mathbf{k} \times \mathbf{E} = i(\omega/c)\mathbf{B}\,, \qquad i\mathbf{k} \times \mathbf{B} = (4\pi/c)\mathbf{j} - i(\omega/c)\mathbf{E} \quad (3.86)$$

in which all quantities are in the Fourier space. The $\rho$, for example, stands for $\rho(t, \mathbf{k})$ which is the spatial Fourier transform of $\rho(t, \mathbf{x})$ and similarly for all other variables. We will assume that, to the first order of approximation, most of the current is produced due to the motion of electrons relative to ions with the latter treated as stationary. Then the equation of motion for the free electron, $m\dot{\mathbf{v}} = -q\mathbf{E}$, has the solution given by $\mathbf{v} = q\mathbf{E}/(i\omega m)$. (Remember that we are working in Fourier space and — at the end — we will have to take the real part of all the variables; a relative $i$ factor merely indicates that two physical quantities are out of phase.) This will correspond to a current $\mathbf{j} = -nq\mathbf{v} = \sigma\mathbf{E}$ where $\sigma = inq^2/\omega m$. This expresses the current in terms of the electric field. Using the charge conservation in Fourier space, we have the relation $\omega\rho = \mathbf{k} \cdot \mathbf{j}$ allowing us to express the charge density as $\rho = (\sigma/\omega)\mathbf{k} \cdot \mathbf{E}$. We can now substitute for $\rho$ and $\mathbf{j}$ in the Maxwell's equation Eq. (3.86) and express them entirely in terms of electric and magnetic field as

$$i\mathbf{k} \cdot \epsilon\, \mathbf{E} = 0\,, \qquad\qquad i\mathbf{k} \cdot \mathbf{B} = 0\,,$$
$$i\mathbf{k} \times \mathbf{E} = i(\omega/c)\mathbf{B}\,, \qquad i\mathbf{k} \times \mathbf{B} = -i(\omega/c)\epsilon\mathbf{E} \quad (3.87)$$

where

$$\epsilon = 1 - (\omega_p/\omega)^2 \qquad (3.88)$$

with $\omega_p$ being the plasma frequency introduced earlier in Eq. (3.83); numerically,

$$\omega_p = \left(\frac{4\pi n q^2}{m}\right)^{1/2} = 56.3 \times \sqrt{n} \text{ kHz} \qquad (3.89)$$

if $n$ is in CGS units. By taking the cross product of the third equation in Eq. (3.87) with $\mathbf{k}$ and using the other equations, it is easy to obtain

$$c^2 k^2 = \omega^2 - \omega_p^2. \qquad (3.90)$$

This is the modified dispersion relation for electromagnetic radiation in a plasma in the simplest possible context. There are two important conclusions which arises from this result.

First, note that $k$ becomes purely imaginary for low frequency radiation with $\omega < \omega_p$. Suppose an electromagnetic wave with a frequency $\omega < \omega_p$ impinges on a slab of plasma. The wave mode inside the plasma will decay exponentially with distance because $k$ is pure imaginary. In other words, low frequency radiation cannot penetrate into plasma and will be reflected back completely. It is this feature which allows short wave radio broadcasts to be reflected back by the ionospheric plasma allowing radio reception beyond the line of sight horizon. The characteristic cut-off frequency in the ionosphere is about $\nu \sim (10-15)$ MHz. By the same token, one cannot receive low frequency radiation in radio band from distant astronomical sources — which is the result we mentioned in page 80.

The physical reason for this phenomenon is easy to obtain. When electrons are displaced relative to ions because of the action of the external electromagnetic field, it will generate a field which opposes the applied electric field thereby attempting to reduce the net field to zero. When this process is effective, the electromagnetic radiation cannot propagate through the region. If we imagine all the electrons in a fictitious area $A$ to be displaced by a distance $\delta x$ normal to the area element, then it will generate an electric field $4\pi\sigma$ where $\sigma = n_e \delta x$ is the surface charge density in the area element. The equation of motion for an electron in this electric field will be $m(d^2/dt^2)\delta x = -4\pi q\sigma = -4\pi n_e q^2 \delta x$. This leads to oscillations of displaced charge densities at the plasma frequency $\omega_p$ showing that this is the characteristic time scale of plasma oscillations. On the other hand, the characteristic time scale at which electric field varies is $\omega^{-1}$. When $\omega_p^{-1} < \omega^{-1}$, the electrons can quickly readjust themselves opposing the electric field and preventing propagation. This is why, for $\omega_p > \omega$ the waves are exponentially damped in a plasma.

Let us now move on to another effect is related to the group velocity of those (high frequency) waves which *do* manage to propagate through the plasma. Using Eq. (3.90) for $\omega > \omega_p$, we find that the group velocity is now given by $v_g \equiv (\partial \omega / \partial k) = c(1 - \omega_p^2/\omega^2)^{1/2}$. If such a group of waves are traveling from a cosmic source towards earth, through a plasma, the travel time to cover a distance $r$ will be

$$t_\omega = \int_0^r \frac{ds}{v_g} = \int_0^r \frac{ds}{c} \left(1 - \frac{\omega_p^2}{\omega^2}\right)^{-1/2}. \tag{3.91}$$

For $\omega \gg \omega_p$, we can use the Taylor expansion of the square root and the definition of $\omega_p^2$ to get

$$t_\omega = \frac{r}{c} + \frac{2\pi q^2}{m_e \omega^2}(DM), \tag{3.92}$$

where the quantity $DM$, which astronomers call the *dispersion measure* of the plasma, is defined to be

$$DM \equiv \int_0^r n_e ds. \tag{3.93}$$

This quantity gives the column density of electrons along the line of sight. Using Eq. (3.92) we can relate the difference in the time of arrival, $\Delta \tau_D$, of a signal at two different frequencies to the dispersion measure. Numerically, in the usual units of the astronomer, ($n$ in cm$^{-3}$ and $l$ in parsecs), we have:

$$\frac{DM}{\text{cm}^{-3}\ \text{pc}} = 2.4 \times 10^{-4} \left(\frac{\Delta \tau_D}{\mu s}\right) \left[\left(\frac{\nu_1}{1\ \text{MHz}}\right)^{-2} - \left(\frac{\nu_2}{1\ \text{MHz}}\right)^{-2}\right]^{-1}. \tag{3.94}$$

If the time delay at the two frequencies can be measured then this equation allows for the dispersion measure to be determined along any given direction.

As an example, consider a source at a distance 1 kpc from which one is receiving radiation that is propagating through a plasma with electron density 0.05 cm$^{-3}$. The dispersion measure will be $DM \approx 50$ pc cm$^{-3}$. Radiation at the two different frequencies $\nu_1 = 1.7$ GHz and $\nu_2 = 1.5$ GHz (in the radio band) will show a time delay of about $t_1 - t_2 = -0.021$ seconds which is easily measurable.

We next consider another phenomenon related to the propagation of waves in *magnetized* plasma (called *Faraday rotation*). This arises when a plane polarized electromagnetic wave propagates along the direction of a magnetic field $\mathbf{B_0}$ in a cold plasma. The equation of motion for an electron in the plasma is now given by

$$m\frac{d\mathbf{v}}{dt} = -q\mathbf{E} - \frac{q}{c}\,\mathbf{v}\times\mathbf{B_0}. \qquad (3.95)$$

Taking the magnetic field to be along the third axis $\hat{\mathbf{n}}_3$ and the electromagnetic wave to be circularly polarized, we have

$$\mathbf{E}(t) = Ee^{-i\omega t}\left(\hat{\mathbf{n}}_1 \mp i\hat{\mathbf{n}}_2\right). \qquad (3.96)$$

From Eq. (3.95) we find that the steady state velocity induced on the electrons will be

$$\mathbf{v}(t) = \frac{-iq}{m(\omega \pm \omega_B)}\,\mathbf{E}(t) \qquad (3.97)$$

where $\omega_B = qB_0/mc$ is the cyclotron frequency of the non relativistic electron. You can now go through the same analysis as before by finding the current, charge density *etc.* and you will find that the plasma behaves as a system with two different dielectric constants $\epsilon_R$ and $\epsilon_L$ for the right and left circularly polarized waves with

$$\epsilon_{R,L} = 1 - \frac{\omega_p^2}{\omega(\omega \pm \omega_B)}. \qquad (3.98)$$

Any linearly polarized wave can be represented as the sum of two circularly polarized waves. Since the speeds of propagation of the two circular polarizations are different, the plane of polarization will be rotated as the wave propagates along the magnetic field. For $\omega \gg \omega_p, \omega \gg \omega_B$, the net rotation while propagating a distance $L$ along the magnetic field can be expressed as

$$\psi = \int_0^L \Delta k\, dz = \frac{2\pi e^3}{m_e^2 c^2 \omega^2}\int_0^L n_e B_0\, dz \equiv \frac{2\pi e^3}{m_e^2 c^2 \omega^2}\,\langle RM\rangle \qquad (3.99)$$

where the last equality defines the quantity, RM, called *rotation measure*. If the propagation is not strictly along the magnetic field, it is the component of the magnetic field, $B_\parallel$, along the direction of propagation, which is the relevant quantity in defining the rotation measure. Numerically, in conventional astronomical units

$$\frac{\Delta\psi}{\text{rad}} = 8.1\times 10^5 \left(\frac{\lambda}{10^2\ \text{cm}}\right)^2 \int_0^{L/\text{pc}} \left(\frac{B_\parallel}{\text{Gauss}}\right)\left(\frac{N_e}{\text{cm}^{-3}}\right) d\left(\frac{z}{\text{pc}}\right). \qquad (3.100)$$

As in the case of dispersion measure, observations of Faraday rotation at two different frequencies can be used to determine the rotation measure. This serves as a diagnostic of astrophysical plasmas.

Finally, we want to consider the behaviour of magnetic field in a moving plasma, and show that — in a highly conducting fluid — the magnetic field is "frozen" to the fluid and moves along with it. To do this, consider a closed curve $C$ which is 'flowing along' with the plasma in the sense that every point in the curve is shifted by an amount $\mathbf{v}dt$ in space in a small time interval $dt$ where $\mathbf{v}$ is the velocity of the fluid element at that point. In general, the shape of the closed curve will change with time and we are interested in the rate of change of magnetic flux

$$\Phi = \int_S \mathbf{B} \cdot d\mathbf{S} \tag{3.101}$$

through a surface $S$ bounded by the curve $C$. This flux will change both because of intrinsic time variation of the magnetic field as well as due to the fact that the curve $C$ is moving in space. The first effect is easy to compute and is given by

$$\left(\frac{d\Phi}{dt}\right)_I = \int_S \frac{\partial \mathbf{B}}{\partial t} \cdot d\mathbf{S} = -\int_S (\nabla \times \mathbf{E}) \cdot d\mathbf{S} = -\oint_C \mathbf{E} \cdot d\mathbf{l}. \tag{3.102}$$

To obtain the change due to the motion of the loop, notice that the area covered by a line element $d\mathbf{l}$, moving with velocity $\mathbf{v}$, in a small time interval $dt$ is $d\mathbf{A} = \mathbf{v}\, dt \times d\mathbf{l}$. This leads to the change of flux through the curve $C$ given by

$$d\Phi = \mathbf{B} \cdot d\mathbf{A} = \mathbf{B} \cdot (\mathbf{v}\, dt \times d\mathbf{l}) = (\mathbf{B} \times \mathbf{v}) \cdot d\mathbf{l}\, dt. \tag{3.103}$$

Adding up all the infinitesimal line elements $d\mathbf{l}$ of the curve and dividing by $dt$, we get the flux change due to the second process to be

$$\left(\frac{d\Phi}{dt}\right)_{II} = -\oint_C (\mathbf{v} \times \mathbf{B}) \cdot d\mathbf{l}. \tag{3.104}$$

Adding the effects in Eq. (3.102) and Eq. (3.104), the net change of flux through the loop will be

$$\frac{d\Phi}{dt} = \left(\frac{d\Phi}{dt}\right)_I + \left(\frac{d\Phi}{dt}\right)_{II} = -\oint_C (\mathbf{E} + \mathbf{v} \times \mathbf{B}) \cdot d\mathbf{l}. \tag{3.105}$$

But notice that, in the non relativistic limit, $v \ll c$, the quantity $\mathbf{E}' = \mathbf{E} + (\mathbf{v} \times \mathbf{B})$ is the electric field in the rest frame of the plasma. If the plasma is highly conducting, we must have $\mathbf{E}' \approx 0$. (This is because even a small electric field will induce a large current if the conductivity is large and the current will redistribute the charges making $\mathbf{E}' \approx 0$.) Hence we find that $(d\Phi/dt) = 0$. That is, the magnetic flux through every closed curve in a conducting plasma remains constant as the plasma moves.

It is also possible to show from this result that any two elements of plasma which were on the same magnetic field lines at some instant of time, will remain glued to the same magnetic field line at all times. This allows one to think of magnetic field lines as frozen into the plasma and following the motion of the plasma.

## Exercises

**Exercise 3.1**   *Getting degenerate:*

(a) How degenerate are you? (b) What about electrons in sodium metal at room temperature?

**Exercise 3.2**   *Jupiter and Saturn:*

The internal structure of planets like Jupiter or Saturn is thought to be described reasonably well by an equation of state $P = K\rho^2$ corresponding to $n = 1$ polytrope which can be exactly solved. Use the known masses and radii of the planets to compute other physical parameters. Is the observed $M, R$ values for Jupiter and Saturn consistent with this equation of state?

**Exercise 3.3**   *One dimensional isothermal profile:*

The one dimensional analogue of an isothermal sphere will satisfy the equation $(d^2\phi/dz^2) = 4\pi G\rho_0 \exp[-(\phi/\sigma^2)]$. Show that this equation has the exact solution $\rho = \rho_0 \operatorname{sech}^2(z/2z_0)$ where $z_0^2 = (\sigma^2/8\pi G\rho_0)$.

**Exercise 3.4**   *What fixes the accretion rate?*

The result we obtained in Eq. (3.46) shows that the flow rate $\dot{M}$ is uniquely fixed at a value dependent only on $M, c_s(\infty), \rho(\infty)$ implying a gas can flow to a compact object only at a specific rate. The purpose of this exercise is to provide an explanation for this result which should have surprised you in the first place. (a) Show that when $\dot{M}$ changes slightly, the velocity and density at any fixed radius $r$ changes by the amount

$$\frac{\delta\dot{M}}{\dot{M}} = \left(1 - \frac{v^2}{c_s^2}\right)\frac{\delta v}{v}; \qquad \frac{\delta\rho}{\rho} = -\frac{v^2}{c_s^2}\frac{\delta v}{v}. \tag{3.106}$$

Argue from this result that at any given $r$, there is a maximum permitted flow rate which occurs when $v^2 = c_s^2$. (b) Determine the maximum value $|\dot{M}_{max}(r)|$ in terms of other parameters in the problem. (c) Show that, for $1 < \gamma \leq 5/3$, the $|\dot{M}_{max}(r)|$ decreases as $r^{-(5-3\gamma)/2(\gamma-1)}$ for small $r$, reaches a minimum value $\dot{M}_{crit}$ at $r = r_c$ and increases as $r^2$ for large $r$. (d) Argue from the above conclusion that, in order to have a flow which makes the transition from sub-sonic to supersonic, we must have exactly $\dot{M} = \dot{M}_{crit}$.

### Exercise 3.5 *Shocking motion:*

Ionized hydrogen, which is getting accreted on to the surface of a white dwarf of radius $R$, is moving through a shock located close to the surface. Show that the shock temperature can be written in a suggestive form as $k_B T \approx m_e c^2 (3m_p/32m_e)(2GM/c^2R)$. Assume that the pre-shocked matter is cold.

### Exercise 3.6 *Getting it totally wrong:*

(a) At the center of the Sun, $T = 1.5 \times 10^7$ K and $\rho = 100$ gm cm$^{-3}$. Estimate the ionisation fraction using Saha's equation. You will find that the center of the Sun should be nearly 25 per cent neutral. Do you believe this? (b) Intergalactic medium has $T = 2.7$ K and $\rho = 10^{-29}$ gm cm$^{-3}$. It should be almost completely neutral while observations suggest that it is highly ionized. What is going on? [Hint: (a) Compute the pressure and inter-particle separation at the center of the Sun. (b) Remember that Saha's equation was obtained using the assumption of thermodynamic equilibrium.]

### Exercise 3.7 *Polarization of CMBR:*

Recently observations at 1 GHz has revealed polarization of the cosmic microwave background radiation at approximately the expected strength. If the universe had a strong magnetic field, the direction of which varies significantly at angular scales smaller than the beam of the detector, then this signal would have been destroyed by Faraday rotation. This allows us to put a bound on the magnetic field strength. Assume the universe was fully ionized with a number density of $2 \times 10^{-8}$ cm$^{-3}$ and the relevant size of the universe is $10^{28}$ cm. Obtain a rough upper limit on the net parallel field strength at scales smaller than the beam size.

### Exercise 3.8 *Relaxation in real life:*

A gaseous region of ionized hydrogen, usually called a HII region, has an electron number density $n_e = 100$ cm$^{-3}$ and kinetic temperature of $T = 10^4$ K. The age of such a region is typically $10^7$ yrs. Estimate the time scale for electrons and protons to achieve Maxwellian distribution individually and the time scale for the temperatures to be equalized. Assume that the logarithmic factor is $L \approx 25$. Repeat the same analysis for a supernova remnant with $n_e = 1$ cm$^{-3}$ and $T = 5 \times 10^7$ K and an age of about 400 yrs. [Ans: For the HII region, $t_{eq,e-e} \approx 4 \times 10^{-6}$ yrs; $t_{eq,i-i} \approx 1.6 \times 10^{-4}$ yrs; $t_{eq,i-e} \approx 7 \times 10^{-3}$ yrs. In the case of supernova remnant, $t_{\mathrm{eq,e-e}} \approx 10^2$ yrs; $t_{\mathrm{eq,i-i}} \approx 4 \times 10^3$ yrs; $t_{\mathrm{eq,i-e}} \approx 2 \times 10^5$ yrs.]

## Notes and References

A good text on fluid mechanics is Landau and Lifshitz (1975b). Astrophysical fluid dynamics is covered in, for example, Shore (1992). Figure 3.3 is adapted from chapter 8, Padmanabhan (2000).

　　1. *Section 3.2:* Detailed discussion of isothermal sphere and its stability can be found in Padmanabhan (2000), chapter 10 and in Padmanabhan (1990).

　　2. *Section 3.3.4:* For a more rigorous discussion of fluid turbulence, see for example, Landau and Lifshitz (1975b)

# Chapter 4

# Stars and Stellar Evolution

From this chapter onwards, we shall use the concepts developed in chapters 1-3 to study different astrophysical systems, starting with the physics of stars. Historically, stellar astronomy pioneered the study of astrophysical systems and it is no surprise that stars are the astrophysical objects about which we have the most comprehensive understanding. To a great extent, this is facilitated by the fact that a large sample of stars can be studied with reasonable accuracy in our own galaxy.

As a consequence, stellar physics also determined the terminology and jargon (often unfortunate and confusing) which is used in the rest of astrophysics. For example, the division of stars visible to naked eye into a set of bins called *magnitudes*, led to brightness of any celestial object being measured by this unit; all elements heavier than hydrogen and helium are traditionally called *metals* by stellar astronomers and this convention (in which carbon and nitrogen are termed metals!) still continues; stars are classified by a random sequence of alphabets *etc.* While these traditions are unlikely to be replaced by more rational ones in the near future, it cannot be denied that stellar physics also set the tone for understanding astrophysical systems based on concepts from basic physics. It is this theme with which we will begin our exploration.

## 4.1 When is gravity important?

Let us begin by considering the electromagnetic and gravitational binding energies of a body of mass $M$ and radius $R$. The atomic binding energy of normal matter is purely electromagnetic and is about $(q^2/a_0)$ per particle where $q$ is the electronic charge and $a_0$ is the atomic radius. If the atoms are closely packed (with the number density $n$ satisfying the relation $na_0^3 \simeq 1$), the atomic binding energy (per particle) of a system is about $\epsilon_a \approx (q^2/a_0) \approx q^2 n^{1/3}$. On the other hand, the gravitational energy per particle is

$$\epsilon_g = \frac{GM^2}{NR} \simeq \frac{Gm_p^2 N}{N^{1/3}a} \simeq Gm_p^2 N^{2/3}n^{1/3} \tag{4.1}$$

where $N \simeq (M/m_p)$ is the total number of particles in the body, and we have again assumed the atoms of mass $m_p$ are closely packed with $N = na^3$. You can immediately see a crucial difference between gravitational binding energy per particle and atomic binding energy per particle. The latter $\epsilon_a \propto n^{1/3}$ depends only on the density and is independent of the *total* number of particles in the system. But in the case of gravitational binding energy $\epsilon_g \propto N^{2/3}n^{1/3}$ showing the binding energy *per particle* itself depends on the total number of particles $N$. The *total* energy of the system at given density varies as $N\epsilon_a \propto N$ in the case of atomic binding (which is extensive, so that two litres of petrol give you twice as many miles as one litre of petrol) while $N\epsilon_g \propto N^{5/3}$ in the case of gravity — which is clearly non-extensive. The ratio of the two binding energies is given by

$$\frac{\epsilon_a}{\epsilon_g} \approx \left(\frac{\alpha}{\alpha_G}\right)\left(\frac{1}{N^{2/3}}\right) \equiv \left(\frac{N_G}{N}\right)^{2/3} \approx \left(\frac{10^{54}}{N}\right)^{2/3} \tag{4.2}$$

where we have defined the 'gravitational fine structure constant' $\alpha_G = Gm_p^2/\hbar c \simeq 10^{-38}$ (by replacing $q^2$ by $Gm_p^2$ in the fine structure constant $\alpha = q^2/\hbar c$). The laboratory systems, with $N \ll 10^{54}$, have negligible gravitational potential energy compared to their atomic binding energy. Clearly, the number $N_G \equiv \alpha^{3/2}\alpha_G^{-3/2} \approx 10^{54}$, arising out of fundamental constants, sets the smallest scale at which the gravitational binding energy becomes as important as electromagnetic binding energy of matter. The corresponding mass and length scales are

$$M_{\text{planet}} \simeq N_G m_p \simeq 10^{30} \text{ gm}, \qquad R_{\text{planet}} \simeq N_G^{1/3} a_0 \simeq 10^{10} \text{ cm}, \tag{4.3}$$

and are comparable to those of a large planet. For larger masses, gravitational interaction changes the structure significantly while for smaller masses gravity is ignorable and matter is homogeneous with constant density so that $M \propto R^3$.

Since we used Newtonian gravity to arrive at this conclusion, it is necessary to verify that the ratio of the gravitational energy to rest mass energy, $(\epsilon_g/Mc^2)$ is small for this scale; this ratio for $M \simeq M_{\text{planet}}, R \simeq R_{\text{planet}}$ is about $\alpha_G^{1/2}\alpha^{3/2}(m_e/m_p) \ll 1$. You see that the smallness of this ratio

— allowing planets to be described by Newtonian gravity — follows purely from the values of fundamental constants.

Most of the astrophysically interesting systems have masses larger than $M_{\text{planet}}$ and require the gravitational force to be balanced by forces other than normal solid state forces. In general such systems can be classified into two categories. The first set has the gravity balanced by the classical motion (random or orbital) of the particles; while the second one has the gravitational pressure balanced by degeneracy pressure. For a compactly packed system with $na_0^3 \approx 1$, the non-relativistic Fermi energy of electrons, $\epsilon_F \simeq (\hbar^2/2m_e)n^{2/3} \simeq (1/2m_e)(\hbar/a_0)^2$, is clearly comparable to the atomic binding energy and one can compare either $\epsilon_a$ or $\epsilon_F$ with $\epsilon_g$ to see how the system is supported against gravity. (Remember that pressure is proportional to energy density; so balancing the pressures is the same as balancing energy densities.) This is meaningful as long as the temperature of the system (as measured by random velocities) is low and $k_B T \ll \epsilon_F \simeq \epsilon_a \approx 10$ eV. If the temperature is significantly higher, then it is the thermal energy $k_B T$ which should be compared to $\epsilon_g$.

When the mass of the system increases, the gravitational pressure increases and — to balance it — both the degeneracy pressure and the thermal pressure will increase. The dynamics of the system will then depend on the relative significance of these two quantities. To take into account both thermal and quantum degeneracy contributions, we shall assume the matter pressure can be approximated by $P \approx nk_B T + n\epsilon_F$ which — while not exact — is an adequate interpolation between the two limits. This pressure can balance the gravitational pressure if the corresponding energy densities are comparable: $(k_B T + \epsilon_F) \simeq G m_p^2 N^{2/3} n^{1/3}$. Using Eq. (3.12) for $\epsilon_F$ for the *non-relativistic* electrons we get

$$k_B T \simeq G m_p^2 N^{2/3} n^{1/3} - \frac{(3\pi^2)^{2/3}}{2} \frac{\hbar^2}{m_e} n^{2/3}. \tag{4.4}$$

For a classical system, the first term on the right hand side dominates and we see that the gravitational potential energy and kinetic energy (corresponding to the temperature $T$) are comparable; this is merely a restatement of the *virial theorem* we proved in chapter 1. If the system contracts under the self gravity and the radius $R$ of the system is reduced, then the density $n$ will increase. But note that the second term ($\propto n^{2/3}$) on the right hand side of Eq. (4.4) grows faster than the first ($\propto n^{1/3}$). So, as the system contracts, the temperature will increase, reach a maximum value $T_{\max}$ and decrease again; equilibrium is possible for any value of the temperature

$T \leq T_{\text{max}}$ with gravity balanced by thermal and degeneracy pressure. The maximum temperature, $T_{\text{max}}$, which the body can acquire, is reached when $n = n_c$, with

$$n_c^{1/3} \cong \frac{\alpha_G}{(3\pi^2)^{2/3}} \left( \frac{N^{2/3}}{\lambda_e} \right); \qquad k_B T_{\text{max}} \simeq \frac{\alpha_G^2}{2(3\pi^2)^{2/3}} \left( N^{4/3} m_e c^2 \right) \quad (4.5)$$

where $\lambda_e \equiv (\hbar/m_e c)$. An interesting phenomenon arises if the maximum temperature $T_{\text{max}}$ is sufficiently high to trigger nuclear fusion at the center of the system; then we obtain a gravitationally bound, self sustained, nuclear reactor usually called a *star*. Let us consider the condition for this to occur.

For two protons to fuse together undergoing nuclear reaction, it is necessary that they are brought within the range of attractive nuclear force which is about $l \approx (h/m_p c)$. Since this requires overcoming the Coloumb repulsion, such direct interaction can take place only if the kinetic energy of colliding particles is at least of the order of the electrostatic potential energy at the separation $l$. This requires energies of the order of $\epsilon \approx (q^2/l) = (\alpha/2\pi) m_p c^2 \approx 1$ MeV. It is, however, possible for nuclear reactions to occur at even lower energies by quantum mechanical tunneling through the Coloumb barrier. This happens when the de Broglie wavelength $\lambda_{\text{deB}} \equiv (h/m_p v) = l(c/v)$ of the two protons (which is clearly larger than $l$) overlap. This occurs when the energy of the protons is about $\epsilon_{\text{nucl}} \approx (\alpha^2/2\pi^2) m_p c^2 \approx 1$ keV. It is conventional to write this expression as $\epsilon_{\text{nucl}} \approx \eta \alpha^2 m_p c^2$ with $\eta \simeq 0.1$. This quantity $\epsilon_{\text{nucl}}$ sets the scale for triggering nuclear reactions in astrophysical contexts. The energy $k_B T_{\text{max}}$ corresponding to the maximum temperature obtained in Eq. (4.5) will be larger than $\epsilon_{\text{nucl}}$ when

$$N > (2\eta)^{3/4} (3\pi^2)^{1/2} \left( \frac{m_p}{m_e} \right)^{3/4} \left( \frac{\alpha}{\alpha_G} \right)^{3/2} \approx 4 \times 10^{56} \quad (4.6)$$

for $\eta \simeq 0.1$. The corresponding condition on mass is $M > M_*$ with

$$M_* \approx (2\eta)^{3/4} (3\pi^2)^{1/2} \left( \frac{m_p}{m_e} \right)^{3/4} \left( \frac{\alpha}{\alpha_G} \right)^{3/2} m_p \approx 4 \times 10^{32} \text{ gm} \quad (4.7)$$

which is comparable to the mass of the smallest stars observed in our universe. The mass of the Sun, for example, is $M_\odot = 2 \times 10^{33}$ gm. Again, you see that the mass of a typical star is completely determined in terms of fundamental constants. We shall now discuss how much of the details regarding a star can be obtained from basic physical considerations.

## 4.2 Stellar magnitudes and colours

Since a contracting body will usually have a higher density at the central region, nuclear reactions will be first triggered at the central region and the heat that is generated will be transported outwards through different physical processes. Once the nuclear reactions occur in the hot central region of the gas cloud, its structure changes significantly. If the transport of this energy to the outer regions is through photon diffusion then the opacity of matter will play a vital role in determining the stellar structure. In particular, you will see that the opacities determine the relation between the luminosity and mass of the star.

A photon with mean free path $l = (n\sigma)^{-1}$, random walking through the hot plasma in the interior of the star, will undergo $N_{\text{coll}} \simeq (R/l)^2$ collisions in traversing the radius $R$. Therefore, it will take the time $t_{\text{esc}} \simeq (lN_{\text{coll}}/c) \simeq (R/c)(R/l)$ for the photon to escape. [This is, of course, significantly larger than the time taken for *collisionless* propagation $(R/c)$ by a large factor $(R/l)$.] The luminosity of a star $L$ will be proportional to the ratio between the radiant energy content of the star, $E_\gamma$, and $t_{\text{esc}}$. The radiation energy content is $E_\gamma \simeq (aT^4) R^3 \propto T^4 R^3$ where $T$ is essentially the core temperature. Then

$$L \propto \frac{E_\gamma}{t_{\text{esc}}} \propto \frac{R^3 T^4 l}{R^2} \propto RT^4 l \propto \frac{RT^4}{n\sigma}. \tag{4.8}$$

For a wide class of stars, we may assume that the central temperature is reasonably constant. This is because the nuclear reactions — which depend *strongly* on $T$ — act as an effective thermostat. When the temperature increases, the nuclear reaction rate increases thereby producing more energy. Since the system is in local virial equilibrium, $E_{\text{tot}} = (1/2)E_{\text{grav}}$ and when the total energy becomes more positive, the *magnitude* of the gravitational energy, $|E_{\text{grav}}| \propto M^2/R$ decreases. This makes $R$ increase which, in turn, decreases the density $\rho \simeq M/R^3$. This decreases the nuclear reaction rate thereby acting as a thermostat. Further, virial theorem gives $k_B T \simeq (GMm_p/R)$ so that $TR \propto M$. If Thomson scattering dominates then $\sigma = \sigma_T \equiv [(8\pi/3)(e^2/m_e c^2)^2]$ and we get

$$L \propto \frac{RT^4}{\sigma_T n} \propto \frac{T^4 R^4}{\sigma_T N} \propto \frac{M^4}{M} \propto M^3. \tag{4.9}$$

The situation is different if the opacity is provided by the interaction of photons with partially ionized atoms rather than by free electron scattering. The cross section for bound-free and bound-bound opacity in thermal equilibrium has been obtained in chapter 2, Sec. 2.5, where it was shown that $l \propto T^{7/2}n^{-2} \propto T^{7/2}R^6 M^{-2}$. In this case, we have

$$L \propto RT^4 l \propto R^7 T^{15/2} M^{-2} \propto M^{11/2} R^{-1/2}. \qquad (4.10)$$

Taking $(GM/R) \approx$ constant so that $R \propto M$ now gives $L \propto M^5$. Taken along with Eq. (4.9), we expect the luminosity of a star to be related to its mass by $L \propto M^\alpha$ with $\alpha \simeq 3 - 5$.

The discussion so far dealt with *core* temperature which governs the nuclear reactions and energy generation. The effective *surface* temperature $T_s$ of a star, of course, will be much lower. To determine $T_s$, we note that the total blackbody luminosity from the star will be $(4\pi R^2)(ac/4)T_s^4$ (see the discussion after Eq. (2.22)). It is convenient to use this relation to actually *define* the surface temperature $T_s$ of the star with a given luminosity $L$ and radius $R$ by $L = (ac)\pi R^2 T_s^4$.

If the radiation from the star is approximated as that of a blackbody, then the intensity $I_\nu$ (which, as you will recall, is the energy per unit area per unit time per solid angle per frequency) of thermal radiation emitted by the star will be

$$I_\nu = \frac{dE}{dAdtd\Omega d\nu} = B_\nu \equiv \frac{2h\nu^3}{c^2}\frac{1}{e^{h\nu/k_B T_s} - 1}. \qquad (4.11)$$

We saw in chapter 2 that the quantity $\nu B_\nu$ (which gives the intensity per logarithmic band in frequency) reaches a maximum value around $h\nu \approx 4k_B T$ which translates to the fact that a a star with surface temperature of 6000 K will have the maximum for $\nu B_\nu$ at 6000 Å. Observationally, it is found that stellar surface temperatures vary from $3 \times 10^3$ K to about $3 \times 10^4$ K as the mass varies from about $0.1 M_\odot$ to $60 M_\odot$. The corresponding variation in the radius is in the range of $(0.8 - 70) \times 10^{10}$ cm while the luminosity ranges from $(10^{-3} - 10^{5.7})L_\odot$ where $L_\odot = 3.8 \times 10^{33}$ erg s$^{-1}$ is the luminosity of the Sun.

Using this description we can obtain a few more characteristics of a star. In principle, one can fit the spectrum of a star to a blackbody spectrum (approximately) and thus measure $T_s$. The total energy flux $l$ received from the star on earth (called the *apparent luminosity*) can also be measured directly. Since $l = L/(4\pi d^2)$ where $d$ is the distance to the star, we can determine the *absolute luminosity* $L$ of the star if the distance to the star is known. Assuming that the distance can be independently measured, we will be able to determine $L$ and plot the location of the stars in a two dimensional $(L - T_s)$ plane. From our definition, it follows that $T_s \propto$ $L^{1/4}R^{-1/2} \propto L^{1/4}M^{-1/2}$. Combining with $M \propto L^{1/5}$, valid when the interior is only partially ionized, we get $T_s \propto L^{1/4}L^{-1/10} \propto L^{3/20}$. On the other hand, if Thomson scattering dominates with $L \propto M^3$, we get $T_s \propto L^{1/12}$. Thus, if the stars are plotted in a log $T_s$ − log $L$ plane, (called *Hertzsprung-Russel diagram* or *H-R diagram*, for short; see Fig. 4.6) we expect them to lie within the lines with slopes $(3/20) = 0.15$ and $(1/12) \simeq$ 0.08. The observed slope is about 0.13 giving reasonable support to the basic ideas outlined above. The diagonal thick line from left top to bottom right in Fig. 4.6 shows an accurate numerical fit to the observed main sequence band.

What we described above is a "theoretician's" H-R diagram which plots the luminosity $L$ of the stars against its effective surface temperature $T_s$. Observational astronomers, however, like to plot a quantity called *visual magnitude* against the *colour* of the star measured in terms of the difference of the magnitudes in blue and visual bands. You need to understand what these terms mean, which we shall now discuss.

For totally crazy (that is, historical) reasons, optical astronomers measure the luminosity in terms of a quantity called *magnitude* $(\mathcal{M})$ which is related logarithmically to the physical luminosity $L$ by

$$\mathcal{M} = -2.5 \log \left( \frac{L}{L_\odot} \right) + \mathcal{M}_\odot. \qquad (4.12)$$

Here $\mathcal{M}$ stands for absolute magnitude (not to be confused with the symbol for mass) and you will notice that the brighter an object is, the *lower* is its magnitude. The logarithm is taken to base ten. One can define a corresponding quantity (called *apparent magnitude, m*) using the measured flux $F = (L/4\pi d^2)$. To take care of the dependence on $d$, the numbers are chosen such that the absolute luminosity of an object is the same as the apparent magnitude it would have if it is placed at a distance of 10 parsecs; that is, $m - \mathcal{M} = 5\log(d/10 \text{ pc})$. To complicate the matters further, the luminosity is usually measured within a given frequency range (usually called $U, V, B$ etc. for ultra-violet, visible, blue ....) and one can have blue magnitude $m_B$, visible magnitude $m_V$ etc. for the same object. The absolute magnitude, determined by integrating over all frequencies, is called *bolometric* magnitude. The relation between magnitudes and more sensible units can be expressed in the form

$$F_X = K_X 10^{-0.4 m_X} \text{erg cm}^{-2} \text{ s}^{-1} \qquad (4.13)$$

where $X$ denotes the band (Ultraviolet, Visible, Blue ...). For the standard cases you will encounter (called U, B, V, R, I, J, K; the last few being in infra-red), the constants are $K_X = (2.90, 6.47, 3.38, 3.82, 2.00, 1.26, 0.19) \times 10^{-6}$. In the case of total integrated flux and *bolometric* magnitude, the corresponding relation is

$$F = 10^{-0.4m} \times 2.52 \times 10^{-5} \text{ ergs cm}^{-2}\text{s}^{-1}. \tag{4.14}$$

The primary effect of all these definitions is to scare physicists away from astronomy and you just need to overcome this fear to work in this area!

Incidentally, it turns out (by accident, not by design) that $10^{-0.4} = 0.398$ which is quite close to $e^{-1} = 0.367$ so that magnitudes have a correspondence with e-folding. This allows you to use standard formulas to calculate magnitude differences: For example, since $10^{+0.4 \times 0.01} \approx e^{0.01} \approx 1.01$ a magnitude difference of 0.01 actually corresponds to 1 per cent flux difference.

These results can be used to estimate some relevant quantities. For example, consider the contribution of starlight to the background light in the sky in the optical band. A solid angle $d\Omega$ will intercept a volume $(1/3)R^3 d\Omega$ of our galaxy if $R$ is the radius of the galaxy. If the number density of bright stars with luminosity of $L \simeq L_\odot$ is $n \simeq 0.1 \text{ pc}^{-3}$, then the flux per steradian is

$$\mathcal{F} \simeq \frac{1}{3}R^3(nL_\odot)\frac{1}{4\pi R^2} \simeq \frac{1}{12\pi}nL_\odot R \simeq 9 \times 10^{-6} \text{ W m}^{-2} \text{ rad}^{-2} \tag{4.15}$$

if $R = 10$ kpc. Using $1 \text{ rad}^2 \simeq 4.25 \times 10^{10} \text{ arcsec}^2$, we get a sky brightness due to integrated starlight of about $2.1 \times 10^{-16} \text{ W m}^{-2} \text{ arcsec}^{-2}$. You should convince yourself that this works out to about 21 mag arcsec$^{-2}$.

For comparison, note that a galaxy consists of about $10^{11}$ stars and could be located at distances ranging from 1 to 4000 Mpc. At 10 Mpc, such a galaxy will subtend an angle of about $\theta \simeq (2R/d) \simeq 400''$ and will have a flux of about $3 \times 10^{-11} \text{ W m}^{-2}$ if the size of the galaxy $R_{\text{gal}} \simeq 10$ kpc. The surface brightness of the galaxy will be about $10^{-16} \text{ W m}^{-2} \text{ arcsec}^{-2} \simeq 21$ mag arcsec$^{-2}$. These galaxies also contribute to the background light in the optical band. Repeating the above analysis we did for stars with $L = 10^{11}L_\odot, R \simeq 4000$ Mpc, $n \simeq 1 \text{ Mpc}^{-3}$, we get a background of $5 \times 10^{-18}$ W m$^{-2}$ arcsec$^{-2} \simeq 24$ mag arcsec$^{-2}$.

Another quantity of great observational importance, closely related to magnitude, is *colour*. To understand this concept, let us consider some source that emits radiation at three wavelengths $\lambda_1, \lambda_2$ and $\lambda_3$. The absolute luminosities of the source at these wavelengths can only be determined if we know the distance to the source. However, one can determine the ratios $L(\lambda_1)/L(\lambda_2)$ and $L(\lambda_2)/L(\lambda_3)$ without knowing the distance to the source. These ratios, of course, contain important information about the spectral characteristics of the source. Since the magnitudes are related to the luminosities by logarithms, the ratio of luminosities gets translated to differences in the magnitudes. For example, the difference $(m_B - m_V)$ between the apparent magnitudes in blue and visual bands will measure the ratio of the total fluxes received from a source in the wavelengths corresponding to B and V. (Of course, no instrument can measure the flux exactly at one wavelength; what is observed is the integral of $F(\lambda)W(\lambda)$ over the wavelength where $W(\lambda)$ is a filter window which is nonzero in a small wave length band $\Delta\lambda$ around a given wavelength.) This quantity $(m_B - m_V)$ is called the B-V *colour index* and is often just denoted as B-V. More precisely, the U-B and B-V indices are defined through the relations

$$U - B = -2.5 \log \left( \frac{F_{3650}\Delta\lambda_U}{F_{4400}\Delta\lambda_U} \right) - 0.87, \qquad (4.16)$$

$$B - V = -2.5 \log \left( \frac{F_{4400}\Delta\lambda_B}{F_{5500}\Delta\lambda_V} \right) + 0.66 \qquad (4.17)$$

where $F_\lambda(\text{erg cm}^{-2} \text{ s}^{-1} \text{ Å}^{-1})$ is the flux density at the specified wavelength $\lambda$ and $\Delta\lambda_U = 680$ Å, $\Delta\lambda_B = 980$ Å, $\Delta\lambda_V = 890$ Å denote the width of the colour filters. The constants are a matter of convention. Clearly these refer to some kind of a 'colour' in the sense that if B-V is large and positive, you expect the source to be bluer than some other source which is peaked in the visible band. Colour indices are of great practical value because they can be obtained by just taking photometric data of the source with different filters and can still provide spectral information. Getting the actual spectrum of the source can be a lot tougher in many cases. Figure 4.1 shows the colour-colour plot for a thermal spectrum and for a power law spectrum with $F_\nu \propto \nu^{-\alpha}$. The curves are marked by the temperature for the thermal spectrum and the index $\alpha$ for the power law spectrum. It is obvious that colours alone can provide fair amount of spectral information.

Let us next consider another important feature of stellar spectrum viz. absorption lines. Since the temperature decreases as one goes from the center to the surface of the star, the radiation emitted from the hot interior can be absorbed by the cooler gas in the stellar atmosphere producing absorption lines characteristic of the local conditions prevailing the outer regions.

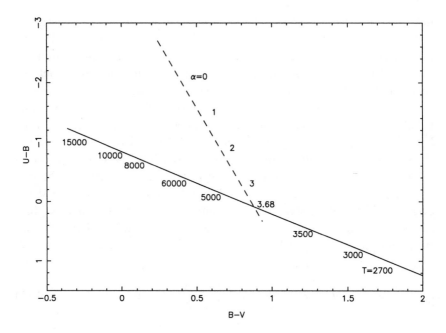

Fig. 4.1  The colour-colour plot for Planck spectrum at different temperatures and a class of power law spectra $F_\nu \propto \nu^{-\alpha}$ parameterized by the index $\alpha$.

A given absorption line will be produced by a particular element, in a particular ionisation state, when an electron makes the transition between two energy levels. The probability for such an absorption by partially ionized elements in their excited states will depend sensitively on the temperature. At high temperatures, atoms tend to be completely ionized and hence will not exhibit the discrete energy levels corresponding to bound states. At low temperatures, on the other hand, there will be very few atoms in the excited state which can produce the given absorption line. It follows that absorption due to atoms in their excited states will be maximum at some intermediate temperature — which can be estimated if the atomic properties are known. Further, absorption lines due to different elements will peak at different temperatures. Hence by choosing different atomic line transitions, one can probe the stellar atmospheres effectively. Such a spectroscopic analysis has led to a classification of stars, usually denoted by the sequence of letters O,B,A,F,...., *etc.* which are roughly in the order of decreasing surface temperature. (Again, these letters are not of any particular significance and the ordering is of historical origin. One favorite astronomical pastime is to find mnemonics to remember this sequence.)

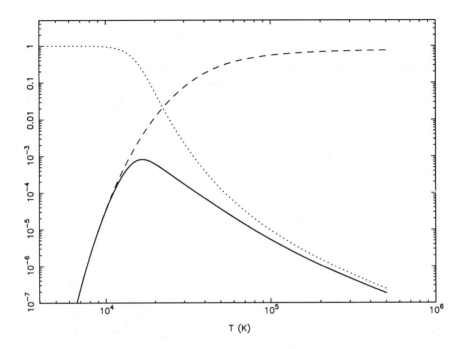

Fig. 4.2   Figure showing why Balmer absorption lines (which are caused by the upward electronic transitions from $n = 2$ state of hydrogen) are seen most prominently in an intermediate range of temperature. The dotted line shows the neutral fraction of hydrogen and the dashed line shows the fraction of these neutral atoms which are in $n = 2$ state. Their product, shown by the thick unbroken line has a clear maximum.

To illustrate this phenomena in a simple context, let us consider the Balmer absorption line due to hydrogen from near the stellar surface. This absorption occurs when an electron in the state $n = 2$ of the hydrogen atom absorbs a photon and makes a transition to any of the higher states. Hence the strength of the absorption line will be proportional to the fraction of hydrogen atoms which are in the $n = 2$ state. This fraction, in turn, is the product of (i) the fraction of hydrogen atoms which are neutral and (ii) the fraction of the neutral hydrogen atoms which are in the $n = 2$ state. To simplify the calculations, let us ignore other excited states and assume that the hydrogen atom is either in the $n = 2$ state or in the $n = 1$ state. The neutral fraction of the hydrogen atoms is $\mathcal{N} = (1 - y)$ where $y$ is the ionisation fraction given in Eq. (3.73). If we denote the right hand side of Eq. (3.73) by $F(T)$ then it is easy to solve the quadratic equation and obtain $\mathcal{N}$ as

$$\mathcal{N} = \frac{1}{2}\left[(2 + F) - (F^2 + 4F)^{1/2}\right].  \qquad (4.18)$$

This quantity is plotted by a dotted curve in Fig. 4.2. As expected, hydrogen is mostly neutral at low temperatures and makes rapid transition to ionized state at high temperatures. The relative population of neutral atoms in the $n = 2$ and $n = 1$ states is given by:

$$\mathcal{R} = \frac{N(n=2)}{N(n=1)} \equiv \frac{N_2}{N_1} = 4e^{-\Delta E/k_B T} = 4e^{-10.2\text{eV}/k_B T}.  \qquad (4.19)$$

The factor 4 comes from the fact that the quantum state $n$ has a degeneracy $g_n = 2n^2$ giving a relative weight of 8/2 to the two states. The fraction of the total *neutral* atoms which is in $n = 2$ state is given by

$$\frac{N_2}{N_2 + N_1} = \frac{\mathcal{R}}{1 + \mathcal{R}}.  \qquad (4.20)$$

The dashed curve in Fig. 4.2 shows this fraction as a function of temperature. Obviously when the temperature is low, most atoms are in the ground state while at high temperatures, some of the atoms can get to $n = 2$ state. The net fraction of atoms which will be contributing to the Balmer line absorption is given by the product

$$\frac{N_2}{N_\text{tot}} \approx \frac{N_2}{N_1 + N_2} \frac{N_1 + N_2}{N_\text{tot}} = \left(\frac{\mathcal{R}}{1 + \mathcal{R}}\right)\mathcal{N}.  \qquad (4.21)$$

This product is also shown by a thick line in Fig. 4.2. It is clear that there is a distinct maxima as a function of temperature due to two competing effects. Obviously, stars which show strong balmer absorption will have a surface temperature near this maxima while stars with either lower or higher surface temperature will not show this signature.

Figure 4.3 shows the actual spectra of different types of stars (with different surface temperatures) and the standard classification symbols which are used. One can clearly see that Balmer absorption is dominant at some intermediate temperature. Similar effects also explain other features — which could at first be surprising — in the Fig. 4.3. For example, there are stars in which Ca II absorption is more prominent than the hydrogen absorption in spite of the fact that hydrogen is about $5 \times 10^5$ times more abundant than calcium. The point is that at the relevant range of temperatures, only a fraction $5 \times 10^{-9}$ of the hydrogen atoms are at the $n = 2$ state

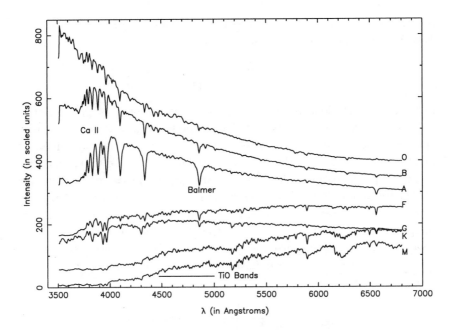

Fig. 4.3   Spectra of different types of stars and the corresponding absorption lines (Figure courtesy: R. Gupta.)

capable of producing Balmer absorption. On the other hand, nearly 99 per cent of all the calcium atoms are in the ground state and can produce the Ca II K line at wavelength $\lambda = 3933$ Å. In other words, there are nearly 400 times more Ca II ions capable of producing the absorption lines compared to the hydrogen atoms capable of producing Balmer lines. Similar features explain why you find exotic stuff like TiO in stellar surface at appropriate temperatures.

## 4.3   Modeling stellar structure

We will next try to get a feel for the detailed structure of the star. The first step will be to obtain the relevant equations which govern the structure of the star. To simplify matters, we shall assume that the star is in steady state, spherically symmetric and does not have any magnetic field. Then the first equation one could write down relates the mass $M(r) \equiv M_r$, contained within a sphere of radius $r$, to the density $\rho(r)$ by $(dM/dr) = 4\pi r^2 \rho$. The second equation we have is that of hydrostatic equilibrium equating the gravitational force to the pressure gradient: $(dP/dr) = -GM\rho/r^2$. Together, we now have *two* equations

$$\frac{dP}{dM_r} = -\frac{GM_r}{4\pi r^4} \; ; \qquad \frac{dr}{dM_r} = \frac{1}{4\pi r^2 \rho}, \qquad (4.22)$$

relating *three* functions $M(r), P(r), \rho(r)$. We can, of course, supplement this by an equation of state which will give the pressure in terms of the density and temperature: $P = P(\rho, T)$ but that will bring in the new unknown function, $T(r)$. (This is the general situation. Of course, in special case of barotropic fluids for which the pressure can be expressed as a function of density alone, $P = P(\rho)$, the system of equations closes. This case was discussed in Sec. 3.2.) So we next need an equation to determine the temperature distribution $T(r)$ inside the star, which in turn depends on how the energy is transported across the star. If the energy transport is radiative, we can relate the radiative flux $\mathcal{F}$ to the temperature gradient using Eq. (2.36):

$$\frac{dT}{dr} = -\frac{3}{4ac}\frac{\kappa\rho}{T^3}\frac{L_r}{4\pi r^2} \qquad \text{(radiation)}. \qquad (4.23)$$

This result for $(dT/dr)$, of course, is valid only if the entire flux of energy $\mathcal{F}(r)$ is transfered through radiative processes. In general, convection could also provide a means of transferring the energy and we need to determine when it is important.

One simple condition for the *absence* of convection can be obtained as follows: Let us consider a fluid column located in a gravitational field in which a blob of matter with specific volume $V(P, s)$ (which is volume of unit mass of matter) moves from a height $z$ to a height $z + \xi$ where the ambient pressure is $P'$. On displacement, its specific volume becomes $V(P', s)$ if the motion is adiabatic and $s$ does not change. This motion of the blob will be stable and the resulting force will drive it back to the original location provided the fluid element is heavier than the fluid it displaces in its new position. The specific volume of the latter is $V(P', s')$ where $s'$ is the equilibrium entropy at the height $z + \xi$. For stability we must have $V(P', s') - V(P', s) > 0$. Writing, $s' - s = \xi(ds/dz)$ and using the thermodynamic[1] relation $(\partial V/\partial s)_P = (T/C_P)(\partial V/\partial T)_P > 0$ we get the condition for stability to be $(ds/dz) > 0$. That is, the entropy must increase with height for convective stability.

---

[1] This is easy to prove from the definition $C_P = T(\partial s/\partial T)_P$ and the relation $(\partial s/\partial T)_P = (\partial s/\partial V)_P(\partial V/\partial T)_P$.

This result can be expressed in terms of $T$ and $P$ by noting that adiabatic evolution ($s = $ constant) implies $T \propto P^{(\gamma-1)/\gamma}$ (see Eq. (3.10)). Hence, if $s = $ constant, the temperature gradient is $(d \ln T / d \ln P) \equiv \nabla_{ad} = (\gamma - 1)/\gamma$. So, in the case of stellar interior, this condition for convective instability can be expressed as $\nabla > \nabla_{ad}$ where

$$\nabla \equiv \frac{d \ln T}{d \ln P}; \qquad \nabla_{ad} \equiv \frac{\gamma - 1}{\gamma}. \tag{4.24}$$

Using Eq. (4.23) and Eq. (4.22) we can find the gradient in the case of the radiative transfer of energy to be

$$\nabla_{\text{rad}} \equiv \left( \frac{d \ln T}{d \ln P} \right)_{\text{rad}} = \frac{3\kappa}{16\pi ac} \left( \frac{P}{T^4} \right) \left( \frac{L_r}{GM_r} \right). \tag{4.25}$$

If the gradient $\nabla_{\text{rad}}$ arising out of radiative transfer is less than $\nabla_{ad}$, then the fluid is stable against convection and we can use Eq. (4.23). On the other hand, if $\nabla_{\text{rad}}$ is greater than $\nabla_{ad}$, we have to take into account the energy transported due to convection. (Note that $\nabla_{\text{rad}}$ increases with $\kappa$; high opacity tends to cause convective instability.) This is a complicated problem and we do not yet have a fundamental theory describing convective energy transfer. However, if we assume very efficient condition, we may approximate the gradient to be that of the adiabatic value and assume that $\nabla = \nabla_{ad}$. This gives

$$\frac{dT}{dr} = -\nabla_{ad} \left( \frac{GM_r}{r^2} \right) \left( \frac{\rho T}{P} \right) = -\left( 1 - \frac{1}{\gamma} \right) \left( \frac{\rho T}{P} \right) \left( \frac{GM_r}{r^2} \right). \tag{4.26}$$

One crucial feature regarding convection is that it is a very fast process compared to radiative transfer. The acceleration induced on a blob of gas moving under convection is about $(GM_r/r^2)(\Delta T/T)$ where $\Delta T$ is the temperature difference. If we take $\Delta T \approx 1$ K, then a blob of gas can travel from $0.5 R_\odot$ to $0.1 R_\odot$ in about 3 weeks time which is considerably shorter than the time scale for radiative diffusion.

We now have three differential equations (two in Eq. (4.22) and either Eq. (4.23) or Eq. (4.26) depending on whether energy transport is radiative or convective) and one equation of state relating the five variables, $P, M, T, L$ and $\rho$. To close the system, we need one more equation which relates the luminosity $L(r)$ to the energy production rate $\epsilon(\rho, T)$ per unit mass. This is given by

$$\frac{dL_r}{dM_r} = \epsilon(\rho, T). \tag{4.27}$$

In the simplest context, $\epsilon$ denotes the energy production due to hydrogen fusing to form helium. The hydrogen fusion leading to helium can occur through either of the two sequences of reactions (called *p-p reactions* or *CNO cycle*) depending on the temperature. The form of $\epsilon$ for these two nuclear reactions cycles are given by the fitting functions:

$$\epsilon_{\text{pp}}(\rho, T) \approx \frac{2.4 \times 10^4 \rho X^2}{T_9^{2/3}} \, e^{-3.380/T_9^{1/3}} \quad \text{erg g}^{-1} \text{ s}^{-1}, \qquad (4.28)$$

$$\epsilon_{\text{CNO}}(\rho, T) \approx \frac{4.4 \times 10^{25} \rho X Z}{T_9^{2/3}} \, e^{-15.228/T_9^{1/3}} \quad \text{erg g}^{-1} \text{ s}^{-1}. \qquad (4.29)$$

where $T_9 = (T/10^9 \text{ K})$ *etc.*. Stars with $M \gtrsim 1.3 M_\odot$ have CNO cycle as the dominant reaction channel while stars with $M \lesssim 1.3 M_\odot$ has p-p reactions as the dominant channel. Of these, the CNO cycle has significantly higher temperature sensitivity, resulting in a larger temperature gradient in the core region, which — in turn — can lead to convection. Hence stars with $M \gtrsim 1.3 M_\odot$ have a convective core while stars with $M \lesssim 1.3 M_\odot$ have a radiative core. When hydrogen burning occurs in a convective environment the ashes of burning will be uniformly distributed over the core region by convection in stars with $M \gtrsim 1.3 M_\odot$. It is obvious that stars with $M \lesssim 1.3 M_\odot$ will be structurally different from those with $M \gtrsim 1.3 M_\odot$.

The equations of stellar structure thus require three input functions $P(\rho, T), \kappa(\rho, T)$ and $\epsilon(\rho, T)$ for their integration. In general, these functions have fairly complicated form since they incorporate several different effects. The pressure, for example, can come from an ideal gas plus radiation plus even degeneracy pressure for some range of stellar parameters; the opacity can come from free electron scattering as well as free-free and bound-free absorption. We have already seen in Eq. (4.28), Eq. (4.29) that the nuclear energy generation can have a complicated density and temperature dependence. Given such complexities, it is necessary to integrate the relevant equations numerically to obtain the properties of stars. (The integration is fairly straightforward; what is difficult is getting the correct input physics as well as modeling convention.) The results of such numerical integration is reasonably well fitted by the following fitting formula which gives the luminosity and the radius of the star in terms of its mass $m = M/M_\odot$ at the start of the main sequence (usually called Zero Age Main Sequence,

ZAMS for short)

$$
\frac{L}{L_\odot} =
\begin{cases}
\dfrac{1.107m^3 + 240.7m^9}{1 + 281.9m^4}, & \text{if } m \leq 1.093; \\[4mm]
\dfrac{13990m^5}{m^4 + 2151m^2 + 3908m + 9536}, & \text{if } m \geq 1.093;
\end{cases}
\tag{4.30}
$$

$$
\frac{R}{R_\odot} =
\begin{cases}
\dfrac{0.1148m^{1.25} + 0.8604m^{3.25}}{0.04651 + m^2}, & \text{if } m \leq 1.334; \\[4mm]
\dfrac{1.968m^{2.887} - 0.7388m^{1.679}}{1.821m^{2.337} - 1}, & \text{if } m \geq 1.334.
\end{cases}
\tag{4.31}
$$

(Of course, the right hand sides will *not* lead to unity for $m = 1$ and they should not; try to figure out why.) These relations as well as the resulting surface temperature $T_s$ obtained by $\sigma T_s^4 = L/(4\pi R^2)$, are plotted in Fig. 4.4. The solid curve refers to the luminosity which is given in the left vertical axis; the dashed curve gives the radius and the dot-dash curve gives the temperature and these are referred in right hand side vertical axis. It is clear that different physical phenomena contribute in different stars making these relations fairly complex.

If you don't care for such complexities, but just want to get a feel for which effects lead to which results, then one can make a fairly drastic approximation to obtain the relevant scalings. This requires assuming that the three input functions $P, \kappa, \epsilon$ are power laws in $\rho$ and $T$. It is conventional to take this in the form

$$
\rho = \rho_0 P^a T^{-b}; \quad \kappa = \kappa_0 \rho^n T^{-s}; \quad \epsilon = \epsilon_0 \rho^\lambda T^\nu.
\tag{4.32}
$$

Such an approximation for $\rho$ and $\kappa$ will be valid if any single process makes the dominant contribution to pressure and opacity. The approximation of $\epsilon$ by a power law is valid in limited ranges depending on which nuclear reaction dominates the energy production. Both the reactions in Eq. (4.28) and Eq. (4.29) — and, in fact, more complicated helium burning, silicon burning and oxygen burning reactions — can be approximated by local power law $\epsilon = \epsilon_0 \, \rho^\lambda T^\nu$. For the p-p chain $\lambda = 1, \nu = 4$; for CNO cycle, $\lambda = 1, \nu = 15$ to 18. Similarly, for reactions involving helium burning, $\lambda = 2, \nu = 40$ *etc.*

If we make the power law approximation in Eq. (4.32) then we expect homologous solutions with power law dependence for all the parameters. While such a system can be easily written down and solved, the net effect

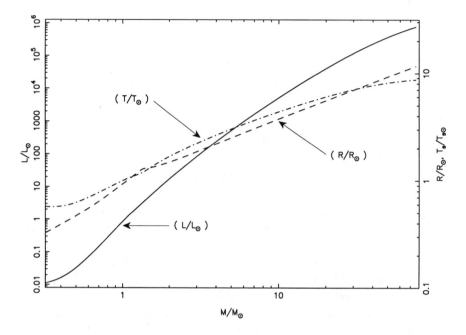

Fig. 4.4 Luminosity, radius and surface temperature of the zero age main sequence stars as a function of mass. The solid curve refers to the luminosity which is given in the left vertical axis; the dashed curve gives the radius and the dot-dash curve gives the temperature and these are referred in right hand side vertical axis.

is easy to mimic by the following trick: You just replace all derivatives by ratios like $dP/dr \to P/R$ etc. (It works for all homologous systems; you will recall that we used the procedure in the case of Lane-Emden equation; see the discussion after Eq. (3.29).) In this case, Eq. (4.22), coupled with $P \propto \rho T$ reduces to $P \propto M^2/R^4$, $\rho \propto M/R^3$, $T \propto M/R$. Substituting these into Eq. (4.23) and using $L \propto \epsilon M$, we get the simple relation $\epsilon \kappa = M^2$. Using Eq. (4.32) in which $\epsilon$ and $\kappa$ are given by power laws and expressing $\rho$ and $T$ in terms of $M$ and $R$, we will get a simple relation between $M$ and $R$, which can be expressed as $R \propto M^{n_R}$. Substituting back in the previous relations, allows you to express everything else as power laws in $M$: like $T \propto M^{n_T}$, $L \propto M^{n_L}$, $T_s \propto M^{n_{HR}}$ etc. You can play around with different assumptions, leading to different power law indices in Eq. (4.32) and compare the results with the more exact fitting formula in Eq. (4.30), Eq. (4.31). The key results of such an analysis can be summarized as follows.

First, for the high mass stars with CNO cycle being the dominant energy production mechanism and electron scattering providing the opacity, homology relations give $n_R \approx (0.78 - 0.81), n_L = 3$ and $n_{\mathrm{HR}} \approx 0.12$ depending on how you model pressure. The observed values are closer to $n_R = 0.56, n_L \approx 3.5$ and $n_{\mathrm{HR}} \approx 0.17$. These are in the same ballpark and the difference is essentially due to the convective regions in the stars.

Second, for low mass stars dominated by p-p chain, we have to consider both electron scattering opacity and Kramer's opacity. For Kramer's opacity, homology models give $n_R \approx (0.1 - 0.2), n_L \approx (5.4 - 5.5), n_{\mathrm{HR}} \approx (0.23 - 0.24)$. With electron scattering opacity, we get $n_R \approx (0.43 - 0.5), n_L \approx 3, n_{\mathrm{HR}} \approx (0.17 - 0.18)$. The observed values are closer to $n_R \approx (0.85 - 0.96), n_L \approx 4.8, n_{\mathrm{HR}} \approx 0.13$. It is clear that, while the broad trends are captured by these models, the details cannot be reproduced. In fact, numerical integration shows (see Fig. 4.4) that the relevant parameters of the stars show different power law indices in different ranges and cannot be reproduced by simple homology arguments.

In all the above discussion we did not say anything about the chemical composition of the star, which, of course, is an important issue. We will see in chapter 6 that, when the universe was just a few minutes old, some amount of nucleons combined together to form Helium which amounted to about 23 percent by weight. (We will also see that none of the heavier elements were produced in significant quantities.) As the universe evolves and the first generation of stars (usually called *Pop I*) form, they will have essentially such a composition: $X \approx 0.77, Y \approx 0.23, z \approx 0$ where $X, Y, Z$ denotes the fraction of hydrogen, helium and heavier elements (metals, as they are called!) respectively. As these stars evolve, heavier elements will be synthesized in them and — as we shall see later — these elements will be spewed out into interstellar medium through supernova explosions. Later generations of stars (called *Pop II*) will have traces (say, $Z \approx 0.02$) of heavier elements. In several contexts, it is useful to know how the various stellar parameters like luminosity, radius *etc.* depend on the metallicity. Similarly, as stellar nuclear reactions produce heavier elements, the mean atomic weight $\mu$ of the stellar material will change and it is also relevant to ask how the results depend on $\mu$. These are important in determining, for example, the track of the star in HR diagram as the evolution proceeds and $\mu$ and $Z$ change. In these two cases, the homology relations can be more reliable.

The mean atomic weight $\mu$ enters the equations mainly through the ideal gas law (and somewhat more weakly through the opacity). Keeping the leading dependence which arises through the ideal gas laws, one can easily show that the radius, temperature and the luminosity of a star of a given mass scales as $R \propto \mu^{m_R}, T \propto \mu^{m_T}, L \propto \mu^{m_L}$. For CNO cycle ($\nu = 15 - 18$) and electron scattering opacity ($n = s = 0$) and ideal gas equation of state, we get

$$m_L = 4.0, \qquad m_R = 0.58 - 0.63, \qquad m_T = 0.42 - 0.36. \qquad (4.33)$$

This shows that the radius and temperature vary rather weakly with the mean atomic weight while the luminosity is a strongly increasing function. (For p-p chain, with ideal gas equation of state and Kramer's opacity, the dependence is even steeper and we get $m_L \approx 8$.) For the first case, the slope in H-R diagram will be

$$m_{HR} = \frac{1}{4}\left(1 - \frac{2m_R}{m_L}\right) \approx 0.17 \qquad (4.34)$$

which is higher than the corresponding value $n_{HR} \approx 0.12$. This implies that as the atomic weight increases, say due to the conversion of hydrogen into helium, the stars will have higher $L$ but will evolve below the main sequence line in the $L - T_{\text{eff}}$ plane.

Since $Z$ is usually a small number, the heavy element abundance does not directly influence $\mu$; however, the opacity has a dependence on $Z$ since the metals contribute significantly to opacity. Using the power law approximations for $\mu$ and $\kappa$ given by $\mu \simeq 0.5 X^{-0.57}$ and $\kappa \propto Z(1 + X)\rho T^{-3.5}$ for low mass stars[2] (with p-p chain reaction as the main source of energy and Kramer's opacity as the transport process), it can be easily shown that

$$R \propto Z^{0.15} X^{0.68}; \quad L \propto Z^{-1.1} X^{-5.0}; \quad T_{eff} \propto Z^{-0.35} X^{-1.6} \qquad (4.35)$$

approximately. These combine to give $L \propto Z^{0.35} X^{1.55}$. The actual $Z$ dependence is weaker than the one indicated above, since the electron scattering opacity is not taken into account; but the relation shows the correct trend. Both luminosity and the effective temperature increase with decreasing $Z$. For a star of given mass but varying $Z$ we get the relation $T_{eff} \propto L^{0.33}$. In other words, the main sequence for metal poor stars, will

---

[2]The $Z$ in this expression for $\kappa$ corresponds to the mass fraction of metals; so naively this expression will give $k = 0$ for, say, pure hydrogen. Of course, this is not true in the rigorous sense; but at the range of temperatures we are interested in, hydrogen will be fully ionized and the bound-free opacity indeed arises only from metals. Thus the scaling $k \propto Z$ is correct as far as its applicability to the stellar interior is concerned.

lie below and to the left of the main sequence for metal rich stars in the $L - T_{eff}$ plane.

The scaling relations obtained above refers mostly to the bulk of the star. The situation in the outermost region of the star (usually called the stellar envelope or *stellar atmosphere*), is somewhat different. Here the radiative processes are still important in determining the passage of energy but there is no energy *production* or significant self gravity. The variation of different physical quantities in the stellar envelope can, therefore, be obtained by assuming that the stellar envelope is located in the (external) gravitational field of the rest of the star. Let the energy flux of radiation through the envelope be $F_{\rm rad}$. Using $F_{\rm rad} = -(ac/3\kappa\rho)\nabla T^4$ and $\nabla P = -g\rho$, where $g$ is the local acceleration due to gravity at the envelope due to the rest of the star, we have

$$F_{\rm rad} = \frac{gca}{3\kappa}\frac{dT^4}{dP} \approx \frac{gca}{3\kappa}\frac{T^4}{P} \qquad (4.36)$$

where we have approximated $(dT^4/dP)$ by $(T^4/P)$ in the spirit of homology relations. Taking $g \propto (M/R^2), \kappa \propto \rho T^{-3.5}, \rho \propto (P/T)$ and using the fact that $F \propto (L/R^2)$ is a constant in the envelope, we get the relations:

$$\frac{L}{R^2} \propto \left(\frac{M}{R^2}\right)\left(\frac{T^{8.5}}{P^2}\right) \; ; \quad P \propto \left(\frac{M}{L}\right)^{1/2} T^{4.25} \; ; \quad \rho \propto \left(\frac{M}{L}\right)^{1/2} T^{3.25}.$$
$$(4.37)$$

The radial dependence of temperature follows from the equation $(dP/dr) \approx (P/r) \approx -g\rho \propto -(M/r^2)(P/T)$; this gives $T \propto (1/r)$ in the envelope and fixes the variation of all relevant quantities in the stellar envelope.

The entire discussion of stellar structure given above was based on the star remaining in a steady state. There is, however, one important physical phenomena which can take place in stars which are in 'almost' steady state. This corresponds to oscillations of the star around its equilibrium configuration, leading to important observable effects. While most stars have a relatively stable luminosity during the significant part of their life, some stars show striking variability in their energy outputs. These intrinsically variable stars have their light output changing in a periodic or semi-periodic fashion due to internal processes. The simplest of the variations arises due to radial pulsations of the stars at some period $\mathcal{P}$ determined by the star's internal structure. An intuitive description of this result can be obtained as follows. We start with the equation of motion for a shell of mass $m$ at radius $r$:

$$m\ddot{r} = 4\pi r^2 P - \frac{GMm}{r^2}. \tag{4.38}$$

In equilibrium, at $r = r_{eq}$, we have the condition $4\pi r_{eq}^2 P_{eq} = (GMm/r_{eq}^2)$. Perturbing the equation around this equilibrium value, by taking $r = r_{eq} + \delta r$, we get

$$m\frac{d^2\delta r}{dt^2} = 4\pi r_{eq}^2 P_{eq}\left(\frac{2\delta r}{r} + \frac{\delta P}{P}\right) + \frac{GMm}{r_{eq}^2}\left(\frac{2\delta r}{r}\right) \tag{4.39}$$

which governs the radial displacement of a shell of matter confined by $r = r_{eq}$ originally. Writing $(\delta P/P) = \gamma(\delta\rho/\rho) = -3\gamma(\delta r/r)$ (where the last equality follows from the mass conservation $\rho r^3 = $ constant) we get

$$m\frac{d^2\delta r}{dt^2} = -\frac{GMm}{r_{eq}^3}[3\gamma - 4]\,\delta r. \tag{4.40}$$

This equation shows that $\delta r$ is oscillatory with a local frequency $\omega^2 = (GM/r_{eq}^3)(3\gamma - 4) \propto \rho(3\gamma - 4)$. The frequency scales as $\sqrt{\rho}$ and the stability requires $\gamma > (4/3)$. (This is something we noticed earlier; see Eq. (1.57).)

The period of oscillation is directly related to the mean density $\bar{\rho}$ of the stars and hence can be connected to the luminosity and surface temperature. A more rigorous calculation confirms the result that the period of oscillation $\mathcal{P} = (2\pi/\omega)$ satisfies the relation $\mathcal{P}\sqrt{\bar{\rho}} = $ constant. For a given mode, the right hand side depends only on the polytropic index $n$ (introduced in chapter 3; Sec. 3.2) and $\gamma_{ad}$. This relation is called the *period-luminosity* relation A reasonable fit to the observations is given by

$$\log\left(\frac{L}{L_\odot}\right) = -17.1 + 1.49\log\left(\frac{\mathcal{P}}{1\text{ day}}\right) + 5.15\log T_{eff}. \tag{4.41}$$

Since both the surface temperature $T_{eff}$ and $\mathcal{P}$ can be measured without knowing the distance to the star, this relation allows us to determine the absolute luminosity of a variable star without knowing its distance. By measuring the apparent luminosity and comparing it with absolute luminosity, we can estimate the distance to the star. In particular, if such a variable star is spotted in another galaxy, this technique will provide an estimate of the distance to the galaxy — which is of considerable significance in cosmology. This is one of the examples in which detailed modeling of a particular class of stars helps in the understanding of issues which are important in a wider context.

## 4.4    The Sun as a star

We shall now illustrate several aspects of the above description of stellar structure using the Sun as a prototype main sequence star. Numerical integration of the relevant equations (leading to what is usually called *standard solar model*) provides the following information about the Sun. (i) The central pressure and central density are $P_c = 2.4 \times 10^{17}$ dynes cm$^{-2}$ and $\rho_c = 154$ gm cm$^{-3}$. (ii) The pressure falls to half its central value at $0.12R_\odot$; the temperature falls to half its central value at $0.26R_\odot$; the luminosity falls to half its value at $0.11R_\odot$ and half the hydrogen is consumed virtually at the origin. Here $R_\odot = 6.96 \times 10^{10}$ cm, is the radius of the visible disk of the Sun. We shall first try to understand these values.

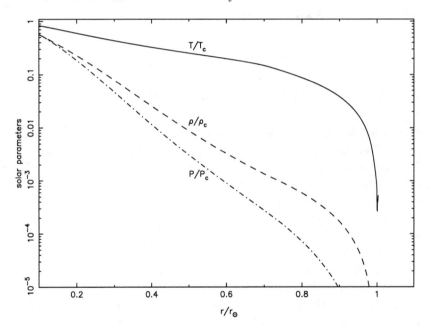

Fig. 4.5    The temperature, pressure and density inside the Sun scaled with respect to the central values. (Figure courtesy: H.M.Antia.)

Since density can be taken to be approximately constant near the center, $(dP/dr) \approx -G(4\pi/3)\rho_c^2 r$ near the center. Integration gives $P(r) \approx P_c - (2\pi/3)G\rho_c^2 r^2$ leading to $P = P_c/2$ at $r_{1/2} = (3P_c/4\pi G\rho_c^2)^{1/2} \approx 0.1R_\odot$. This is pretty close to the value of $0.12\ R_\odot$ obtained from numerical integration. Next, since Sun is radiative near the core, $\nabla = (d\ln T/d\ln P)$ is less than

$\nabla_{\text{ad}} = (1 - \gamma^{-1}) = 0.4$ (for $\gamma = 5/3$); thus the temperature must vary more shallowly than $T \propto P^{0.4}$ and at the radius where $T$ has dropped by a factor 2, the pressure would have dropped *at least* by $2^{2.5} = 5.7$. In reality, the ratio is more like 13 which suggests a scaling law $T \propto P^{0.27}$. This has to be expected, since numerical work also shows that the gradient $\nabla$ varies from 0.34 at center to 0.23 at $r = 0.26R_\odot$ and to 0.19 at $0.5R_\odot$.

The energy generation in the Sun is primarily due to p–p chain and the thermonuclear reactions take place in the core region with $r \lesssim 0.2R_\odot$. In the region $0.2R_\odot \lesssim r \lesssim 0.7R_\odot$, the energy transport is radiative while in most of the outer regions $0.7R_\odot \lesssim R_\odot$, the energy transport is due to convection. (This is seen from the fact that the adiabatic and radiative temperature gradients inside the Sun cross over around 0.71 $R_\odot$.) The p–p reaction rate which generates solar energy, varies as

$$\rho\epsilon_{\text{pp}} \propto X^2 T^4 \rho^2 \propto X^2 P^2 T^2 \propto X^2 P^{2.5} \qquad (4.42)$$

where the last relation follows from the scaling $T \propto P^{0.27}$ mentioned above. If $q \equiv r/(\sqrt{2}r_{1/2})$, then the pressure varies near the center as $P \propto (1 - q^2)$ leading to an energy generation $\epsilon_{\text{pp}} \propto (1 - q^2)^{5/2}$. Using this it is easy to find the value of $q$ at which half the total luminosity is generated. We get $q_{1/2} = 0.52$, predicting that half the luminosity should come from within $0.074R_\odot$. This is smaller than the true value of $0.1R_\odot$ obtained from numerical integration because we have neglected the variation of $X$.

Since half the hydrogen is depleted at the center, more energy generation is occurring at larger radii. Since Sun is $t_\odot = 4.5$ Gyr old and the efficiency of pp chain is 0.007, the amount of hydrogen which must have fused is about $(L_\odot t_\odot / 0.007 c^2) = 0.046 M_\odot$. The mass within half luminosity radius is $0.097 M_\odot$ and thus originally contained $0.097 X M_\odot = 0.068 M_\odot$ of hydrogen. In this region, $(0.046 M_\odot / 2) = 0.023 M_\odot$ of hydrogen must have been burnt. This is a fraction $(0.023/0.068) \approx (1/3)$ of the total mass of hydrogen in this region.

At the center of the Sun, $\mu = 0.829, \rho = 154$ gm cm$^{-3}$ and $T = 1.54 \times 10^7$ K. The gas energy density is $\epsilon_{\text{gas}} = (3/2)(\rho/\mu m_p)k_B T \approx 3.6 \times 10^{17}$ erg cm$^{-3}$ while the radiation energy density is $\epsilon_{\text{rad}} = aT^4 \approx 4.25 \times 10^{14}$ erg cm$^{-3}$. The ratio between them is about $1.2 \times 10^{-3}$. At half the solar radius, $\mu = 0.605, \rho = 1.43$ and $T = 3.9 \times 10^6$ K, giving $\epsilon_{\text{gas}} = 1.16 \times 10^{15}$ erg cm$^{-3}$ and $\epsilon_{\text{rad}} \approx 1.75 \times 10^{12}$ erg cm$^{-3}$ giving a ratio of $1.5 \times 10^{-3}$. In fact, this ratio is nearly constant in the Sun throughout the range $0 < r < 0.8R_\odot$.

This constancy implies that

$$\mathcal{R} \equiv \left( \frac{\rho}{1 \text{ gm cm}^{-3}} \right) \left( \frac{T}{10^6 \text{ K}} \right)^{-3} = \text{constant} \qquad (4.43)$$

giving $p \propto \rho T \propto \rho^{4/3}$; thus, most of the Sun can be described by a $n = 3$ polytrope. The gravitational binding energy of a polytrope of index $n = 3$ is [see Eq. (3.28)]

$$U_{\text{grav}} = -\frac{3}{(5-n)} \left( \frac{GM^2}{R} \right) \approx -\frac{3}{2} \left( \frac{GM_\odot}{R_\odot} \right). \qquad (4.44)$$

Virial theorem, on the other hand, gives $U_{\text{kinetic}} = -U_{\text{grav}}/2$. To relate thermal energy $U_{\text{th}}$ to $U_{\text{kinetic}}$, note that

$$\frac{U_{\text{kinetic}}}{U_{\text{th}}} = \frac{(3/2)Nk_BT}{C_VT} = \frac{(3/2)[C_P - C_V]}{C_V} = \frac{3}{2}(\gamma - 1). \qquad (4.45)$$

Since Sun is mostly ionized particles, $\gamma = (5/3)$ and $U_{\text{th}} \approx U_{\text{kinetic}}$. This gives the thermal time scale (also called *Kelvin-Helmholtz time scale*) to be $t_{\text{KH}} = (U_{\text{th}}/L_\odot) = 7 \times 10^{14}$ s which is a factor $(350 - 700)$ larger than diffusion time scale $t_{\text{diff}}$. The reason is quite simple. All the Sun's radiation leaks out at $t \approx t_{\text{diff}}$ but it is replenished by the thermal energy of matter. So for all the Sun's thermal energy to be carried off in radiation, it has to leak out by a larger factor of $(\epsilon_{\text{gas}}/\epsilon_{\text{rad}}) \approx 700$. This result is a direct consequence of the equation for radiative diffusion $(dP_{\text{rad}}/dr) = -\kappa\rho L/(4\pi cr^2)$. Writing

$$\frac{dP_{\text{rad}}}{dr} \approx \frac{P_{\text{rad}}}{P_{\text{gas}}} \frac{dP_{\text{gas}}}{dr} \approx \frac{P_{\text{rad}}}{P_{\text{gas}}} \frac{2\epsilon_{\text{gas}}}{3r} \qquad (4.46)$$

we find that $L \approx (\epsilon_{\text{rad}}/\epsilon_{\text{gas}})(4\pi r^3/3)\epsilon_{\text{gas}}(c/\kappa\rho r^2)$. Noting that $t_{\text{diff}} \approx (\kappa\rho r^2/c) = r^2/c\lambda$, this equation can be written as $t_{\text{KH}} \approx (\epsilon_{\text{gas}}/\epsilon_{\text{rad}})t_{\text{diff}}$.

For an ideal monotonic gas, the onset of convection takes place when $(d\ln T/d\ln P) > 0.4$ [see Eq. (4.24)] which occurs around $r \approx 0.7R_\odot$ for the Sun. Also note that very close to the surface, radiative conditions will prevail and hence solar atmosphere has to be treated separately.

One key result of modeling the solar interior is that Sun emits a steady flux of neutrinos which should be detectable at Earth. The *mean* solar neutrino flux on Earth can be easily estimated by taking the dominant reaction channel. The cumulative effect of a p-p chain is to convert 4 protons into one helium nucleus, releasing 2 positrons, 2 neutrinos and about 28 MeV of energy, which is eventually radiated as photons. Since

2 neutrinos come out with the release of 28 MeV of energy in the form of radiation, the total number of neutrinos released per second will be about $2(L_\odot/28 \text{ MeV})$. Hence the neutrino flux on Earth, at a distance $r$ from the Sun, will be about

$$F = \frac{2L_\odot}{4\pi r^2 (28 \text{ MeV})} = \frac{2 \times 4 \times 10^{33} \text{ erg } s^{-1}}{4\pi (1.5 \times 10^{13} \text{ cm})^2 \times 28 \text{ MeV}} = 6 \times 10^{10} \text{ cm}^{-2}s^{-1}.$$
(4.47)

Several experiments have been set up to measure the flux of solar neutrinos arriving on Earth. One of these procedures involves allowing the solar neutrinos to react with $^{37}Cl$ thereby producing an electron and $^{37}Ar$. The product $^{37}Ar$ is radioactive with a half life of 35 days. By measuring the amount of Argon nuclei which is produced, one can estimate the number of neutrino captures per second which took place in the target. (This reaction requires a threshold energy of 0.81 MeV for the neutrino and hence is sensitive only at higher energy neutrinos.) With an absorption cross section of about $10^{-44}$ cm$^2$ and an effective flux of about $10^9$ neutrinos cm$^{-2}$ s$^{-1}$ one would expect $10^{-35}$ reactions per second per target atom. It is convenient to introduce a unit called *Solar Neutrino Unit*, abbreviated as SNU, which stands for $10^{-36}$ captures per second per target atom. From Eq. (4.47) one expects a detection at the level of about 10 SNU. A more precise calculation gives $7.7^{+1.2}_{-1.0}$ SNU where the error bar covers uncertainties in nuclear reaction rates, opacities, model building techniques and is fairly conservative. Observations, however, gave a much lower value for the solar neutrino flux, around $2.55 \pm 0.25$ SNU. This discrepancy has been originally called the *solar neutrino problem.*

We now understand that the solution to the solar neutrino problem lies in the phenomenon of neutrino oscillations. In the simplest models describing electro-weak interactions, it is assumed that there are 3 different species of neutrinos, viz. electron neutrino ($\nu_e$), muon neutrino ($\nu_\mu$) and tau neutrino ($\nu_\tau$) all of which are *massless*. In that case, each neutrino will retain its identity as it propagates from Sun to Earth and the solar neutrino experiment will essentially measure the flux of electron neutrino $\nu_e$; the experiments are insensitive to the other two types of neutrinos. There is, however, fair amount of evidence that the neutrinos are actually *massive*. If that is the case, then one can construct particle physics models in which a phenomenon called *neutrino oscillations* can exist. In this phenomenon, the nature of the neutrino changes periodically in time so that the identity of the neutrino can oscillate between, say, $\nu_e$ state and the $\nu_\mu$ state. It is then possible to arrange the masses of the neutrinos in such a way that the deficit of electron neutrinos seen in terrestrial experiments can be explained as due to "conversion" of $\nu_e$ to $\nu_\mu$, say, as the neutrino propagates from the Sun to Earth.

Finally we consider a phenomenon of some terrestrial significance originating from the sun, viz. the solar wind. There is a steady flow of hot material from the sun, which can be modeled as the time reversed solution to accretion (see Sec.3.3.1) viz. a wind flowing from a central object. The solar wind can be modeled (to the lowest order of approximation) as a spherically symmetric flow of a non dissipative fluid in steady state. In this case, the equation of state is best approximated as $P = \rho k_B T$ for which Eq. (3.42) is not valid. However, it is easy to integrate the original Eq. (3.40) and obtain

$$\frac{v^2}{c_s^2} - \log \frac{v^2}{c_s^2} = 4 \log \frac{r}{r_c} + \frac{2GM}{rc_s^2} + \text{constant} \qquad (4.48)$$

where $r_c \equiv (GM/2c_s^2)$ is the critical radius at which the velocity of flow is equal to the isothermal sound speed $c_s \equiv (k_B T)^{1/2}$, which is now a constant independent of $r$. The constant of integration can be evaluated by using $v = c_s$ at $r = r_c$ and we get its numerical value to be 3. Substituting back and rearranging the terms, we can write the solution as

$$\frac{\mathcal{M}^2}{2} - \ln \mathcal{M} = \frac{2r_c}{r} + 2 \ln \frac{r}{r_c} - \frac{3}{2} \qquad (4.49)$$

where $\mathcal{M} = (v/c_s)$ is the Mach number. The relevant solution is the one in which $v$ is small near the solar radius $r = R_\odot$ and increases to supersonic values for $r \gg r_c$. At such large radii, the first and third terms of Eq. (4.49) balance each other giving

$$v \approx 2c_s \sqrt{(\log(r/r_c))}. \qquad (4.50)$$

The corresponding pressure and the density falls as

$$P \propto \rho \propto \frac{1}{r^2 \sqrt{(\log r)}} \qquad (4.51)$$

at large distances. On the other hand, close to the solar radius, $v \ll c_s$ and we can ignore the first term on the left hand side of Eq. (4.49) and obtain

$$\dot{M} = 4\pi R_\odot^2 \rho(R_\odot)v(R_\odot) = 4\pi R_\odot^2 \rho(R_\odot)c_s \mathcal{M}(R_\odot)$$

$$\approx 4\pi r_c^2 \rho(R_\odot)c_s \exp\left(\frac{3}{2} - \frac{2r_c}{R_\odot}\right). \tag{4.52}$$

To make numerical estimates relevant to solar wind, we note that, near the surface of the Sun, $\rho \approx 10^{-15}$ gm cm$^{-3}$, $c_s \approx 1.3 \times 10^7$cm s$^{-1}$, giving $\dot{M} \approx 1.4 \times 10^{12}$ gm s$^{-1} \approx 2 \times 10^{-14} M_\odot$ per year. From the solution we can now estimate $v$ and $\rho$ at the location of earth at $r = 1$ AU $= 1.5 \times 10^{13}$ cm. We get $\mathcal{M} \approx 3.8$ and $v \approx 500$ km s$^{-1}$. From the conservation of $\dot{M}$, it follows that the density of particles in solar wind at the location of Earth is about $\rho \approx 10^{-23}$ gm cm$^{-3}$. These values match reasonably well with the observed features of solar wind.

## 4.5   Overview of stellar evolution

So far the discussion concentrated on the time independent, equilibrium, structure of stars powered by nuclear reactions. To have a complete picture, this needs to be supplemented by several aspects of stellar evolution, starting from the formation of the stars and proceeding to its late stages. Being a time dependent problem, this is inherently more complicated than what we have discussed so far. In this section, we shall provide a simplified discussion of the key features of stellar evolution, to give you a flavor of what is involved.

### 4.5.1   *Formation of a main sequence star*

We will see in later chapters that when a galaxy forms in an expanding universe, it will be clumpy and contain fair amount of smaller sub-structures. Our own galaxy contains large molecular clouds with masses in the range $(10^5 - 10^6)M_\odot$, with temperatures (10-100) K and densities $(10 - 10^4)$ cm$^{-3}$. To form stars by collapsing such gas clouds requires an increase in density by a factor of about $10^{24}$ and an increase in temperature by a factor of about $10^6$. This is an inherently complex process and many of the details are still unclear. We shall provide a simplified discussion of a possible scenario.

Let us analyse the collapse of a gas cloud under self gravity, ignoring the effects of rotation and magnetic field and treating it as spherically symmetric. The first question we need to address is an estimate of the mass which can collapse under self gravity. The gravitational potential energy of a cloud of mass $M$ and radius $R$ is $U = -(3/5)GM^2/R$ (see Exercise 1.1 of Chapter 1) while the thermal kinetic energy is $(3/2)Nk_BT$ where $N = (M/\mu m_H)$ and $\mu$ is the mean molecular weight. It was shown in chapter 1 that such a cloud is unstable to collapse if the magnitude of the gravitational potential energy is larger than (about) twice the kinetic energy. This gives the condition for collapse to be

$$\frac{3Mk_BT}{\mu m_H} < \frac{3}{5}\frac{GM^2}{R}. \tag{4.53}$$

Eliminating the radius of the cloud in terms of the mean density $\rho_0 = (3M/4\pi R^3)$, this criterion can be expressed as $M > M_J$ where

$$M_J \simeq \left(\frac{5k_BT}{G\mu m_H}\right)^{3/2}\left(\frac{3}{4\pi\rho_0}\right)^{1/2} \tag{4.54}$$

is called the *Jeans mass*. Numerically, one finds that

$$M_J \cong 1.2 \times 10^5 M_\odot \left(\frac{T}{100\ K}\right)^{3/2}\left(\frac{\rho_0}{10^{-24}\ \text{gm cm}^{-3}}\right)^{-1/2}\mu^{-3/2}. \tag{4.55}$$

The density $\rho \approx 10^{-24}$ gm cm$^{-3}$ and temperature $T \approx 100$ K are reasonable for clouds in interstellar medium. In such a case, the above result shows that only objects with $M \gtrsim 10^5 M_\odot$ can collapse initially. You see that this value is much higher than typical stellar masses.

When such a large cloud starts collapsing under self gravity, its physical parameters will change during the course of the collapse, thereby changing the Jeans mass itself. If the Jeans mass decreases as the cloud collapses, then subregions inside the original cloud can become locally unstable and start collapsing themselves. This process will continue with smaller and smaller inhomogeneities undergoing gravitational collapse, as long as the Jeans mass continues to decrease. Therefore, it is quite possible that lower mass objects, which should act as pregenitors of stars, arise due to such a secondary phenomena — namely, fragmentation inside the collapsing cloud.

The Jeans mass in Eq. (4.55) depends on the temperature and density as $M_J \propto T^{3/2}\rho^{-1/2}$. For a collapsing cloud, $\rho$ keeps increasing; whether $M_J$ increases or decreases will depend on the evolution of the temperature $T$. As the cloud collapses, it releases gravitational potential energy which can either: (a) increase the internal energy of the material and cause pressure readjustment or, (b) be radiated away if the time scale for cooling, $(t_{cool})$, is shorter than the time scale for free fall collapse $(t_{ff})$. If the cloud manages to radiate away the gravitational energy which is released, then it can contract almost isothermally. Equation (4.54) shows that, for isothermal contraction, $M_J$ varies as $\rho^{-1/2}$. As the star contracts, $\rho$ increases and the Jeans mass decreases allowing smaller regions inside the gas cloud, containing fragments of smaller mass, to collapse on their own gravitationally. Under most of the interstellar conditions, the collapse time scale $t_{ff} \propto (G\rho)^{-1/2}$, is much larger than the cooling time scale of the gas. Hence it is reasonable to assume that collapse of the cloud will be isothermal in the initial stages of the contraction.

This process will cease when the condition for isothermality is violated and Jeans mass starts increasing. This will occur when, at some stage in the evolution, the cloud becomes optically thick to its own radiation. After this stage, the collapse will be adiabatic rather than isothermal. For an ideal, mono-atomic, gas with $\nabla_{ad} = (2/5)$, the temperature scales as $T \propto P^{2/5} \propto \rho^{2/3}$ during adiabatic compression. Then $M_J \propto T^{3/2}\rho^{-1/2} \propto \rho^{1/2}$, which means that the Jeans mass will grow with time. Hence the smallest mass scale which can form due to fragmentation will correspond to the Jeans mass at the moment when the cloud makes a transition from isothermal to adiabatic evolution.

We can estimate this lower limit to the mass as follows. During the free fall time of a fragment, $t_{ff} \simeq (G\rho)^{-1/2}$, the total energy radiated away is of the order of $E \approx (GM^2/R)$. In order to maintain isothermality, the rate of radiation of energy has to be

$$A \approx \frac{E}{t_{ff}} \simeq \frac{GM^2}{R}\,(G\rho)^{1/2} = \left(\frac{3}{4\pi}\right)^{1/2} \frac{G^{3/2}M^{5/2}}{R^{5/2}}. \qquad (4.56)$$

However, we have seen in chapter 2 that a body in thermal equilibrium at temperature $T$ cannot radiate at a rate higher than that of a blackbody at the same temperature. Hence the rate of radiation loss by a cloud fragment can be written as

$$B = (4\pi R^2)\,(\sigma T^4)\,f \qquad (4.57)$$

where $f$ is a factor less than unity. For isothermal collapse, $B \gg A$ and a transition to adiabaticity occurs when $B \approx A$. This happens when

$$M^5 = \left[\frac{64\pi^3}{3}\right] \left[\frac{\sigma^2 f^2 T^8 R^9}{G^3}\right]. \qquad (4.58)$$

The fragmentation would have reached its limit when the Jeans mass is equal to this value: $M_J = M$. Replacing $M$ by $M_J$ and expressing $R$ in terms of the density, we can estimate the numerical value to be

$$M_J = 0.02\, M_\odot \, \frac{T^{1/4}}{f^{1/2} \mu^{9/4}} \qquad (4.59)$$

where $T$ is in Kelvin. For $T \simeq 10^3$ K and $f \approx 0.1$, we get $M \approx 0.36 M_\odot$; this result does not change much if the parameters are varied within reasonable range, because of the weak dependence on $T$ and $f$. Thus collapse of a cloud can lead to fragments with masses of the order of solar mass or above, but not significantly below.

It must be stressed that the existence of rotation and magnetic field — the two factors we have ignored — can have significant effects on the collapse of such a protocloud. For example, if a cloud with an initial radius of 50 pc and a slow rotation of $10^{-8}$ rad yr$^{-1}$ collapses to a final radius of about 100 AU, its angular velocity will go up to about 100 rad yr$^{-1}$ if the angular momentum is conserved. This corresponds to a period of a few days and the cloud would literally fly apart. Obviously, the cloud must find a way of losing its angular momentum if the collapse scenario has to work. Similarly, the interstellar magnetic field of about $10^{-5}$ G will be amplified to more than about $10^4$ G if the collapse proceeds with the magnetic flux being frozen during the collapse. Such high fields are not observed in the stars and hence there must also exist some mechanism to dissipate the magnetic field as the cloud collapses.

Further evolution of collapsing cloud leading to formation of protostar is quite complicated and many of the details are still uncertain. Eventually, several of the collapsing fragments reach a sufficiently hot core in which nuclear reactions can take place. In general, lower mass cloud fragments will be more abundant inside a given interstellar cloud. Hence the number of stars that form per unit volume per unit mass interval will be a strong function of the mass. Low mass stars are formed in larger numbers and last longer, thereby exhibiting greater abundance. Since the process of star formation is extremely complex, it is not easy to obtain from first principles a distribution function for the number density of stars of different masses that is formed in the galaxy. One popular fit to the initial mass function (or IMF) which is extensively used by the astronomers is called the *Salpeter IMF*. This is given by $\xi(M) \propto M^{-2.35}$ which is a simple power law. With appropriate normalization, Salpeter IMF becomes:

$$\xi(m)dm = 2 \times 10^{-12} m^{-2.35} dm \text{ pc}^{-3} \text{ yr}^{-1}; \qquad m \equiv (M/M_\odot) \qquad (4.60)$$

which gives the birth rate of stars of different masses in the range $0.4 \lesssim (M/M_\odot) \lesssim 10$ reasonably accurately. (A more detailed fit is given by *Scalo IMF* which has the form $\xi(M) \propto M^{-n}$ with three different values of $n$ in three different ranges.) We will have occasion to use mass function in the later chapters.

### 4.5.2 Life history of a star

The evolution of a star, once it reaches the main sequence, with nuclear reactions triggered in the central region, depends essentially on its mass. Figure 4.6 gives the tracks of the stars in the $L - T_s$ plane as they evolve. The diagonal branch running across (from top left to bottom right) gives the ZAMS based on Eqs. (4.30), (4.31). The following discussion highlights the key points and will walk you through the stellar evolution.

To begin with, if $M < 0.08 M_\odot$, nuclear reactions cannot be sustained in the contracting cloud for it to become a star. Such objects end up as *planets* or *brown dwarfs*, which are configurations in which the electron degeneracy pressure is significant and the material is fully convective. (To be precise, brown dwarfs have some nuclear reactions in the core but these are not the ones we are interested in for stellar structure.) Similarly, systems with masses higher than about $(60 - 100) M_\odot$ are unstable and cannot last for significant period of time. Since radiation and thermal energies in a star scale as $E_\gamma \propto T^4 R^3$, $E_{th} \propto MT$, it follows that $E_\gamma/E_{th} \propto T^3 R^3/M \propto M^2$ (where we have used the result $T \propto M/R$). In a sufficiently high mass star, the radiation energy will dominate over thermal energy and blow it apart. This sets the upper and lower limits for viable stellar mass.

In the allowed mass range, stellar structure and evolution are characterized by some key values for the masses. Based on some features in the evolution which will be described below, we can divide the stars into low mass, intermediate mass and high mass stars. You will also recall that stars with $M \lesssim 1.3 M_\odot$ have a radiative core while stars with $M \gtrsim 1.3 M_\odot$ have a convective core.

The first — and the longest — phase of stellar evolution occurs when stars act as gravitationally bound systems in which nuclear reactions fusing hydrogen into helium are taking place in the center. The process of combining four protons into a helium nuclei, releases about $0.03 m_p c^2$ of energy,

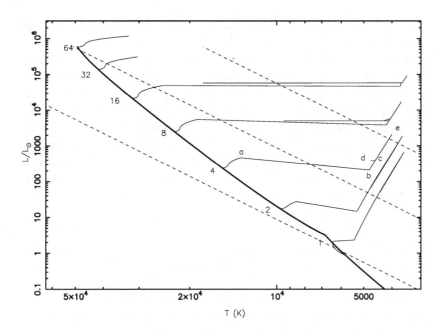

Fig. 4.6 The *approximate* tracks of the stars, as they evolve, in the $L - T_s$ plane. The thick diagonal line from left top to right bottom denotes the stars in the ZAMS. The evolution of stars of masses $(1 - 64)M_\odot$ are shown above and to the right of this diagonal line by zig-zag lines. One of them with $M = 4M_\odot$ has a few phases of evolution marked by the letters a to e discussed in the text. The dashed lines are loci of constant radii at $R/R_\odot = 1, 10, 80$ increasing from the bottom.

which is about a fraction 0.007 of the original energy $4m_p c^2$. Assuming that a fraction $\epsilon \approx 0.01$ of the total rest mass energy of the star can be made available for this nuclear reaction, the lifetime of the nuclear burning phase of the star will be about

$$t_{\rm ms} = \frac{\epsilon M}{L} \approx 3 \times 10^9 \text{ yr} \left( \frac{\epsilon}{0.01} \right), \qquad (4.61)$$

for $M \simeq M_\odot$ and assuming the opacity is due to Thomson scattering so that Eq. (4.9) is applicable. This defines the key time scale in stellar evolution. A more precise fitting formula for this time scale is given by

$$\left( \frac{t_{\rm ms}}{10^6 \text{ yr}} \right) = \frac{2550 + 669m^{2.5} + m^{4.5}}{0.0327m^{1.5} + 0.346m^{4.5}}, \quad m = \frac{M}{M_\odot}. \qquad (4.62)$$

In fact $t_{\rm ms}$ represents the longest stretch of time in a star's life spent in steady state nuclear burning. Most of the scaling relations obtained in Sec. 4.3 are applicable to such stars and they will form a band in the H-R diagram called the *main sequence band*. Since any given star spends maximum amount of time in the main sequence, it follows that a significant fraction of the stars that we observe will be found along the main sequence. Stars of different masses, of course, spend widely different amounts of time in the main sequence with the high mass stars spending less time.

When the star is in main sequence, it is essentially converting hydrogen into helium in its core and maintains balance between the energy produced and energy radiated out. Even in the main sequence, its radius and luminosity are varying slightly. For example, the $4M_\odot$ star shown in Fig. 4.6 will move from the ZAMS to the point marked 'a' while burning hydrogen in the core. Eventually, at $t \gtrsim t_{\mathrm{ms}}$, the hydrogen in the core runs out and the stellar evolution moves to the next stage. Since the electrostatic potential barrier between the helium nuclei is higher, further reactions which fuse the helium nuclei cannot take place immediately and the core begins to shrink under its own weight, in thermal time scale, and heats up. The hydrogen surrounding the core follows suit and will be compressed to high enough densities so that nuclear reactions are again ignited in the hydrogen *shell*. This is a somewhat peculiar situation in which the core is collapsing, taking the surrounding hydrogen shell along with it, leading to the energy production from the *shell*. The luminosity of the star increases and the increased pressure pushes the outer layers of the star further from the center making the star move into a *red giant* phase. (It is 'red' because the surface temperature $T_s \propto (L/R^2)^{1/4}$ decreases with increasing $R$, making the black body peak shift to the redder side.) In massive stars ($M \gtrsim M_\odot$) there is a distinct evolutionary track from 'a' to 'b' which is crossed very rapidly. The time scale for this is very short and is well fitted by

$$\frac{t_{\mathrm{HG}}}{t_{\mathrm{ms}}} = \frac{0.543}{m^2 - 2.1\,m + 23.3}; \quad m = \frac{M}{M_\odot}. \qquad (4.63)$$

The subscript HG stands for *Hertzsprung Gap*, as it is called. Since this time scale is very short, very few stars can be caught observationally in this phase thereby leading to a gap in the HR diagram in this region. Low mass stars do not go through this phase of the evolution does not have a distinct Hertzsprung gap.

The next stage of evolution in moderate to high mass stars is usually called the *giant phase* during which the star essentially moves along the track marked 'b-e' (and similar tracks for other stars). The point 'b' is called the base of the *giant branch* and the track 'b-e' is called the giant branch. The luminosity of the star at the base of the giant branch, $L_{\text{bgb}}$, is essentially determined by its mass:

$$\frac{L_{\text{bgb}}}{L_\odot} = \frac{2.15\,m^2 + 0.22\,m^5}{1.0 + 1.4 \times 10^{-2}\,m^2 + 5.0 \times 10^{-6}\,m^4}, \quad m = \frac{M}{M_\odot}. \qquad (4.64)$$

The radius and the luminosity of the star while on this track is reasonably well fitted by

$$\frac{R}{R_\odot} = (0.25\,l^{0.4} + 0.8\,l^{0.67})\,m^{-0.27}; \quad m = \frac{M}{M_\odot}; \quad l = \frac{L}{L_\odot}. \qquad (4.65)$$

The collapse of the core to a sufficiently high density will trigger the next set of nuclear reactions producing carbon from helium. This ignition usually takes place when the luminosity is about $L = L_{\text{bgb}} + 2000 L_\odot$. When the helium ignition occurs, the core expands slightly and the envelope contracts (which is just the reverse of the original evolution) causing the star to move down to 'c'. (These fine tracks are not visible in the Fig. 4.6 due to lack of resolution.) The star now settles to a new equilibrium configuration with a helium burning core and a hydrogen burning shell. During this period, the luminosity of the stars remain roughly constant (earning them the name *horizontal branch* stars) at

$$\frac{L_{\text{He}}}{L_\odot} = 0.763\,m^{0.46}\frac{L_0}{L_\odot} + 50\,m^{-0.1}; \quad m = \frac{M}{M_\odot}. \qquad (4.66)$$

Its radius, however, decreases making the star evolve along the horizontal line from 'c' to 'd'. During its evolution in the horizontal branch, the stars convert the mass in the helium core (about 0.45 $M_\odot$) into oxygen and carbon in almost equal proportion. These reactions typically release a fraction $7.2 \times 10^{-4}$ of the rest mass energy. Taking the luminosity of a horizontal branch star to be about 50 $L_\odot$, we find that the lifetime of a star in the horizontal branch is about $t_{\text{HB}} \approx 7.2 \times 10^{-4} \times (0.45 M_\odot c^2 / 50 L_\odot) = 0.1$ Gyr which is considerably shorter than the time spent in the main sequence. A more precise fitting function for the time scale for core helium burning is given by

$$\frac{t_{\text{He}}}{t_{\text{ms}}} = \begin{cases} 0.0886 + 4.748 \times 10^{-4}\,m + 1.186 \times 10^{-6}m^2 & (\text{if } m \geq 10) \\ (37.58 - 23.5\,m + 4.406\,m^2)^{-1} & (\text{if } m < 10). \end{cases} \qquad (4.67)$$

Eventually, the helium in the core will get exhausted and a similar course of events take place again. The star will burn helium to carbon in an inner shell around the collapsing inert carbon core and, of course, hydrogen to helium in another outer shell. In a sufficiently high mass star, there could be onion-like structure with different nuclear reactions taking place at different shells. A high mass star ($M > 8M_\odot$) can continue to burn hydrogen (for about $7 \times 10^6$ yr), helium ($5 \times 10^5$ yr), carbon (600 yr), neon (1 yr), oxygen (0.5 yr) and silicon (1 day) leading to an iron core with the time scale spent in each phase being progressively shorter as indicated in the bracket. The overall giant phase of evolution lasts for about $0.15\, t_{\rm ms}$. It is conventional to split this giant evolution into a *red giant* phase (before helium ignition) and a *super giant* phase (after core helium has been exhausted).

The outer regions of the star can expand to very large radii (of several AU) at this stage. The temperature in the star's envelope can get sufficiently low so that solid particles can condense out of vapour phase producing "dust". (This is exactly how you get soot particles up in the chimney.) The dust particles will be blown away from the star by radiation pressure thereby leading to a stellar wind which can result in a mass loss of up to several times $10^{-5} M_\odot$ yr$^{-1}$.

The evolution is different for low mass stars which do not get hot enough to ignite anything beyond helium. During the later part of evolution, the outer regions of the star undergo violent instabilities leading to an eventual ejection of significant fraction of material in the form of a *planetary nebula*. The core region will become a *white dwarf* supported by the degeneracy pressure of electrons. To achieve this balance, the Fermi energy $\epsilon_F$ of the electrons must be larger than the gravitational potential energy $\epsilon_g \cong Gm_p^2 N^{2/3}n^{1/3}$ [see Eq. (4.1)]. When the particles are non relativistic, $\epsilon_F = (\hbar^2/2m_e)(3\pi^2)^{2/3}n^{2/3}$ [see Eq. (3.12)] and the condition $\epsilon_F \geq \epsilon_g$ can be satisfied (at equality) if

$$n^{1/3} = \frac{2}{(3\pi^2)^{2/3}} \left( \frac{G\,m_p^2 m_e}{\hbar^2} \right) N^{2/3}. \tag{4.68}$$

Using $n = (3N/4\pi R^3)$ and $N = (M/m_p)$, this reduces to the following mass-radius relation:

$$RM^{1/3} \simeq \alpha_G^{-1} \lambda_e m_p^{1/3} \simeq 8.7 \times 10^{-3}\, R_\odot M_\odot^{1/3}. \tag{4.69}$$

A white dwarf with $M \simeq M_\odot$ will have $R \simeq 10^{-2} R_\odot$ and density $\rho \simeq 10^6 \rho_\odot$. The evolution of low mass stars usually leads to helium or carbon cores to survive as white dwarfs. We will discuss the structure of white dwarfs in detail in the next chapter.

The end stages of high mass stars can be more dramatic. At the end, the core collapses catastrophically and one is led to a *supernova explosion* and a neutron star as a remnant. In the late stages of the evolution, about 1.4 $M_\odot$ of original hydrogen gets converted into iron before the star explodes as a supernova. This reaction effectively releases a fraction $8.5 \times 10^{-3}$ of the rest mass energy. If we take the luminosity to be about $10^3 L_\odot$, the life time turns out to be about $0.18 \, (L/10^3 L_\odot)^{-1}$ Gyr.

Several high energy processes become important in the core region of the star at this stage. For example, neutrino interactions with nuclei and cooling of the core through neutrino emission can be significant even at the silicon burning phase. Neutrinos interact only very weakly with matter and are characterized by a very small scattering cross section $\sigma$. For example, the reaction $\bar{\nu}_e(p, n)e^+$ (the compact notation $A(B, C)D$ stands for the reaction $A+B \rightarrow C+D$) has a cross-section $\sigma = 10^{-43}$ cm$^2(\epsilon_\nu/1$ MeV$)^2$ while the reaction $\nu_e(e^-, e^-)\nu_e$ has a cross-section $\sigma = 10^{-48}$ cm$^2(\epsilon_\nu/1$ MeV$)$. The mean free path for neutrinos of energy $\epsilon_\nu \approx 1$ MeV, corresponding to the first reaction is about $\lambda = (n\sigma)^{-1} = (m_p/\rho\sigma) \simeq 10^{19}$ cm $\rho^{-1}$ in c.g.s. units. This is about $10^6 R_\odot$ for typical stars and about $(10^3 - 10^4)R_{\text{wd}}$ for white dwarfs. Therefore neutrinos can escape freely from stars and white dwarfs carrying energy with them. The main drain of energy arises through the reactions

$$e^+ + (Z, A) \rightarrow (Z - 1, A) + \nu \rightarrow (Z, A) + e + \bar{\nu} \qquad (4.70)$$

in which the kinetic energy of the electron is transfered to a $\nu\bar{\nu}$ pair. These reactions are significant at $T > 5 \times 10^9$ K. At these temperatures, $e^+e^-$ annihilations will also result in $\nu\bar{\nu}$ pair production in 1 out of 20 cases and $\gamma$ production in the remaining 19 out 20 cases. The energy production per volume in this process is about

$$\mathcal{E} \simeq (n_e n_{\bar{e}}) \, \sigma c \, (k_B T) \simeq \left(\frac{aT^4}{k_B T}\right)^2 \sigma c \, (k_B T). \qquad (4.71)$$

Using $\sigma \approx 10^{-44}$ cm$^2(T/3 \times 10^9$ K$)^2$ we get this energy production rate to be $\mathcal{E} \simeq 10^{20}$ erg cm$^{-3}(T/3 \times 10^9$ K$)^9$. The corresponding cooling time at this temperature $t_{\text{cool}} \simeq (aT^4/\mathcal{E})$ is about 2 hours, showing that this can be a significant drain of the energy. For a 20 $M_\odot$ star, the photon luminosity during this phase is about $4.4 \times 10^{38}$ erg s$^{-1}$ while the neutrino luminosity can be as high as $3.1 \times 10^{45}$ erg s$^{-1}$.

At this stage, the photo disintegration of nuclei due to collisions with energetic photons (which are present at high temperatures), as well as inverse beta decay, reduces the pressure support of the core thereby triggering a rapid core collapse. The inverse beta decay also converts a significant fraction of electrons and protons into neutrons making the core neutron rich. As the core radius decreases and the density increases, the neutrinos produced by the weak interactions will get trapped inside. An order of magnitude estimate for $\nu$−trapping, which illustrates the scalings involved, can be obtained as follows: When the collapse of the core (mostly made of iron) begins, the physical parameters are about $\rho_{\text{core}} \approx 10^{(9-10)}$ gm cm$^{-3}$, $t_{\text{collapse}} \approx (G\rho)^{-1/2} \approx 0.1$ s and $\epsilon_F \approx 50$ MeV $(\rho_{12}/\mu_e)^{1/3}$. The neutrinos get trapped when their diffusive time scale becomes less than the collapse time scale. Taking the nuclear composition to be that of a $^{56}$Fe core, the mean free path is about $\lambda_\nu \simeq 0.3 \times 10^5 \rho_{12}^{-5/3}$ cm. If the size of the region is $R$ and the number of scatterings are $N_{\text{scat}}$ before the neutrino escapes, then the random walk argument gives $R \simeq N_{\text{scat}}^{1/2} \lambda_\nu$ and the diffusion time scale is

$$\tau_{\text{diff}} \simeq N_{\text{scat}} \left(\frac{\lambda_\nu}{c}\right) \simeq \left(\frac{R}{\lambda_\nu}\right)^2 \left(\frac{\lambda_\nu}{c}\right) \simeq \frac{R^2}{c\lambda_\nu} \propto \frac{\rho^{-2/3}}{\rho^{-5/3}} \propto \rho. \qquad (4.72)$$

A more exact calculation (taking into account the spherical geometry by solving the diffusion equation in the spherical coordinates) changes the formula only by a factor $(3/\pi^2)$ and we get

$$\tau_{\text{diff}} \simeq \frac{3R^2(\rho)}{\pi^2 c\lambda_\nu(\rho)} \simeq 0.02\ \rho_{12}\ \text{sec}. \qquad (4.73)$$

This becomes larger than the free fall collapse time

$$t_{\text{coll}} = \left(\frac{8\pi G\rho}{3}\right)^{-1/2} = 1.3 \times 10^{-3} \rho_{12}^{-1/2} \text{sec} \qquad (4.74)$$

when the density of matter is higher than $\rho_{\text{trap}}$ with

$$\rho_{\text{trap}} \simeq 1.5 \times 10^{11}\ \text{gm cm}^{-3}. \qquad (4.75)$$

At higher densities, neutrinos are trapped inside the star for the duration of the collapse. The neutrino density increases very fast and they soon become degenerate. Collapse proceeds further until nuclear densities $\rho_{\text{nucl}} \approx [3\,m_N/4\pi(1.2\ \text{fm})^3] \approx 10^{14}$ gm cm$^{-3}$ are reached.

When the nuclear forces prevent further contraction of the core, a shock wave could be set up from the center to the surface of the star which is capable of throwing out the outer mantle of the star in a supernova explosion. The total kinetic energy of the outgoing shock is about $10^{51}$ erg which is roughly one percent of the energy liberated in neutrinos. When the outer material expands to about $10^{15}$ cm and becomes optically thin, an impressive optical display arises, releasing about $10^{49}$ erg of energy in photons with a peak luminosity of about $10^{43}$ erg s$^{-1} \simeq 10^9 L_\odot$. Such a supernova is one of the brightest phenomena in the universe in the optical band. The outgoing gas ploughs through the interstellar medium with ever decreasing velocity and eventually will appear as a shell like supernova remnant with a typical radius of about 40 parsec. The central remnant is a *neutron star* supported essentially by the degeneracy pressure of the neutrons. We shall discuss these remnants in detail in the next chapter.

The neutrino trapping also has the effect of reducing their luminosity significantly. When the star collapses, from some radius $R$ to a radius $R_{\text{nuc}}$ at which the density is of the order of nuclear densities, the amount of gravitational potential energy released is of the order of $(GM^2/R_{\text{nuc}})$ since $R_{\text{nuc}} \ll R$. Writing $\rho_{\text{nuc}} \approx (3M/4\pi R_{\text{nuc}}^3)$, we get $R_{\text{nuc}} \approx 12$ km for $M \approx M_\odot$. If this energy was released in the form of neutrinos, within a *collapse* time scale, then the luminosity can be easily estimated to be about

$$L_{\nu,\text{max}} \simeq \frac{(GM^2/R_{\text{nuc}})}{t_{\text{coll}}} \simeq 10^{57} \text{ erg s}^{-1}. \qquad (4.76)$$

On the other hand, neutrino trapping leads to the same energy being liberated over the *diffusive* time scale with $\tau_{\text{diff}} \gg t_{\text{coll}}$ when $\rho \approx \rho_{\text{nuc}}$. As a result, the actual neutrino luminosity turns out to be more like

$$L_\nu \simeq \frac{(GM^2/R_{\text{nuc}})}{\tau_{\text{diff}}} \simeq 10^{52} \text{ erg s}^{-1}. \qquad (4.77)$$

Hence the bulk of the liberated gravitational energy is actually transformed internally to other forms of energy like thermal energy, excitation energy of nuclear state *etc.* rather than being lost to the system. This suggests that after the neutrino trapping occurs, the collapse proceeds almost adiabatically in the system.

Equation (4.69) is still applicable to the remnant of a supernova explosion with $m_e$ replaced by the mass of the neutron, $m_n$; correspondingly the right hand side of Eq. (4.69) is reduced by $(\lambda_n/\lambda_e) = (m_e/m_n) \simeq 10^{-3}$. Such objects — called *neutron stars* — will have a radius of $R \simeq 10^{-5} R_\odot$ and density of $\rho \simeq 10^{15} \rho_\odot$ if $M \simeq M_\odot$. For such values $(GM/c^2 R) \simeq 1$ and general relativistic effects are beginning to be important. (See chapter 5 for a more detailed discussion.)

   The nature and mass of the final remnant produced in the stellar evolution is broadly correlated with the initial mass of the star. By and large, white dwarfs are formed at the end of the evolution of stars with the initial mass of $(1 - 8)M_\odot$ and neutron stars are formed from the stars with the initial mass of $(8 - 50)M_\odot$. The formation rate of any stellar remnant can be easily estimated if the initial mass function for stars is known. If one uses the empirically determined initial mass function given by Eq. (4.60), then the birth rate $\mathcal{B}$ of stars in the range of masses $(M_1, M_2)$ in the galaxy is given by

$$\mathcal{B} = V_{\text{disc}} \int_{(M_1/M_\odot)}^{(M_2/M_\odot)} \psi_s dm = 0.19 m^{-1.35}\Big|_{(M_2/M_\odot)}^{(M_1/M_\odot)} \text{ yr}^{-1} \qquad (4.78)$$

where

$$V_{\text{disc}} = \pi r^2 \ (2H) = 1.3 \times 10^{11} \text{ pc}^3 \qquad (4.79)$$

is the volume of the disk of the galaxy with vertical height $H = 90$ pc and radius $r = 15$ kpc which is appropriate for stars more massive than $1 \ M_\odot$. The number density of compact objects formed from these parent stars will be

$$n = T_0 \int_{(M_1/M_\odot)}^{(M_2/M_\odot)} \psi_s dm = 0.018 m^{-1.35}\Big|_{(M_2/M_\odot)}^{(M_1/M_\odot)} \text{ pc}^{-3} \qquad (4.80)$$

where $T_0 \approx 12 \times 10^9$ yr is the age of the galaxy. For white dwarfs, taking the mass range of parent star to be $(1-8)M_\odot$, we get an integrated galactic birth rate of 0.18 per year with the number density $n_{\text{wd}} \simeq 1.7 \times 10^{-2} \text{ pc}^{-3}$. Neutron stars, on the other hand, originate from stars with initial mass $(8-50)M_\odot$ and we get an integrated galactic birth rate of 0.01 per year with the number density $n_{\text{ns}} \simeq 10^{-3} \text{ pc}^{-3}$. The mean distance $(n^{-1/3})$ between the white dwarfs should be about 3.8 pc while that between neutron stars should be about 10 pc.

   To determine the mass contributed by these objects, we have to take into account the fact that much of the initial mass will be lost during the evolutionary process. The white dwarfs and neutron stars have average masses of about $< M >_{\text{wd}} = 0.65 M_\odot$ and $< M >_{\text{ns}} = 1.4 M_\odot$ so that

$$\frac{\rho}{\rho_{\text{gal}}} = \frac{n < M >}{\rho_{\text{gal}}} \qquad (4.81)$$

where $\rho_{\text{gal}} \approx 0.14 M_\odot \text{ pc}^{-3}$ is the mass density of the galaxy. This gives $(\rho/\rho_{\text{gal}}) \simeq 0.08$ for white dwarfs and $(\rho/\rho_{\text{gal}}) \simeq 0.01$ for neutron stars.

## Exercises

**Exercise 4.1**     *Simplest exercise in this book:*

Astronomers have discovered an unusual *square* object in the sky which appears to be 4 arcsec × 4 arcsec in angular size and has a magnitude of 10. (a) What is the surface brightness of this object in magnitude per square arcsec? [Hint: If you think it is (10/16) mag arcsec$^{-2}$, you haven't figured out the guiles of the astronomers!] (b) If the object is moved to a distance which is twice as far, what will be its magnitude? (c) What will be its surface brightness now? (d) If Sun is a perfect blackbody at $T = 5800$ K, what is the flux (in W m$^{-2}$) you will measure if you were right on the surface of the Sun? This is the *brightness of the surface* of Sun. (e) From Earth, Sun appears as a disk of diameter 0.5 degree and the total luminosity of the Sun is $3.86 \times 10^{26}$ W. What is the flux per square arcsec of the Sun's disk (in W m$^{-2}$ arcsec$^{-2}$) which we measure for the Sun? This is the *surface brightness* of the Sun. (f) Express the surface brightness of the Sun in magnitude per square arcsec.

**Exercise 4.2**     *Practice with magnitudes:*

In a device which counts photons statistically, there will be a fluctuation around the mean number of photons observed. Assume that when you expect to observe $N$ photons on the average, the actual value can fluctuate by an amount $\pm\sqrt{N}$. (a) How many photons need to be collected if the magnitude of a star has to be measured to an accuracy $\pm 0.02$? How long an exposure will one require with an one meter telescope to measure the B magnitude of an $M_B = 20$ star with the same accuracy? (b) Assume that the night sky has the same brightness as that produced by one 22.5 B magnitude star per square arc second over a bandwidth of 940 Å. Express the night sky brightness in erg cm$^{-2}$ s$^{-1}$ ster$^{-1}$Å$^{-1}$. (c) Consider an attempt to measure the magnitude of a $M_B = 24$ star within an accuracy of $\pm 0.02$ magnitude. Assume that the starlight is spread out over 1 square arc second. How many photons are required to be collected to achieve this and what should be the exposure time of a 4 meter telescope for this measurement? [Ans: (a) $N \approx 2.9 \times 10^3$ photons; about 26 seconds. (b) $F = 3.5 \times 10^{-6}$ erg cm$^{-2}$ s$^{-1}$ ster$^{-1}$Å$^{-1}$. ]

**Exercise 4.3**     *Scaling relation and molecular weight:*

Use the scaling arguments to determine the dependence of the luminosity of a star on the mean molecular weight, if the opacity is given by Kramer's formula; i.e., prove Eqs. (4.35) and related results. Normalize this result using the Sun for which $X = 0.70$, $Y = 0.28$ and $Z = 0.02$. Hence determine how different the luminosity will be for a star with $X = 0.98$ and $Z = 0.02$ and no helium.

### Exercise 4.4    *Stars as polytropes:*

(a) The low mass stars ($M < 0.3 M_\odot$) can be approximated as a $\gamma = 5/3$ polytrope. One can then integrate the Lane-Emden equation to obtain various parameters. Observations suggest $R/R_\odot \cong (M/M_\odot)^{0.94}$ in this range. Use this to determine the central density, pressure and temperature in terms of $M/M_\odot$. (b) High mass stars ($M > 50 M_\odot$) can be approximated as a polytrope with $\gamma = 4/3$ and has the observed relation $R/R_\odot = 5.5(M/100 M_\odot)^{0.45}$. Once again estimate the central parameters in terms of $M/M_\odot$. (c) Assume that the opacity is of the form $\kappa = \kappa_0 \rho T^{-s}$. Integrate the equation for radiative equilibrium and find a relation between temperature and pressure as a function of radius inside the star. (Ignore radiation pressure compared to gas pressure.) Use the results of (a) and (b) to determine $\kappa_0$ in terms of the mass and luminosity of star.

### Exercise 4.5    *Degenerate to the core:*

The electron number density at the core of the Sun is about $4.4 \times 10^{25}$ cm$^{-3}$. Are these electrons degenerate? What about the helium nuclei at the core of the Sun?

### Exercise 4.6    *This won't do:*

If the Sun's energy output is purely due to gravitational contraction, how long would it last? If the Sun were powered by chemical reactions (which releases, say, $10^4$ cal g$^{-1}$), how long will it last?

### Exercise 4.7    *Radioactive decay and light curves:*

The $^{56}$Ni decays to $^{56}$Co with a half-life of 6.1 days which, in turn, decays $^{56}$Fe with a half-life of 77.1 days. Estimate the form of the luminosity of a supernova as a function of time when the cobalt decay is dominant. For supernova 1987A take $L \approx 10^{42}$ erg s$^{-1}$ at $t = 0$ for the purpose of cobalt decay extrapolation. What is the total amount of energy released and the mass of nickel synthesized?

### Exercise 4.8    *Bound on the neutrino mass:*

Neutrinos with energies (7.5 – 35) MeV were detected over a time interval of about 12.4 seconds from supernova 1987A which occurred at a distance of about 50 kpc from Earth. Assuming that all these neutrinos were emitted simultaneously and the delay in arrival time is due to neutrinos of different energies having different speeds put an upper bound on the possible mass of the neutrino. [Ans. $m_\nu c^2 \lesssim 17$ eV.]

### Exercise 4.9    *Betelgeuse going boom:*

Betelgeuse is a super giant star of mass 20 $M_\odot$ located at a distance of 200 pc from us. It is very likely to explode as a supernova within the next million years. Assume that this results in an iron core of mass 1.4 $M_\odot$, density of $10^9$ gm cm$^{-3}$

and that the core will collapse to a radius of 10 km. What is the amount of gravitational binding energy that is released? What is the free fall collapse time? If 1 per cent of the binding energy comes out as visible light in 100 days, what is the average luminosity of the supernova? What is the energy flux at Earth? Do you expect this supernova to out shine the Sun?

### Exercise 4.10   *Decreasing star formation:*

Suppose that the star formation rate in our galaxy decreases as $\exp[-(t/\tau)]$ where $\tau = 3$ Gyr and that the main sequence life time of a 3 $M_\odot$ star is about 350 Myr. Compute the fraction of all the 3 $M_\odot$ stars ever made which are still on the main sequence. [Ans. About $4.6 \times 10^{-3}$.]

### Exercise 4.11   *Return gift:*

Using the relation between remnant mass and initial mass of the star described in the text and Salpeter IMF, estimate the fraction of the mass that is returned to the interstellar medium after 10 Gyr.

### Exercise 4.12   *Growing old:*

Estimate what fraction of light from an old globular cluster comes from the stars which are still in the main sequence.

### Exercise 4.13   *Infant mortality:*

Star formation occurs in giant molecular clouds which will result in a weakly bound cluster of stars. Consider a cluster with $N = 10^3$ stars and radius $R = 1$ pc. Estimate the time scale over which such a star cluster will be disrupted by the shearing motion in the galaxy. How does it compare with its relaxation time?

### Exercise 4.14   *Real life astronomy – Galactic center:*

Take a look at Dwek *et al.*, Ap.J.,**445**, p. 716 (1995) which has an analysis of the infrared emission from the center of the galaxy. Take a look at figure 1 of this reference. (a) Using this picture, make a rough estimate of the total infrared flux in W m$^{-2}$. (Be careful with the units!) Assuming a distance to this region to be 8.5 kpc, compute the luminosity of the central region in these wavelengths in terms of $L_\odot$. (b) Most of this light comes from red giants. (Why?) If a typical red giant is an M0 type star, look up the stellar classification scheme to find the peak wavelength at which red giants emit. (c) You will find that the peak is not at 2.2 $\mu$m at which this data is taken. Hence the total luminosity needs to be corrected. Estimate the correction factor [Ans: It is about 14.5.] and recompute the total luminosity. If a single M0 red giant has a total luminosity of 400 $L_\odot$, estimate the number of red giants in this region. (d) The number of red giant stars should be a small fraction of total number of stars. Can you devise a method to roughly estimate the total number of stars given the number of red giant stars?

**Exercise 4.15**   *Real life astronomy – HR diagram:*

Get hold of Hipparcos data from the web (or from some friendly source!). (a) Write a program which will compute (i) the distance to each star in the data base from the parallax and then (ii) the absolute magnitude. It is better to stick to those stars which does not have too large a parallax error. (b) Plot the apparent magnitude against the B-V colour as well as the absolute magnitude versus the B-V colour. How do they compare? (c) Plot B-V colour against the V-I colour. Can you explain the nature of this plot?

**Exercise 4.16**   *Colourful astronomy:*

Verify the two curves in Fig. 4.1. It is rather trivial to show that the curve is a straight line for a class of sources with power law spectrum $F_\nu \propto \nu^{-\alpha}$ when $\alpha$ is varied. What is more surprising is that the colour-colour indices for the Planck spectrum seems to be a straight line. Actually it is not strictly straight. Work it out analytically and convince yourself why it looks so straight. Also obtain the slope of this curve in terms of the three wavelengths for B, U and V.

**Notes and References**

Stellar structure and evolution are discussed in detail in several text books like, for example, Padmanabhan (2001), Hansen and Kawaler (1994), Clayton (1983).

    1. *Sections 4.3 and 4.4:* The fitting formulas for the stellar evolution quoted in the text are from Eggleton et al (1989). A specialist text on supernova physics is Arnett (1996).

# Chapter 5

# Relics of Stars

We saw in the last chapter that stellar evolution can lead to distinct types of remnants. The low mass stars end up as white dwarfs while the higher mass stars lead to supernova and neutron stars. All these types of remnants are of astrophysical interest on their own and we will discuss them (though somewhat briefly) in this chapter, beginning with the dynamics of the region around a supernova and its observational signatures.

## 5.1 Supernova remnants

When a supernova explosion takes place and stellar material is ejected, the local region of the interstellar medium, around that event, is significantly disturbed. The ejected material moves through the interstellar medium and will eventually be the source of electromagnetic radiation in different wave bands. The study of such objects, called *supernova remnants*, helps in the understanding of supernova explosions as well as several evolutionary processes in the interstellar medium (ISM).

In a typical supernova explosion about $E \approx 10^{51}$ erg of energy is released over a time scale which is short compared to any other relevant time scale in the problem. The initial velocity of the ejected matter, $v_{\rm ej}$, is related to the energy $E = (1/2)Mv_{\rm ej}^2$ by

$$v_{\rm ej} \approx 10^4 \text{ km s}^{-1}(E_{51}^{1/2}m^{-1/2}) \approx 10^{-2} \text{ pc yr}^{-1}(E_{51}^{1/2}m^{-1/2}) \qquad (5.1)$$

where we use the notation, $E_{51} \equiv (E/10^{51}$ erg$)$ *etc.*, $m \equiv (M/M_\odot)$ and we have used the convenient conversion $10^6$ km s$^{-1} \approx 1$ pc yr$^{-1}$. Such an explosion will lead to a shock front propagating into the surrounding medium.

In the simplest picture for the supersonic expansion of a hot, gaseous sphere, there will be an abrupt discontinuity between the expanding gas and the material swept up by the shock front. A shock wave will run ahead of the contact discontinuity and the region between the 'piston' and the shock wavefront will be heated to a high temperature. In the case of strong shocks with high Mach number $\mathcal{M}_1$, the ratio of densities on either side of shock wave is $\rho_2/\rho_1 = (\gamma+1)/(\gamma-1) = 4$ if $\gamma = (5/3)$ (see chapter 3, Sec. 3.3.3). The temperature ratio, $(T_2/T_1) = 2\gamma(\gamma-1)\mathcal{M}_1^2/(\gamma+1)^2 = (5/16)\mathcal{M}_1^2$ (for $\gamma = (5/3)$), can be very large for strong shocks. Hence we expect the shocked gas in supernova remnants to be intense x-ray emitters.

In the evolution of such supernova remnants, one can distinguish several different phases of evolution each of which has its own observational signature. In the first phase, the material moves as a blast wave with a constant velocity $v_{ej}$ given by Eq. (5.1) so that the radius of the remnant grows linearly with time: $r \simeq v_{ej}t$. There is very little resistance to this blast wave in the ISM initially. As the mass swept up by the shock front increases, the blast wave will slow down and the $r \propto t$ evolution will end (at some radius $r = r_f$, say) when the mass of interstellar medium swept up by the blast wave, $M_{swept} \approx (4\pi/3)\rho_{ISM}r_f^3$, is comparable to the ejected mass $M_{ej}$. This gives

$$r_f \simeq 2 \text{ pc} \left(\frac{M_{ej}}{M_\odot}\right)^{1/3} \left[\frac{\rho_{ISM}}{2 \times 10^{-24} \text{ gm cm}^{-3}}\right]^{-1/3}. \qquad (5.2)$$

The corresponding time scale is

$$t_f \simeq \frac{r_f}{v_{ej}} \simeq 200 \text{ yr} \left(\frac{M_{ej}}{M_\odot}\right)^{5/6} E_{51}^{-1/2}\rho_{24}^{-1/3}. \qquad (5.3)$$

The next phase of expansion is well described by the Sedov solution which we studied in chapter 3, Sec. 3.3.3 in which we found that $r \propto t^{2/5}$. This phase could also be interpreted as the one in which the energy is constant. The total energy of the system can be expressed as

$$E \approx \frac{1}{2}\left(\frac{4\pi}{3}\right)\rho r^3 v^2 + \text{(constant)} (\rho v^2)\left(\frac{4\pi}{3}\right)r^3 \qquad (5.4)$$

where the first term is the kinetic energy of bulk motion and the second term is the internal energy of the system. Constancy of $E$ implies that $r^3v^2 = \text{constant}$. Using $v = \dot{r}$ and integrating, we obtain $r \propto (E/\rho)^{1/5}t^{2/5}$. As we mentioned in chapter 3, Sec. 3.3.3 the constant of proportionality is of order unity for $\gamma = 5/3$. So this phase is characterized by the scaling:

$$r \simeq \left(\frac{E}{\rho}\right)^{1/5} t^{2/5} \simeq 0.3 \text{ pc } E_{51}^{1/5} n_H^{-1/5} t_{yr}^{2/5}. \tag{5.5}$$

The corresponding velocity and temperature are:

$$v = \dot{r} \simeq 5000 \text{ km s}^{-1} \left(\frac{r}{2 \text{ pc}}\right)^{-3/2} E_{51}^{1/2} n_H^{-1/2} \tag{5.6}$$

$$T \simeq 6 \times 10^8 \text{ K} \left(\frac{r}{2 \text{ pc}}\right)^{-3} E_{51} n_H^{-1} \simeq 10^6 \text{ K } E_{51}^{2/5} n_H^{-2/5} \left(\frac{t}{3 \times 10^4 \text{ yr}}\right)^{-6/5}.$$

These scalings predict a temperature of about $3 \times 10^6$ K for a supernova remnant with an age of about $10^4$ yrs. Plasma at such temperatures will radiate strongly in x-rays. Also note that Eq. (5.6) defines a characteristic time scale

$$t_{\text{sedov}} \simeq 3 \times 10^4 \text{ yr } T_6^{-5/6} E_{51}^{1/3} n_H^{-1/3} \tag{5.7}$$

as a function of the temperature. For example, the supernova remnant Cygnus Loop has reached the end of this phase right now with $R_{\text{now}} \approx 20$ pc and $v_{\text{now}} \approx 115$ km s$^{-1}$. The age of this remnant, obtained from the above relation is about $t \simeq 6.5 \times 10^4$ yr.

The Sedov phase, which conserves the energy, ends when the energy loss due to radiative cooling becomes significant. As the temperature falls below $10^6$ K, the ions of C,N and O will acquire electrons and form atomic systems. The recombination, followed by the cascading down to the ground state through the emission of photons, will cool the gas. We saw in Chapter 2 that the rate of cooling at these temperatures is given by $n_H^2 \Lambda(T) \approx 10^{-22}$ erg cm$^3$ s$^{-1} n_H^2 T_6^{-1/2}$ (see Eq. (2.112)). The third phase will be the one in which the cooling losses are important. The transition to this phase can be estimated by the condition $t_{\text{cool}} \lesssim t_{\text{sedov}}$ where

$$t_{\text{cool}} \simeq \frac{nk_B T}{n^2 \Lambda(T)} \simeq 4 \times 10^4 \text{ yr} \left(\frac{T_6^{3/2}}{n_H}\right) \tag{5.8}$$

and $t_{\text{sedov}}$ is defined in Eq. (5.7). The condition $t_{\text{cool}} \lesssim t_{\text{sedov}}$ leads to $T_6 \lesssim E_{51}^{1/7} n^{2/7}$ or equivalently,

$$\dot{r} \propto T^{1/2} \lesssim 200 \text{ km s}^{-1} (E_{51} n_H^2)^{1/14}. \tag{5.9}$$

This relation shows that the velocity at the end of sedov phase is about 200 km s$^{-1}$ and is only weakly dependent on $E$ and $n_H$.

At this stage, the shock becomes isothermal. There will be a hot interior region [with radius $r(t)$] surrounded by a cool dense shell. The shell moves with approximately constant radial momentum piling up interstellar gas like a snow plough. (Since the shell is thin, we can describe it by a single radius $r(t)$.) The cooling takes away most of the shock energy and the motion of the shell forward will be described approximately by the equation for momentum conservation:

$$\frac{d}{dt}[(4\pi/3)\rho r^3 \dot{r}] \approx 0. \tag{5.10}$$

If the thin shell is formed at a time $t_0$ with radius $r = r_0$ and velocity $v_0$, then the first integral to the above equation is:

$$(4\pi/3)\rho r^3 \dot{r} = (4\pi/3)\rho r_0^3 v_0. \tag{5.11}$$

This integrates to:

$$r = r_0 \left(1 + 4\frac{v_0}{r_0}(t - t_0)\right)^{1/4}; \qquad \dot{r} = v_0\left(1 + 4\frac{v_0}{r_0}(t - t_0)\right)^{-3/4}. \tag{5.12}$$

For large $t$, we get $r \propto t^{1/4}$ and

$$v = \dot{r} \propto t^{-3/4} \simeq 200 \text{ km s}^{-1} \left(\frac{t}{3 \times 10^4 \text{ yr}}\right)^{-3/4}. \tag{5.13}$$

The constant factor in the above equation is fixed by equating the the sedov phase velocity of Eq. (5.6) to 200 km s$^{-1}$.

In the final phase, the speed of the shell drops below the sound velocity of the gas in the ISM, which is about $(10 - 100)$ km s$^{-1}$, in a time scale of about $t \approx (1 - 5) \times 10^5$ yrs. In about this time scale, the remnant loses its identity and it is dispersed by random motions in the interstellar medium. The complete evolution is shown schematically in Fig. 5.1 and is a nice example of different physical processes operating at different time scales in the same system.

Supernova emits x-rays profusely during the first two phases from the hot material located behind the shock. The key emission process is thermal bremsstrahlung from the plasma at a temperature of $10^6$ K or higher. The bright optical filaments are formed during phase 3 and their emission comes in the form of line radiation in the material with a temperature of $10^4$ K. This emission is clearly characteristic of the radiating atoms.

Supernova explosions and the eventual dispersion of ejected material have the effect of enriching the interstellar medium with the material pro-

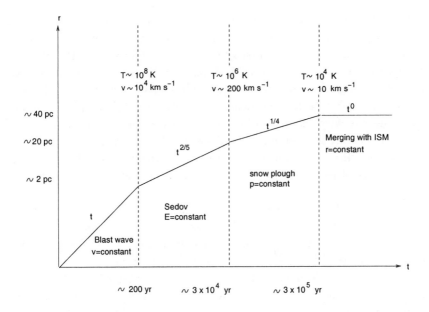

Fig. 5.1 Different stages in the evolution of a supernova remnant, illustrated in terms of the behaviour of its radius, as a function of time.

cessed in stellar interiors. In particular, the heavy elements synthesized inside a star reach the interstellar medium through this process. To make a simple estimate, let us assume that the supernova explosion can be approximated as an instantaneous appearance of a remnant of radius $R = 100$ pc which survives for a time of $t \simeq 10^6$ yr. If the galaxy is modeled as a disk of radius 15 kpc and thickness of 200 pc and if the supernova rate for our galaxy is $(1/30)$ per year (occurring randomly in the disc), then the volume averaged supernova rate is

$$\mathcal{R}_{\mathrm{snr}} \approx 2.3 \times 10^{-13} \ \mathrm{pc}^{-3} \ \mathrm{yr}^{-1}. \tag{5.14}$$

Assuming the supernova explosion is a Poisson process, we can estimate the fraction of the volume filled by the supernova remnants to be

$$f = 1 - \exp\left(-\left(\frac{4\pi}{3}\right) R^3 \mathcal{R}_{\mathrm{snr}} t\right) \approx 0.55. \tag{5.15}$$

In other words, more than half the volume of the galaxy would have been run through a supernova remnant. Any given point of the galactic disc will participate in this process, once in a period of about $10^6$ yrs. Thus the entire

ISM is affected by the occurrence of supernova explosions. Since massive stars evolve at shorter time scale and are also more likely to end up as supernovas, the evolution of first generation of massive stars (Pop II stars) changes the character of the ISM. Second and later generations of stars (Pop I) condense out of this enriched interstellar medium and will have higher proportion of heavier elements. Since the older stars are created from the gas which is yet to be metal enriched by supernova while the younger stars are formed out of metal enriched gas, it is obvious that there should be a direct relation between the age of the star and the fraction of all elements other than hydrogen and helium present in the star, usually called metallicity.

Also note that the total amount of energy released in supernova explosions over the life time of the universe $H_0^{-1} \approx 10h^{-1}\text{Gyr}$ (where $H_0 = 100h\text{km s}^{-1}\text{Mpc}^{-1}$ is the current expansion rate of the universe and $h \approx 0.7$; see chapter 6) is about 4000 $h^2 E_{51}$; for comparison, the binding energy of visible part of the galaxies is about 1300 $h^2$ if the galaxies contain about ten times the dark matter as visible matter (see Chapter 8). Supernova explosions *are* energetic events even globally.

In addition to these x-ray and optical emissions, supernova remnants are also strong sources of radio waves. This is due to the synchrotron radiation from the electrons spiraling in the magnetic fields. We saw in chapter 2, Sec. 2.4.2 that, if the differential number density of relativistic electrons per unit volume is taken to be

$$n(E)\,dE = KE^{-p}\,dE \qquad (5.16)$$

then the total synchrotron flux at frequency $\nu$, from an optically thin source at a distance $d$, can be expressed as

$$S_\nu = \frac{V}{d^2}G\,K\,B^{(1+p)/2}\,\nu^{-(p-1)/2}; \qquad G \simeq \frac{1}{4\pi}\frac{e^3}{mc^2}\left(\frac{3e}{4\pi m^3 c^5}\right)^{(p-1)/2}$$

$$(5.17)$$

where $V$ is the volume of the source and $B$ is the average magnetic field [see Eq. (2.58); note that $S_\nu = Vj_\nu/(4\pi d^2)$]. We will first determine the dependence of the flux $S_\nu$ on the radius $r$ of the remnant, so that one can determine $p$ from the observations. To do this, we need to first fix the dependence of $B$ and $K$ on $r$. In the case of a supernova remnant, the gas at $T > 10^4$ is strongly ionized during the Sedov phase and hence the magnetic field will be frozen to the plasma fluid (see chapter 3, Sec. 3.4.2).

The conservation of magnetic flux implies

$$B(r) = B_0 \left(\frac{r_0}{r}\right)^2 \tag{5.18}$$

showing $B \propto r^{-2}$ as the remnant expands. Next, to determine the radial dependence of $K$, we will use the conservation of the number of relativistic electrons. If the energy density of relativistic electrons decay due to adiabatic expansion of the volume, the energy density $\epsilon$ varies as $d(\epsilon V) = -PdV$. Taking the pressure of relativistic electrons to be $P \simeq (1/3)\epsilon$, this equation integrates to give $\epsilon \propto r^{-4}$. The total energy of the electrons $E = \epsilon V$ decays as $r^{-1}$ allowing us to write

$$E(r) = E_0 \left(\frac{r_0}{r}\right). \tag{5.19}$$

This, in turn, implies that the condition for the conservation of total number of electrons

$$V_0 \int_{E_1}^{E_2} K_0 E^{-p} dE = V_0 \left(\frac{r}{r_0}\right)^3 \int_{E_1 r_0/r}^{E_2 r_0/r} K(r) E^{-p} dE \tag{5.20}$$

can only be satisfied if $(K(r)/K_0) = (r/r_0)^{-(2+p)}$ which fixes the scaling of $K(r)$. Using this result and Eq. (5.18) in Eq. (5.17), we get $S_\nu(r) = S_\nu(r_0)(r/r_0)^{-2p} = S_\nu(t_0)(t/t_0)^{-4p/5}$ where we have used the Sedov relation $r \propto t^{2/5}$. The surface brightness $\Sigma_\nu = (S_\nu/\pi r^2)$ of the remnant scales as

$$\Sigma_\nu(r) = \Sigma_\nu(r_0) \left(\frac{r}{r_0}\right)^{-2(p+1)} \tag{5.21}$$

which relates observable quantities to the index $p$ for electron distribution.

While the scalings are easy, the actual numerical estimate of the synchrotron emission would require an estimate of the magnetic field in the remnant which is difficult to determine. One way of addressing this problem is as follows: It is possible to show that the total energy content of the system (emitting certain amount of synchrotron radiation) has a minimum as a function of the magnetic field strength. If we further assume that the system *is* in the minimum energy configuration, we can determine the magnetic field.

The total energy density of a source containing electrons of number density $n_e$ and a magnetic field $B$ is given by

$$\rho_{\text{tot}} = \rho_e + \rho_B = n_e \gamma m c^2 + \frac{B^2}{8\pi}. \tag{5.22}$$

We will now show that, for a given measured value of the synchrotron flux $j_\nu$, the first term varies as $B^{-3/2}$; that is, $\rho_e \propto \rho_B^{-3/4}$. Hence, for a given flux, the total energy density has a minimum as a function of the magnetic field. To obtain this scaling, note that the synchrotron emissivity is proportional to $j_\nu \propto n_e P/\nu_c$ where $P \propto \gamma^2 B^2$ is the power emitted by a single electron and $\nu_c \propto \gamma^2 B$ is the frequency at which synchrotron peaks (see Eq. (2.52)). This gives $j_\nu \propto n_e B$ which can be equivalently expressed as

$$j_\nu \approx C\rho_e\rho_B^{3/4}; \quad C = \frac{2\sigma_T}{3(8\pi)^{1/4}\sqrt{2\pi qmc}} = 6 \times 10^{-13} \text{ c.g.s.} \quad (5.23)$$

Hence, for a fixed $j_\nu$, we have $\rho_e \propto \rho_B^{-3/4}$ leading to a distinct minimum for $(\rho_e + \rho_B)$. Minimizing $\rho_{\text{tot}}$ now gives the minimum condition to be the one in which the energy densities are essentially in equipartition, $\rho_e \approx \rho_B$ within a numerical factor of order unity which is unimportant. Since energy densities are proportional to the pressure, this condition is equivalent to pressure equilibrium between the magnetic field and the relativistic electrons.

To get a feel for the numbers involved, let us consider the supernova remnant Cas A. This has a flux of $2.72 \times 10^{-23}$ Watts m$^{-2}$ Hz$^{-1}$ and an angular diameter of $4'$. Assuming the source is spherical and is about 2.8 kpc away, we get the luminosity at 1 GHz to be about $2.6 \times 10^{18}$ Watts Hz$^{-1}$ and the volume of the source to be $5.3 \times 10^{50}$ m$^3$. We will approximate the total emission from this system to be about $2.6 \times 10^{27}$ Watts. In that case, a simple calculation gives

$$B_{\text{min}} \approx 2 \times 10^{-4} \text{ Gauss}; \quad \rho_{\text{min}}V \approx 10^{41} \text{ J}. \quad (5.24)$$

These numbers are quite appreciable. For comparison, the kinetic energy of optical filament in the source is about $2 \times 10^{44}$ J and the rest mass energy of $1M_\odot$ is about $2 \times 10^{47}$ J. There must exist a fairly efficient mechanism for converting the explosive energy of supernova into particles and magnetic field to lead to values indicated by Eq. (5.24). The modeling of such remnants is an active area of research.

The magnetic field also plays another — completely different — role in supernova remnants which is worth mentioning. In studying the supernova shock propagation, we have used the fluid description for the matter. A closer look shows that fluid approximation leading to standard shock wave theory (developed in chapter 3, Sec. 3.3.3) may not be applicable! An ejection velocity $10^4$ km s$^{-1}$ corresponds to individual kinetic energy of about 2 MeV for, say, each proton. When such protons hit the hydrogen atom, the latter will be ionized; the cross section for this process is about $\sigma_{\text{ion}} \approx a_0^2 \simeq 10^{-17}$ cm$^2$ and the energy lost per ionisation is about 50 eV. If the medium has a density of about 1 hydrogen atom per cubic centimeter, the stopping length $l$ for a 2 MeV proton is about

$$l \simeq \left( \frac{2 \text{ MeV}}{50 \text{ eV}} \right) \frac{1}{n_l \sigma_{\text{ion}}} \simeq 10^3 \text{ pc.} \tag{5.25}$$

This is larger than any macroscopic length scale of interest in our scenario and hence one suspects that fluid approximation is inapplicable. The situation is, however, saved because the interstellar medium contains magnetic fields which affect the motion of charged particles. A magnetic field strength of, say, $B \approx 3 \times 10^{-6}$ Gauss would result in a Larmor radius of about $R_L \approx 2 \times 10^{10}$ cm $\approx 10^{-8}$ pc. Even though the energy density of such a magnetic field is not of much direct dynamical consequence, $R_L$ will play the role of effective mean free path leading to the formation of a collisionless hydromagnetic shock.

Similar situation arises in the study of many other plasma systems, like e.g., the solar wind in the vicinity of earth's magnetic field. Whenever the Larmor radius is small, it can act as the effective mean free path for the scattering of charged particles. The ratio between the mean free path due to standard Coulomb collisions $[l \propto (T^2/n)]$ and the Larmor radius $[R_L \propto (T^{1/2}/B)]$ varies as $(BT^{3/2}/n)$ and can be large in tenuous high temperature plasmas with strong magnetic fields. This ratio is unity for a critical magnetic field

$$B_c = 10^{-19} \text{ Gauss} \left( \frac{T}{10^5 \text{ K}} \right)^{-3/2} \left( \frac{n}{1 \text{ cm}^{-3}} \right). \tag{5.26}$$

The magnetic field in most astrophysical plasmas will be much larger than $B_c$ and hence the above effect will be important. In the supernova, the magnetic field itself is swept up by this process and is collected along with the gas by the moving shock front. The shock scenario here is thus more complicated than the normal collisional shock and one does expect part of the energy deposited by the shock to appear as relativistic kinetic energy of a fraction of the charged particles. In spite of these differences, the Larmor radius acts as the effective mean free path for the system and renders the fluid approximation at least approximately valid.

## 5.2   White dwarfs

We saw in the last chapter that at the end of stellar evolution, when the
nuclear fuel in the star is exhausted, the gravitational force will start contracting the matter again and the density will increase. Eventually, the
density will be sufficiently high so that the quantum degeneracy pressure
will dominate over thermal pressure. Such astrophysical objects, formed at
the end of stellar evolution, are usually termed as *compact* since — as we
shall see — their sizes are significantly smaller than main sequence stars
of similar mass. Depending on the average density, these compact objects
can have different internal structures which we shall now discuss.

For densities in the range of $10^5$ gm cm$^{-3}$ $\lesssim \rho \lesssim 10^9$ gm cm$^{-3}$, the
remnant is made of a non degenerate ideal gas of ions and a degenerate gas
of electrons. To get a feel for such an object, let us work out the numbers. A
typical example could be a *carbon white dwarf* with temperature $T_{\text{eff}} = 2.7 \times
10^4$ K and mass $M = 1M_\odot$. To begin with, these numbers show that the
white dwarfs are fairly faint objects and are difficult to detect. For example,
a 4 m class telescope can reach a bolometric limiting magnitude of about
24.7. The radius of the white dwarf, estimated as $R = (L/4\pi\sigma T_{\text{eff}}^4)^{1/2}$,
is about 5000 km. Consider such a white dwarf located at a distance $r$
from Earth. The flux received from it will be $F = \sigma T_s^4 (R/r)^2$. Using the
fact that $m_{\text{bol}} = 0$ corresponds to $F_0 = 2.75 \times 10^{-5}$ erg cm$^{-2}$ s$^{-1}$ (see
Eq. (4.14)), we can easily verify that the limiting bolometric magnitude is
reached when $r \approx 2.5$ kpc. Thus only nearby white dwarfs can be detected
by such telescopes. Since they are significantly dimmer than the normal
stars, they will lie to the left and below the main sequence in a HR diagram.

Given the mass and radius, we find that the mean density is $\rho \approx
(3M_\odot/4\pi R^3) \approx 4 \times 10^6$ gm cm$^{-3}$. For this density, one can estimate
the mean distance between the carbon ions as $d_{\text{ii}} \approx (\rho/m_C)^{-1/3} \approx 0.02$
Å where $m_C \approx 12m_H$ is the mass of carbon nucleus. This is smaller
than the radius of the carbon atom $r_C \approx (a_0/Z) \approx (a_0/6) \approx 0.08$ Å.
Hence the electrons will be stripped off the nuclei and will exist as a separate fermionic gas. Further, the mean distance between the electrons will
be $d_{\text{ee}} \approx (Z\rho/m_C)^{-1/3} \approx 10^{-2}$ Å while the thermal wavelength of the
electron is about $\lambda_e = (h^2/m_e k_B T)^{1/2} \approx 10$ Å. Since $\lambda_e \gtrsim d_{\text{ee}}$, quantum mechanical treatment is required. To decide whether the electrons
are relativistic or not, we should compare $m_e c$ with the Fermi momentum
$p_F = (3\pi^2)^{1/3}\hbar n_e^{1/3} = (3\pi^2)^{1/3}\hbar(\rho/\mu_e)^{1/3}$, where $\mu_e = (\rho/n_e m_p)$ is the

number of baryons per electron. The Fermi momentum of the electron will be equal to $m_e c$ at the density $\rho = \rho_c$ where

$$\rho_c \equiv \frac{8\pi}{3} m_p \mu_e \frac{m_e c^3}{h} \approx 10^6 \mu_e \ \text{gm cm}^{-3}. \tag{5.27}$$

In our case the actual density is marginally in relativistic domain but one can just about get away with a non relativistic treatment.

The equilibrium condition for such a system will require the degeneracy pressure of matter to be large enough to balance gravitational pressure. Equivalently, the Fermi energy $\epsilon_F(n)$ must be larger than the gravitational potential energy $\epsilon_g \cong G m_p^2 N^{2/3} n^{1/3}$. When the particles are non relativistic, $\epsilon_F(n) = (\hbar^2/2m_e)(3\pi^2)^{2/3} n^{2/3}$ and this leads to Eq. (4.68). Using $n = (3N/4\pi R^3)$ and $N = (M/m_p)$, one obtains the following mass-radius relation:

$$RM^{1/3} \simeq \alpha_G^{-1} \lambda_e m_p^{1/3} \simeq 8.7 \times 10^{-3} R_\odot M_\odot^{1/3} \tag{5.28}$$

for the white dwarfs. A white dwarf with $M \simeq M_\odot$ will have $R \simeq 10^{-2} R_\odot$ and density $\rho \simeq 10^6 \rho_\odot$.

At lower densities, $(\rho \ll 10^5 \ \text{gm cm}^{-3})$, ions form a crystal lattice structure and the electron density is nonuniform over the scale of the lattice. The coulomb interactions become important in deciding the structure of the matter and the equation of state becomes stiffer; eventually, the electrostatic forces lead to the occurrence of nonzero density at approximately zero pressure, as in the case of normal solids made of non degenerate matter. The Coulomb interaction between electrons is characterized by the energy $E_{\text{coul}} = (Ze^2/d_{ee}) \approx (Ze^2 n_e^{1/3})$ where $n_e$ is the number density of electrons. The ratio between Coulomb energy to Fermi energy is $(E_{\text{coul}}/\epsilon_F) \approx Z^{2/3} \alpha (\rho_c/\rho)^{1/3}$ where $\alpha$ is the fine structure constant. When $E_{\text{coul}} \ll \epsilon_F$, one can treat the gas as ideal. This approximation breaks down at low densities $(\rho \lesssim \rho_c Z^2 \alpha^3 \simeq Z^2 \mu_e)$ and the Coulomb interactions lead to the formation of a crystal structure with finite density at zero pressure. The ratio of electrostatic energy of nearest neighbours to the thermal energy, $\Gamma_c \equiv (Z^2 q^2/a k_B T)$, where $a$ is about the Bohr radius, determines whether Coulomb effects are important or not. $\Gamma = 1$ occurs at $\rho \approx 85(T/10^6 \ \text{K})^3 \ \text{gm cm}^{-3}$. Large values of $\Gamma(\approx 170)$ lead to crystallization.

Since the white dwarfs actually have a nonzero temperature, they will have a finite luminosity due to radiation of heat energy. This will, in turn, make them cool and their location in the HR diagram will change. To study the evolution of the track of white dwarfs in the HR diagram, it is necessary to model its cooling and the resulting change of the temperature. The luminosity will be determined by the nature of the physical processes taking place near the surface of the white dwarf. But the equation of state for degenerate matter will be inapplicable near the surface of a white dwarf since (whatever may be the internal structure) we expect $\rho \to 0$ near the surface. Hence the surface layer of the white dwarf needs to be treated separately along the following lines:

Let us assume that the internal density of a white dwarf decreases from some central value $\rho_c$ at $r = 0$ to $\rho = 0$ at the surface, $r = R$. Then the equation of state will cease to be that of a degenerate gas at some intermediate point $r = r_0$; that is, for $r < r_0$, we shall assume that the equation of state is that of a degenerate gas while at $r > r_0$, the equation of state is that of an ideal gas. Since the degeneracy is decided by the ratio $(\epsilon_F / k_B T)$ — see Eq. (3.13) — it makes sense to determine $r_0$ by the following condition: At $r = r_0$, the Fermi energy $\epsilon_F = (p_F^2 / 2m_e)$ of the system will be equal to the thermal energy $k_B T$. This condition gives

$$\rho_0 = \frac{m_p \mu_e}{3\pi^2} \left( \frac{\hbar}{m_e c} \right)^{-3} \left( \frac{2k_B T_0}{m_e c^2} \right)^{3/2} \tag{5.29}$$

which provides the relation between the density $\rho_0$ and temperature $T_0$ at the transition point. Numerically, in c.g.s units, we have:

$$\rho_0 \approx 6 \times 10^{-9} \, \mu_e \, T_0^{3/2}. \tag{5.30}$$

Further, we will argue that the large thermal conductivity of the degenerate electron gas ensures that the temperature is same $(T = T_0)$ for all $r < r_0$. To see this, we need to estimate the thermal conductivity of the degenerate electron gas — which comes under the general class of parameters known as *transport coefficients*. Since this concept is important, we will take a moment to set up the general formalism and then will apply it to compute conductivity.

The context we want to study is a very common situation in which a system is not in thermal equilibrium globally, but can still be described as in a *local* thermodynamic equilibrium with the parameters like temperature, $T(\mathbf{x})$, varying gradually in space. In this case, the gas will exhibit gradients in temperature, density and velocity fields. When these gradients are small, one can study the effects of these gradients — to the lowest order in their variation — by introducing what are conventionally called transport coefficients. In this description, density gradient leads to diffusion of particles, temperature gradient leads to conductive heat flux and velocity gradient leads to viscosity. The relevant transport coefficients are coefficients of diffusion, thermal conduction and viscosity. We will work it out for diffusion and you can easily extend the idea to find other transport coefficients.

Consider a gaseous system in local thermodynamic equilibrium in which the particles are moving randomly in all directions. Let $\Delta A$ be a small area in some plane perpendicular to $y$-axis which we take to be along the vertical direction. The number of particles crossing this area $\Delta A$ , in unit time, either from above or from below will be $(nv\Delta A/6)$. The factor 6 arises because, on the average, one third of particles will be traveling along the $y$-axis of which half will be traveling downwards. The particles collide, on the average, within one mean-free-path $l = (n\sigma)^{-1}$. Hence the particles coming from above will transport the heat energy at the point $(y + l)$, while the particles going from below will transport the heat energy at the point $(y - l)$ in the opposite direction — thereby leading to thermal conduction. Suppose the density of the particles varies along the $y-$axis. The *net* number $\Delta N$ of particles which crosses the area $\Delta A$ per second will be

$$\Delta N = \left(\frac{1}{6}\right) \Delta A v \left[n(y+l) - n(y-l)\right] \approx \left(\frac{1}{3}\right) (\Delta A) \, v \, l \left(\frac{\partial n}{\partial y}\right). \quad (5.31)$$

The flux of particles along the $y-$axis, $J_y \equiv (\Delta N/\Delta A)$ is conventionally expressed as $J_y = D\,(\partial n/\partial y)$, where $D$ is called the *coefficient of diffusion*. Comparison with Eq. (5.31) shows that

$$D = \left(\frac{1}{3}\right)(vl) \quad (5.32)$$

which is a general result for coefficient of diffusion.

Coming back to the case of white dwarfs, the mean free path is $l = (n_i\sigma)^{-1}$ where $\sigma \simeq (2Ze^2/m_e v^2)^2$ is the relevant cross-section for charged particle interaction. This gives

$$D \cong \frac{1}{12\pi}\left(\frac{m_e}{Ze^2}\right)^2 \frac{v^5}{n_i}. \quad (5.33)$$

If $\rho \simeq 10^6$ gm cm$^{-3}$ and $v \simeq v_F \simeq c$ then we get $D \simeq 200$ cm$^2$ s$^{-1}$; the corresponding diffusion time scale in the white dwarf interior is $t_{\text{cond}} \simeq (R^2/D) \simeq 10^6$ yr which is quite short compared to the age of the galaxy. Hence thermal conduction can quickly equalize the temperature at $r < r_0$

and we may take $T = T_0$ for all $r < r_0$. It is the thin surface layer of non degenerate matter near the surface, $r_0 < r < R$, which prevents efficient cooling of white dwarfs and determines its luminosity.

We can now find the scaling relation, relating the luminosity of the white dwarf to other parameters. We saw earlier (see chapter 4, Eq. (4.37)) that in a radiative envelope $\rho \propto (M/L)^{1/2} T^{3.25}$. At the point of matching ($r = r_0$) between the degenerate core and the envelope, Eq. (5.30) requires that we must have $\rho_0 \propto T_0^{3/2}$. Combining the two relations we get $(M/L)^{1/2} T_0^{3.25} \propto T_0^{3/2}$; or $L \propto MT_0^{7/2} \propto MT_{\text{core}}^{7/2}$ where the last proportionality follows from the fact that core is isothermal. The characteristic cooling time for such a system is

$$t_{\text{cool}} \approx \frac{U}{L} \propto \frac{MT_{\text{core}}}{MT_{\text{core}}^{7/2}} \propto T_{\text{core}}^{-5/2}. \tag{5.34}$$

This leads to the scaling law $L \propto MT_{\text{core}}^{7/2} \propto Mt_{\text{cool}}^{-7/5}$.

Combining the relations $L \propto \dot{T}$ and $L \propto T^{7/2}$, one finds that $\dot{T} \propto T^{7/2}$ or, equivalently, $\dot{L} \propto -L^{12/7}$. Including all the numerical factors and integrating, the luminosity evolution of white dwarfs is found to be

$$t = 6.3 \times 10^6 \, \text{yrs} \left( \frac{A}{12} \right)^{-1} \left( \frac{M}{M_\odot} \right)^{5/7} \left( \frac{\mu}{2} \right)^{-2/7} \left[ \left( \frac{L}{L_\odot} \right)^{-5/7} - \left( \frac{L_0}{L_\odot} \right)^{-5/7} \right]. \tag{5.35}$$

This expression relates the time $t$ to the luminosity $L(t)$ if the initial luminosity (at $t = 0$) was $L_0$. Ignoring the $L_0$ term at late times (when $L \ll L_0$), we get

$$t \approx 3.9 \times 10^6 \, \text{years} \left( \frac{A}{12} \right)^{-1} \left( \frac{M}{M_\odot} \right)^{5/7} \left( \frac{L}{L_\odot} \right)^{-5/7}. \tag{5.36}$$

Let $N(L)$ denotes the number density of white dwarfs in the range of luminosities $(L, L + dL)$; then the rate of change of this number density is given by

$$\dot{N} = \dot{L} \frac{dN}{dL} = \left( \frac{\dot{L}}{L} \right) \left( \frac{dN}{d \ln L} \right) = \left( \frac{1}{t(L)} \right) \left( \frac{dN}{d \ln L} \right) \propto \left( \frac{L}{M} \right)^{5/7} \left( \frac{dN}{d \ln L} \right). \tag{5.37}$$

If the white dwarfs are produced at an approximately constant rate, then we expect $\dot{N} \approx$ constant giving

$$\frac{dN}{d\ln L} \propto \left(\frac{L}{M}\right)^{-5/7}. \tag{5.38}$$

Thus the luminosity function, giving the number density of white dwarfs per logarithmic interval of luminosity, is a power law with slope $(-5/7)$. We do expect this power law to be truncated at the extreme limits. At low luminosities and temperatures, crystallization will modify the structure while at very high luminosities and temperatures, one has to take into account effects due to neutrino cooling. In the intermediate range, we expect the above power law to hold. This simple result agrees reasonably well with observations.

The coolest white dwarfs which have been observed have $L \approx 10^{-4.5} L_\odot$ with a cooling time — given by Eq. (5.36) — of about $10^{10}$ years which is comparable to the age of the galaxy. A dramatic drop in the population of white dwarfs is observed in our galaxy around $L \simeq 10^{-4.5} L_\odot$. This is consistent with an age of $(9 \pm 1.8) \times 10^9$ years for our galaxy. Adding the time spent in the pre white dwarf stage in the stellar evolution would suggest that the star formation in the disk of the galaxy began about $(9.3 \pm 2) \times 10^9$ years ago. This is about 6 billion years shorter than the age of the *globular clusters* which are distributed in a more spherically symmetric manner in our galaxy and are not confined to the disk. This suggests that the disk of the galaxy formed later than the halo — a scenario we will come back to in chapter 7.

## 5.3 Neutron stars and pulsars

At densities $\rho \gtrsim 10^9$ gm cm$^{-3}$ the electrons can combine with the protons in the nuclei and thus change the composition of matter. The equation of state at higher densities needs to be determined by studying the the the minimum energy configuration of matter allowing for nuclear reactions. The details are quite complicated but a simplified analysis — which captures the essential physics — can be based on the idea that, electrons can combine with protons through *inverse beta decay* to form neutrons, which can then provide the degeneracy pressure that will balance gravity. Equation (5.28) is still applicable with $m_e$ replaced by $m_n$; correspondingly the right hand side of Eq. (5.28) is reduced by $(\lambda_n/\lambda_e) = (m_e/m_n) \simeq 10^{-3}$. Such objects — called *neutron stars* — will have a radius of $R \simeq 10^{-5} R_\odot$ and density of $\rho \simeq 10^{15} \rho_\odot$ if $M \simeq M_\odot$. For these values $(GM/c^2 R) \simeq 1$ and general relativistic effects are beginning to be important.

When such a neutron star forms as a result of a collapse, its angular momentum is conserved to a large degree. If the initial angular momentum of the pre-collapse core was $J_{in} \cong M\Omega_{in}R_{in}^2$, then the conservation of angular momentum, $J_{in} = J_{final}$, requires $\Omega_{final} \approx \Omega_{in}(R_{in}/R_{final})^2$. If the ratio between the initial and final radii is about $10^5$ and if we take $\Omega_{in} = (2\pi/20 \text{ days})$, this leads to a fairly rapid rotation for the neutron star with $\Omega_{final} \approx (2\pi/10^{-4} \text{ sec})$.

As the core collapses to form a neutron star, its electrical conductivity becomes very high. In the limit of infinite conductivity, the magnetic flux through the star, $\Phi \propto BR^2$, will also be conserved (see chapter 3, Sec. 3.4.2). This leads to the final magnetic field of $B_{final} = B_{in}(R_{in}/R_{final})^2$. If $B_{in} \approx 100$ Gauss and $(R_{final}/R_{in}) \approx 10^5$, we get $B_{final} \approx 10^{12}$ Gauss and hence the neutron star will be threaded by a very strong magnetic field. Such rotating, magnetic neutron stars emit a characteristic pattern of pulsed radiation with remarkably stable period. These objects, called *pulsars*, are extensively studied and are used as probes of several other astrophysical phenomena. For example, isolated pulsars allow the probing of the interstellar electron density and magnetic field by the processes of plasma dispersion and Faraday rotation described in Sec. 3.4.2. Pulsars in binary systems and clusters allow the study of a host of other phenomena related to evolution of binary stars. They also act as clocks located in strong gravitational fields thereby helping one to test specific predictions of general relativity.

The energy emitted by a pulsar span a wide band of wavelengths — from radio to hard x-rays — with most of the energy being emitted at higher frequencies. In spite of this fact, pulsars are most conveniently observed in radio frequencies and — over years — significant amount of data has accumulated about radio pulsars. There are now more than 1200 pulsars which are believed to be a galactic population of isolated neutron stars whose radiation is powered by rotation. The periods of rotation range from milliseconds to several seconds, with most of the pulsars being confined to periods of 0.2 to 2 seconds. As the rotational speed of the pulsar decreases, the period will increase, with the typical rate of change of period being $\dot{P} \approx 10^{-15}$. But for this slow decay, the period of the pulsar happens to be remarkably stable, making them excellent clocks. One of the most extensively studied pulsars is the Crab pulsar, from which radiation is observed over 60 octaves from 75 MHz to 500 GeV.

To the lowest order of approximation, one can think of pulsars as rotating magnetic dipole moments emitting magnetic dipole radiation. Though this model is too simplistic, it leads to the correct order of magnitude estimate for radiation and hence is worth discussing briefly.

Consider a rotating neutron star with a magnetic dipole moment $\mathbf{m}$, radius $R$ and angular velocity $\Omega$. Let the direction of $\mathbf{m}$ be inclined at an angle $\alpha$ to the rotation axis. In that case, the magnetic moment $\mathbf{m}$ varies with time as

$$\mathbf{m} = \frac{1}{2}BR^3 \left(\mathbf{e}_\parallel \cos\alpha + \mathbf{e}_\perp \sin\alpha \cos\Omega t + \mathbf{e}'_\perp \sin\alpha \sin\Omega t\right) \qquad (5.39)$$

where $\mathbf{e}_\parallel$ is the unit vector along the rotation axis and $\mathbf{e}_\perp, \mathbf{e}'_\perp$ are (fixed) mutually orthogonal unit vectors perpendicular to $\mathbf{e}_\parallel$. Note that the magnitude of the magnetic moment at the pole, $\alpha = 0$, for a magnetized sphere of radius $R$ and magnetic field $B$ must be $|\mathbf{m}| = (BR^3/2)$, which fixes the overall scaling in the above equation. Any time varying magnetic dipole moment emits radiation at a rate

$$\frac{dE}{dt} = -\frac{2}{3c^3}|\ddot{\mathbf{m}}|^2 = -\frac{B^2 R^6 \Omega^4 \sin^2\alpha}{6c^3}. \qquad (5.40)$$

(This is completely analogous to the electric dipole radiation discussed in chapter 2, Sec. 2.1.1. Equation (5.40) is the magnetic counter part of Eq. (2.8).) This radiation is emitted at the frequency of rotation $\Omega$ of the pulsar, and vanishes in the case of an aligned rotator with $\alpha = 0$. Assuming that the energy is supplied by the kinetic energy of rotation, $K = (1/2)I\Omega^2$, it follows that

$$\frac{dE}{dt} = I\Omega\dot{\Omega} = -\frac{B^2 R^6 \Omega^4 \sin^2\alpha}{6c^3}. \qquad (5.41)$$

Using the expression for $(dE/dt)$, we can define a characteristic time scale $T$ by

$$T \equiv -\left(\frac{\Omega}{\dot{\Omega}}\right)_0 = \frac{6Ic^3}{B^2 R^6 \Omega_0^2 \sin^2\alpha} \qquad (5.42)$$

where the subscript '0' denotes the current value. Integrating Eq. (5.41) and fixing the initial conditions correctly, we get

$$\Omega = \Omega_i \left(1 + \frac{2\Omega_i^2}{\Omega_0^2}\frac{t}{T}\right)^{-1/2} \qquad (5.43)$$

where $\Omega_i$ is the initial angular velocity at $t = 0$. This equation gives the angular velocity of the pulsar as a decreasing function of time; clearly, the pulsar is spinning down due to energy loss. Setting $\Omega = \Omega_0$ gives the present age of the pulsar to be

$$t = \frac{T}{2}\left(1 - \frac{\Omega_0^2}{\Omega_i^2}\right) \simeq \frac{T}{2} \qquad \text{(for } \Omega_0 \ll \Omega_i). \tag{5.44}$$

These formulas can be used to make some simple estimates regarding the pulsar emission. The Crab pulsar, for example, has originated as a result of supernova explosion which occurred in the year 1054 AD and hence is about $T \cong 10^3$ years old. Its current period, $P = (2\pi/\Omega_0)$, is about 0.033 seconds. Estimating $|\dot{\Omega}| = (\Omega_0/T) \approx 10^{-8}$ s$^{-2}$, we can calculate the energy loss rate to be about $(dE/dt) = 6.4 \times 10^{38}$ erg s$^{-1}$. If $\sin\alpha \approx 1$, then Eq. (5.42) allows us to determine the magnetic field to be $B \approx 5.2 \times 10^{12}$ Gauss. These values appear to be reasonable.

## 5.4 Black holes

When the density is still higher, the Fermi energy has to be supplied by relativistic particles and $\epsilon_F$ now becomes $\epsilon_F \simeq \hbar c n^{1/3}$. This expression scales as $\epsilon_F \propto n^{1/3}$, just like the gravitational energy density, $\epsilon_g$ in Eq. (4.1). Therefore, the condition $\epsilon_F \geq \epsilon_g$ can be satisfied only if $\hbar c \geq G m_p^2 N^{2/3}$ or, only if $N \leq \alpha_G^{-3/2} \simeq N_G \alpha^{-3/2}$. The corresponding mass bound (called *Chandrasekhar limit*) is $M \lesssim M_{\text{ch}}$ where

$$M_{\text{ch}} \simeq m_p \, \alpha_G^{-3/2} \simeq 1\, M_\odot. \tag{5.45}$$

(A more precise calculation gives a slightly higher value.) Thus, when the mass of the stellar remnant is higher than $M_{\text{ch}} \simeq \alpha_G^{-3/2} m_p$, no physical process can provide support against the gravitational collapse.

The difference in the behaviour of the equilibrium radius $R(M)$ for a body of mass $M$ in the non relativistic and extreme relativistic limits is a simple consequence of the limiting forms of the equation of state. The same result can be interpreted in terms of the pressure balance along the following lines. Pressure balance against gravity requires the ratio $\zeta$ between the gravitational pressure $(GM^2/R^4)$ and the degeneracy pressure $P$ to be some constant of the order of unity. Taking $P \propto \rho^\gamma \propto (M/R^3)^\gamma$, this ratio becomes

$$\zeta = (GM^2/PR^4) \propto M^{2-\gamma} R^{3\gamma-4}. \tag{5.46}$$

For a non relativistic system with $\gamma = 5/3$, we have $\zeta \propto M^{1/3}R$. In equilibrium, $\zeta$ will have some definite numerical value; if $M$ is now increased, equilibrium can be restored by $R$ decreasing so as to maintain the constancy of $M^{1/3}R$. Hence a system supported by non relativistic degeneracy pressure will contract if more mass is added. On the other hand, for the relativistic systems with $\gamma = 4/3$ we have $\zeta \propto M^{2/3} \simeq (M/M_{\rm ch})^{2/3}$, which is independent of the radius. If $M < M_{\rm ch}$, pressure exceeds gravity and the system will expand reducing the density and making the system (eventually) non relativistic; then an equilibrium solution can be found at a suitable value of $M$ and $R$. But if $M > M_{\rm ch}$, the gravitational force exceeds the pressure and the system will contract. But since $\zeta$ is independent of $R$, the contraction will *not* lead to a new equilibrium configuration and the system will continue to collapse under self gravity. In such a case, the star will form a *black hole* and is likely to exert very strong gravitational influence on its surroundings.

Such a body necessarily has to be described by general relativity with a metric in Eq. (1.79)

$$ds^2 = \left(1 - \frac{2GM}{c^2 r}\right)c^2 dt^2 - \left(1 - \frac{2GM}{c^2 r}\right)^{-1} dr^2 - r^2\left(d\theta^2 + \sin^2\theta d\phi^2\right).$$

$$(5.47)$$

This leads to a wide variety of curious phenomena, of which we will discuss a few.

An immediate consequence of this metric is that radiation emitted by a body having such a strong gravitational field will be significantly redshifted as it propagates towards infinity. If $\omega(r)$ is the frequency of radiation emitted by a body of radius $r$ and $\omega_\infty$ is the frequency with which this radiation is observed at large distances, then our discussion after Eq. (1.75) shows that $\omega_\infty = \omega(r)(1 - 2GM/c^2 r)^{1/2}$; that is, radiation from near $r \gtrsim 2GM/c^2$ is redshifted by a large factor.

To explore this in greater detail, consider a wave packet of radiation emitted from a radial distance $r_e$ at time $t_e$ and observed at a large distance $r$ at time $t$. The trajectory of the wave packet is, of course, given by $ds^2 = 0$ in Eq. (5.47) which — when we use $d\theta = d\phi = 0$ — is easy to integrate. We get

$$c(t - t_e) = r - r_e + \frac{2GM}{c^2}\ln\left(\frac{1 - 2GM/c^2 r}{1 - 2GM/c^2 r_e}\right) = r - r_e + \frac{4GM}{c^2}\ln\left(\frac{\omega_e}{\omega(r)}\right).$$

$$(5.48)$$

In the last equality we have expressed the result in terms of the frequency observed at $r$. This result shows that, when the radiation is emitted from a location just outside the Schwarzschild radius (ie., $r_e \gtrsim 2GM/c^2$), and is observed at a large distance (ie., $r \gg 2GM/c^2$) the frequency of radiation measured by an observer at infinity is exponentially redshifted:

$$\omega(t) \propto \exp -(c^3 t/4GM) \equiv K \exp -(\kappa t/c) \qquad (5.49)$$

where $K$ is a constant and we have introduced the parameter

$$\kappa = \frac{c^4}{4GM} = \frac{GM}{(2GM/c^2)^2} \qquad (5.50)$$

which gives the gravitational acceleration $(GM/r^2)$ at the Schwarzschild radius $r = 2GM/c^2$. This quantity, called the *surface gravity*, plays an important role in black hole physics.

Incidentally, this result tells you that the wavefront $\phi(t)$ has a phase which itself is decaying exponentially with time:

$$\phi(t) = \exp[-i \int \omega(t) dt] = \exp[i\Omega e^{-(\kappa t/c)}]; \quad \Omega = cK/\kappa. \qquad (5.51)$$

(Remember that the instantaneous frequency is just the time derivative of the phase of the wave when $\omega$ is not a constant.) This is clearly not a monochromatic wave and has power at all frequencies. The power spectrum of this wave is given by $P(\nu) = |f(\nu)|^2$ where $f(\nu)$ is the Fourier transform of $\phi(t)$ with respect to $t$:

$$\phi(t) = \int_{-\infty}^{\infty} \frac{d\nu}{2\pi} f(\nu) e^{-i\nu t}. \qquad (5.52)$$

Because of the exponential redshift, this power spectrum will *not* vanish for $\nu < 0$. Evaluating this Fourier transform (by changing to the variable $\Omega \exp[-(\kappa t/c)] = z$ and analytically continuing to Im $z$; see exercise 5.8) one gets:

$$f(\nu) = (c/\kappa)(\Omega)^{i\nu\kappa/c} \Gamma(-i\nu c/\kappa) e^{\pi\nu c/2\kappa}. \qquad (5.53)$$

Taking the modulus, we get:

$$\nu |f(-\nu)|^2 = \frac{\beta}{e^{\beta h\nu} - 1}; \quad \beta \equiv \frac{1}{k_B T} = \frac{2\pi c}{\hbar \kappa}. \qquad (5.54)$$

This leads to the the remarkable result that the power, per logarithmic band in frequency, at negative frequencies is a Planck spectrum with temperature

$k_B T = (\hbar\kappa/2\pi c)$; and, more importantly,

$$|f(-\nu)|^2/|f(\nu)|^2 = \exp(-\beta\hbar\nu). \tag{5.55}$$

If the wave $\phi(t)$ is interacting, for example, with a two level system, then you know from standard time dependent perturbation theory that the modes which vary as $e^{\mp i\nu t}$ cause the upward and downward transitions between the two levels. The above result shows that such a two level system — that interacts with the wave — will be driven to a steady state with the two levels populated in accordance with the thermal distribution of temperature $\beta$. (Though $f(\nu)$ in Eq. (5.53) depends on $\Omega$, the power spectrum $|f(\nu)|^2$ is independent of $\Omega$.) An observer detecting the exponentially redshifted radiation in Eq. (5.51) at late times ($t \to \infty$), originating from a region close to $r = 2GM/c^2$, will attribute to this radiation a Planckian power spectrum given by Eq. (5.54). This result lies at the foundation of associating a temperature with a black hole.

Mathematically, such a result will arise for any wave with an exponential redshift. If $\phi(t)$ varies as in Eq. (5.41) it has a power spectrum which is Planckian with temperature $T = \kappa/2\pi$. We will use this result in chapter 6 in the context of exponential expansion of the universe which leads to the redshift of the form in Eq. (5.41).

The general relativistic description of gravity also changes the nature of orbits in the gravitational field of a point mass. To obtain this result in the simplest manner, let us begin with the trajectory of a particle in special relativity under the action of a central force. The angular momentum $\mathbf{L} = \mathbf{r} \times \mathbf{p}$ is still conserved but the momentum is now given by $\mathbf{p} = \gamma m \mathbf{v}$ with $\gamma \equiv (1-v^2)^{-1/2}$. (We will temporarily use units with $c = 1$). So the relevant conserved component of the angular momentum is $L = mr^2(d\theta/ds) = \gamma m r^2 \dot\theta$ rather than $mr^2\dot\theta$. (This, incidentally, means that Kepler's second law regarding areal velocity does not hold in special relativistic motion in a central force.) Consider now the motion of a *free* special relativistic particle described in *polar* coordinates. The standard relation $E^2 = \mathbf{p}^2 + m^2$ can be manipulated to give the equation

$$E^2 \left(\frac{dr}{dt}\right)^2 = E^2 - \left(\frac{L^2}{r^2} + m^2\right). \tag{5.56}$$

(This is still the description of a free particle moving in a straight line but in the polar coordinates!) Since special relativity must hold around any event, we can obtain the corresponding equation for general relativistic motion by simply replacing $dr$, $dt$ by the proper quantities $\sqrt{|g_{11}|}dr$, $\sqrt{|g_{00}|}dt$ and the energy $E$ by the redshifted energy: $E/\sqrt{|g_{00}|}$ (see Eq. (1.75)) in this equation. This gives the equation for the orbit of a particle of mass $m$, energy $E$ and angular momentum $L$ around a mass $M$. With some simple manipulation, this can be written in a suggestive form as:

$$\left(1 - \frac{2GM}{r}\right)^{-1}\frac{dr}{dt} = \frac{1}{E}\left[E^2 - V_{\text{eff}}^2(r)\right]^{1/2} \tag{5.57}$$

with an effective potential:

$$V_{\text{eff}}^2(r) = m^2\left(1 - \frac{2GM}{r}\right)\left(1 + \frac{L^2}{m^2r^2}\right). \tag{5.58}$$

Several important aspects of the motion can be deduced from the nature of this effective potential which is shown in Fig. 5.2.

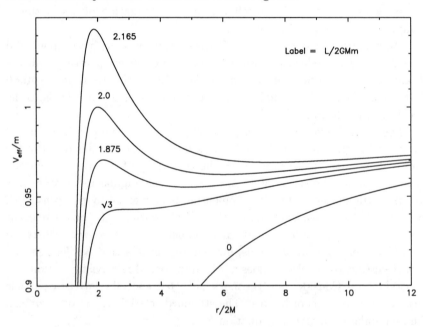

Fig. 5.2 The effective potential governing the motion of a particle in the Schwarzschild metric for different values of the angular momentum.

(i) For a given value of $L$ and $E$ the nature of the orbit will, in general, be governed by the turning points in $r$, determined by the equation $V_{\text{eff}}^2(r) = E^2$. For a given $L > 2GMm$ the function $V_{\text{eff}}(r)$ has one maxima and one minima. Further, $V_{\text{eff}} \to m$ as $r \to \infty$ for all values of $L$. If the energy $E$

of the particle is lower than $m$ then there will be two turning points. The particle will orbit the central body with a perihelion and aphelion. This is similar to elliptic orbits in Newtonian gravity but the perihelion will now precess. (See exercise 5.7.)

(ii) If $m < E < V_{max}(L)$ there will be only one turning point. The particle will approach the central mass from infinity, reach a radius of closest approach and will travel back to infinity. This is similar to the hyperbolic orbits in Newtonian gravity.

(iii) If $E = V_{max}(L)$, then the orbit will be circular at some radius $r_0$ determined by the condition $V'(r_0) = 0, V(r_0) = E$. This gives the radius of the circular orbit and the energy to be

$$\frac{r_0}{2GM} = \frac{L^2}{4G^2M^2m^2} \left[ 1 \pm \sqrt{1 - \frac{12G^2M^2m^2}{L^2}} \right];$$ (5.59)

$$E^2 = \frac{L^2}{GMr_0} \left( 1 - \frac{2GM}{r_0} \right)^2 .$$ (5.60)

The upper sign refers to stable orbit and the lower to the unstable orbit. The stable orbit closest to the center has the parameters, $r_0 = 3GM, L = 2\sqrt{3}GMm$ and $E = \sqrt{(8/9)}m$.

(iv) If $E > V_{max}(L)$, the particle "falls to the center". This behaviour is in sharp contrast to Newtonian gravity in which a particle with nonzero angular momentum can *never* reach $r = 0$.

(v) As the angular momentum is lowered, $V_{max}$ decreases and for $L = 2GMm$ the maximum value of the potential is at $V_{max} = m$. In this case all particles from infinity will fall to the origin. Particles with $E < m$ will still have 2 turning points and will form bound orbits.

(vi) When $L$ is reduced still further, the maxima and minima of $V(r)$ approach each other. For $L \leq 2\sqrt{3}GMm$, there are no turning points in the $V(r)$ curve. Particles with lower angular momentum will fall to the origin irrespective of their energy.

So far, we have thought of black holes as end stages of stellar evolution. More complicated processes can lead to the formation of black holes with masses significantly higher than stellar masses in the centers of galaxies. In the extra galactic domain there are objects called *active galactic nuclei* and *quasars* (see Chapter 9) which have a luminosity of about $10^{44}$ erg s$^{-1}$ which are thought to be powered by accretion discs around very massive black holes.

## 5.5   Compact remnants in stellar binaries

In the description of stellar evolution we have assumed that the star is an isolated system and is not affected by the surrounding, which is certainly not the case if the star is a part of a binary star system. It is then possible for one of the binary stars to evolve into a compact remnant — white dwarf, neutron star or even a black hole — and accrete matter out of the second star leading to a wide variety of new physical phenomena. In such an accretion process, the gravitational potential energy is converted into the kinetic energy of matter and dissipated as thermal radiation. Some of the sources of high energy radiation — both galactic and extra galactic — are generally believed to be powered by such accretion processes. In the galactic scale, accretion discs around stars can be a source of thermalized x-ray emission; in the extra galactic domain there are objects called *active galactic nuclei* and *quasars*.

The basic energetics in the process of accretion is the following: When a mass $M$ falls from infinite distance to a radius $R$, in the gravitational field of a massive object with mass $M_c$, it gains the kinetic energy $E \simeq (GM_cM/R)$. If this kinetic energy is converted into radiation with efficiency $\epsilon$, then the luminosity of the accreting system will be $L = \epsilon\,(dE/dt) = \epsilon\,(GM_c/R)\,(dM/dt)$. In galactic contexts, the compact object will have a mass in the range of $(0.6 - 1.4)M_\odot$ in case of white dwarfs or neutron stars and somewhat higher in case of black holes. For material accreting on to a neutron star of mass $1.4\ M_\odot$ and radius $R = 10^6$ cm, the accretion luminosity will be

$$L_{\text{acc}} = \frac{GM_c\dot{M}}{R} = 2.95\ L_\odot \left(\frac{R}{10^6\ \text{cm}}\right)^{-1}\left(\frac{M_c}{1.4M_\odot}\right)\left(\frac{\dot{M}}{10^{-12}M_\odot\ \text{yr}^{-1}}\right).$$

$$(5.61)$$

The accretion luminosity will therefore increase in proportion to $\dot{M}$.

There is, however, a natural upper bound to the luminosity generated by any accretion process on to a compact remnant of mass $M_c$. This limiting luminosity, called Eddington luminosity, (see Eq. (2.38) with $\kappa = \sigma_T/m_p$ arising from the electron scattering opacity) is

$$L_E = \frac{4\pi G m_p c}{\sigma_T} M_c \simeq 1.3 \times 10^{46}\left(\frac{M_c}{10^8 M_\odot}\right)\text{erg s}^{-1} = 10^{4.5}\left(\frac{M_c}{M_\odot}\right)L_\odot.$$

$$(5.62)$$

The first form of the equation is applicable in the case of accretion on to supermassive black holes and thought to be relevant in the context of active galactic nuclei. The second form of the equality is relevant for galactic x-ray sources powered by stellar mass black holes. The temperature of a spherical system of radius $R$ radiating $L \simeq L_E$ will be determined by $(4\pi R^2)\sigma T^4 = L_E$; that is,

$$T \simeq 1.8 \times 10^8 \text{ K} \left(\frac{M_c}{M_\odot}\right)^{1/4} \left(\frac{R}{1 \text{ km}}\right)^{-1/2}. \tag{5.63}$$

For a solar mass compact ($R \simeq 1$ km) stellar remnant, this radiation will peak in the x-ray band.

Equating the accretion luminosity in Eq. (5.61) to the maximum possible luminosity in Eq. (5.62) we obtain the maximum possible accretion rate to be

$$\dot{M}_{\text{Edd}} = (1.5 \times 10^{-8} M_\odot \text{ yr}^{-1}) \left(\frac{R}{10^6 \text{ cm}}\right). \tag{5.64}$$

For a typical neutron star with $M_c = 1.4 M_\odot, R = 10^6$ cm, the maximum accretion rate that can be sustained in steady state is about $1.5 \times 10^{-8} M_\odot \text{yr}^{-1}$. It is also clear from the above relations that an accretion rate of $(10^{-12} - 10^{-8}) M_\odot \text{yr}^{-1}$ will produce luminosities in the range of $(3 - 10^4) L_\odot$.

These considerations lead to two different classes of accreting sources in the binary systems involving normal stars and a compact neutron star. In the first case, we have a neutron star orbiting around a high mass normal star. High mass stars have outer regions which are loosely bound and hence there is a strong stellar wind from such stars. the orbiting neutron star can accrete a *fraction* of the material in the wind which is flowing past it, thereby leading to a steady $\dot{M}$. For typical stellar wind losses, this accretion rate turns out to be well described by the empirical relation

$$\dot{M}_{\text{acc}} \approx 5 \times 10^{-20} M_\odot \text{ yr}^{-1} \left(\frac{M}{M_\odot}\right)^{5.67} \tag{5.65}$$

where $M$ is the mass of the star. In order to have significant luminosity, we need $\dot{M}_{\text{acc}}$ to be at least about $10^{-12} M_\odot \text{ yr}^{-1}$. The above equation shows that the mass of the star has to be more than about 20 $M_\odot$ for this to be viable. These binary systems are called *High mass X-ray binaries* or HMXB, for short.

An alternative possibility for a binary system is a low mass star and a neutron star orbiting around each other. If the radius of the low mass

star increases significantly during the course of evolution, the strong gravitational field of the neutron star can remove the mass lying in the outer
region of the companion star. (This is essentially same as the tidal disrup-
tion discussed in chapter 1, Sec. 1.2 in connection with the Roche limit;
see Eq. (1.38). In fact, this process is called *Roche limit overflow*.) This
accretion can occur at two different time scales, depending on the mass of
the star. If the mass of the star is more than that of the compact remnant
(i.e, if $M \gtrsim 1.4 M_\odot$), then the transfer can be unstable and occur relatively
rapidly at the thermal time scale of the star:

$$\tau_{\text{th}} = \frac{GM^2}{RL} \simeq 3 \times 10^7 \text{ yr } \frac{(M/M_\odot)^2}{(R/R_\odot)(L/L_\odot)}. \tag{5.66}$$

Using the scaling relations $L \propto M^3, R \propto M$, and assuming that most of
the stellar mass (say, about 0.8 $M_\odot$) is transfered at this time scale, we get
a mass transfer rate of

$$\dot{M} \simeq \frac{M}{\tau_{\text{th}}} \simeq 3 \times 10^{-8} \left(\frac{M}{M_\odot}\right)^3 M_\odot \text{ yr}^{-1}. \tag{5.67}$$

This is, however, *not* a viable scenario because, for $M > 1.4 M_\odot$, this will
lead to an accretion rate higher than the maximum accretion rate allowed
by Eq. (5.64). The second possibility is for stars with masses less than that
of the compact remnant (i.e, if $M \lesssim 1.4 M_\odot$). In this case, mass transfer
takes place at (a much slower) nuclear time scale of the star:

$$\tau_{\text{nucl}} \approx 10^{10} \left(\frac{M}{M_\odot}\right) \left(\frac{L}{L_\odot}\right)^{-1} \text{ yr}. \tag{5.68}$$

If $L \propto M^3$ we have $\dot{M} \approx 10^{-10} M_\odot \text{ yr}^{-1} (M/M_\odot)^3$. This is a possible
accretion scenario for $M \lesssim 1.4 M_\odot$; for $M = 1.4 M_\odot$ this will lead to a
maximum accretion rate of about $3 \times 10^{-10} M_\odot \text{ yr}^{-1}$ which will correspond
to a luminosity of $L_{\text{acc}} \approx 10^3 L_\odot$. Such systems are called *Low Mass X-ray
Binaries* or LMXB, for short. Thus you see that viable accretion powered
x-ray sources can exist only when the companion to the neutron star has a
mass less than $1.4 M_\odot$ or greater than about 20 $M_\odot$.

Accretion to a white dwarf can lead to yet another class of phenomena of which the most important is a supernova explosion called Type I
Supernova — to distinguish it from the supernova occurring at the end
stage of an isolated star, which is called Type II. The Type I supernova can
be modeled as an explosion occurring in a accreting carbon-oxygen white

dwarf which is a member of a closed binary system. When the mass of the white dwarf reaches about 1.3 $M_\odot$, carbon burning begins at the center of the star in the degenerate core and moves towards the surface. This motion occurs subsonically (called *deflagration front* rather than *detonation front*) converting about one half of white dwarf's mass into iron before the degeneracy is removed. Further expansion can lead to a cooling which will eventually dampen the nuclear burning leaving shells of intermediate mass elements around a nickel iron core. The energy released in the process can disrupt the star resulting in the supernova explosion.

Finally, we mention another remarkable result related to compact binaries though it is not directly connected with accretion of matter. This involves an application of general relativity in a compact binary system to verify a result of fundamental significance — that led to an award of the Nobel prize. Considering its importance, we shall briefly describe how this is achieved. Let us consider a binary star system of which one member is a pulsar from which we are receiving periodic signals and the other companion is a neutron star or black hole. In Newtonian gravity, such a binary system will be described by elliptical orbits. Observation of such a system can only determine the combination of masses given in Eq. (1.98) of Ex. 1.7. The situation, however, is different in general relativity. When general relativistic effects are taken into account the major axis of the elliptical orbit will precess in a specific calculable rate. (See Ex. 5.7.) Using this, one can determine individually, the masses $m_1, m_2$, the size of the major axis $A$ and $\sin i$ where $i$ is the inclination of the orbital plane with respect to the line of sight. Further, such an orbiting system will emit gravitational radiation and lose energy at a calculable rate, $\dot{E}$. This loss of energy will lead to a change in the period of the orbit given by $\dot{P}/P = 3(\dot{E}/2E)$. Since all the parameters of the system are known, one can now actually predict the value of $(\dot{P}/P)$ and then compare it with the observations. This exercise has been carried out with remarkable success for a pulsar PSR 1913+16. The predicted rate of change of the period for the star based on emission of gravitational radiation is $\dot{P} = -(2.40258 \pm 0.00004) \times 10^{-12}$ which should be compared with the observed value $\dot{P} = -(2.425 \pm 0.010) \times 10^{-12}$. They match quite well with a ratio being $(1.002 \pm 0.005)$ thereby providing an indirect confirmation of the existence of gravitational radiation.

## Exercises

**Exercise 5.1**   *Supernova in an ISM bubble:*

It is possible that the strong stellar winds preceding the supernova explosion of a star could have led to a density profile in the local ISM which has an approximate form $\rho_{\text{ISM}} \propto r^{-\beta}$ (so that $\rho_{\text{ISM}} \neq$ constant) around the star. Repeat the analysis of supernova blast wave propagation in such a medium and obtain the scaling relations. In particular, show that the Sedov phase now has $R \propto t^{2/(5-\beta)}$. This indicates that the shock wave accelerates for $\beta > 3$ and decelerates for $\beta < 3$. Give a physical reason for this behaviour.

**Exercise 5.2**   *Cutting it off:*

Consider a supernova remnant with an age of about $10^3$ yr and a magnetic field of $10^{-4}$ Gauss. Electrons with an *arbitrary* energy distribution (not necessarily a power law) are contributing to the synchrotron radiation from this remnant. Explain why the electron spectrum should have a sharp cut-off in energy at an energy scale which depends on time.

**Exercise 5.3**   *Dissecting Crab nebula:*

Given the parameters in the text for the Crab nebula: (i) Estimate (a) the energy density of synchrotron photons in the Crab nebula; (b) the ratio of inverse Compton cooling rate to synchrotron cooling rate for a relativistic electron in the Crab nebula. (ii) If the CMB photons undergo inverse Compton scattering with the relativistic electrons in Crab nebula, estimate the spectrum of the resulting gamma rays. Observations of gamma ray spectrum from the Crab suggest that $J \approx 3 \times 10^{-7} (E/1 \text{ TeV})^{-2.5}$ photons m$^{-2}$ s$^{-1}$ TeV$^{-1}$. If this arises due to the above process, what magnetic field strength does this result imply? How is the result modified when we take into account the inverse Compton scattering of the synchrotron photons?

**Exercise 5.4**   *Going too fast:*

If a neutron star spins too fast, it will start losing material from the equatorial region due to centrifugal force. Show that this implies a minimum period $P_{\text{min}} \propto M^{-1/2} R^{3/2}$ where $R$ and $M$ are the radius and mass of the neutron star. Estimate $P_{\text{min}}$ supplying all the numerical factors for $M = 1.4 M_\odot$, $R = 10$ km.

**Exercise 5.5**   *Neutron star collisions:*

Consider the collision of two neutron stars of equal mass obeying the mass radius relationship $RM^{1/3} =$ constant. Compute the binding energy which is released in the process, assuming the same form of the mass radius relationship holds for the final object. What do you expect the final object to be?

**Exercise 5.6** *Separating explosively:*

Supernova explosions often occur in binary systems resulting in substantial amount of mass being lost from the binary. Suppose that the initial mass of the two stars were $M_1$ and $M_2$ and that the star with larger mass lost a mass $\Delta M$ due to supernova explosion. Assume that this mass $\Delta M$ leaves the system instantaneously and exerts no further influence. Show that if the amount of mass lost is greater than half the total mass of the system, then the resulting system will become unbound. [Hint: The hard way to do this problem is to work out what happens to the center of mass after the explosion using momentum conservation and then compute the total energy of the final system to figure out when it becomes positive. Some miraculous algebraic cancellations occur leading to the above result. The cleverer method is to note that the energy of the binary can be written as $E = (1/2)M_{\text{tot}}V^2 + \mu(v^2/2) - (GM_{\text{tot}}\mu/r)$ where $M_{\text{tot}}$ is the total mass, $\mu$ is the reduced mass and $V$ is the velocity of the center of mass. For the initial circular orbit, the second term is $-(1/2)(GM_{\text{tot}}\mu/r)$. After the explosion $V^2$ is instantaneously unchanged while $\mu$ and $M_{\text{tot}}$ change. Take it from there.]

**Exercise 5.7** *Precession of perihelion:*

Since the equation of motion for the particle in Schwarzschild metric is not that in an inverse square potential, we do not expect the Runge-Lenz vector to be conserved. Hence, an elliptical orbit of the Newtonian gravity will precess due to the lowest order general relativistic correction. Obtain the rate of precession along the following lines: (a) Use the arguments given in the text to conclude that $r^2\dot{\theta}/\sqrt{g_{00}}$ is a constant. Hence obtain the equation to the orbit valid in the lowest order of approximation:

$$\frac{d^2u}{d\theta^2} + u = \frac{GMm^2}{L^2} + \frac{3GM}{c^2}u^2 \tag{5.69}$$

where $u = (1/r)$ and $L$ is a constant. Retaining only the first term on the right hand side gives the Newtonian elliptical orbit while the second term leads to the precession. Solving this equation approximately, show that the precession of the major axis for a low eccentricity, nearly circular, orbit is: $\delta\phi \approx 2\pi[1 - (6GM/c^2r_0)]^{-1/2}$ per orbit where $r_0$ is the radius of the original orbit.

**Exercise 5.8** *Black hole temperature:*

Obtain the results in Eq. (5.53) and Eq. (5.54) along the following lines. The first step is to change the integration variable to $z = \Omega \exp[-(\kappa t/c)]$. From the analytic properties of the integrand, show that it is legitimate to rotate the contour of integration along the $\text{Im}-z$ axis. The resulting integral is essentially a gamma function and will lead to Eq. (5.53). To get Eq. (5.54) from this result, you need $|\Gamma(iz)|^2$ which is equal to $(i/z)\Gamma(iz)\Gamma(1 - iz)$. Take it from there.

**Exercise 5.9** *Pseudo Newtonian potential:*

Some aspects of the motion in Schwarzschild metric can be mimicked by using an effective Newtonian potential $\phi_{\text{eff}} = -GM/(r - R_S)$ where $R_S = 2GM/c^2$. Compare the motion of a non relativistic particle under the action of this potential with the exact result in Schwarzschild metric.

**Exercise 5.10** *Spinning up the companion:*

When the disk accretion takes place around a central compact object, the accretion disk can exert a torque on the compact object and spin it up. The rate at which the angular momentum gets transfered from the Alfven radius $r_m$ (see Eq. (3.48)) is given by $\dot{m}r_m^2\omega(r_m) = I\dot{\omega}$. This will change the rotational period of the compact object at the rate $\dot{P}/P = -(\dot{\omega}P/2\pi)$. Show that this leads to a scaling law $\dot{P}/P \propto L^{6/7}P$ where $L$ is the accretion luminosity. The spin up will stop when the rotational velocity of the compact object is equal to the angular velocity of the disk at the Alfven radius. Show that the final period obeys the scaling $P_{\text{final}} \propto B^{6/7}\dot{m}^{-3/7}$.

**Exercise 5.11** *Losing mass is attractive?*

Consider two stars of mass $m_1$ and $m_2$ in which $m_1$ transfers mass to $m_2$ through filling the Roche lobe. This process conserves mass and angular momentum but not energy since the accretion dissipates energy into heat. Assume that the stars are on circular orbit of radius $R$ which is also changing with time due to accretion. From conservation of angular momentum and mass, show that $\dot{R}/R = 2(m_2 - m_1)(\dot{m}_2/m_1m_2)$. Hence show that if the lighter star is losing mass the stars gradually separate away while if the heavier star is losing mass the stars come closer catastrophically.

**Exercise 5.12** *What color is white dwarf?*

Suppose we plot the position of white dwarfs in a diagram giving the V-magnitude against the V-I colour. Estimate the slope of this line. How does it compare with the slope of the main sequence stars?

**Exercise 5.13** *Energy budget:*

Compare the total energy in the magnetic field and particles in Crab today with the original energy released in the supernova explosion and energy lost by the pulsar during its age.

**Exercise 5.14** *Supernova and intra cluster medium:*

The material between the galaxies in a galaxy cluster, usually called intra cluster medium (ICM for short), contains fair amount of metals and gas but, of course, no stars. The metallicity is about $Z_{\text{ICM}} \approx 0.3Z_\odot$ where $Z_\odot = 0.02$. (a) Assume that a typical supernova arises from a $20M_\odot$ star which leaves behind a $1.5M_\odot$

remnant and ejects the rest of its material of which about 2 $M_\odot$ is in the form of metals. Compute the metallicity $Z_{\rm SN}$ of the gas ejected from a single supernova. (b) The total intracluster mass can be written as $M_{\rm ICM} = M_{\rm ej} + M_p$ where $M_{\rm ej}$ is the gas from the supernova ejecta with metallicity $Z_{\rm SN}$ and $M_p$ is the zero metallicity primordial gas which fell in the potential well as the cluster formed. Using the values given so far, estimate the fraction $(M_{\rm ej}/M_{\rm ICM})$ of the gas that came from supernova. (c) The energy released by the supernova $E_{\rm SN} = 10^{51}$ erg will also go to heat up the gas. Compute the resulting temperature of the gas and compare it with the intra cluster temperature estimated from x-ray luminosity which is about (5-10) keV. What do you conclude? (d) If $M_{\rm ICM} \approx 6 \times 10^{13} M_\odot$, find the number of supernova needed to generate the metallicity of $0.3Z_\odot$.

## Notes and References

A text focussed on stellar remnants is Shapiro and Teukolsky (1983). Figure 5.1 is adapted from chapter 4 of Padmanabhan (2001).

1. *Section 5.2:* A detailed discussion of transport coefficients can be found in Landau and Lifshitz (1975c). White dwarfs are discussed in many books on stellar evolution; see, for example, Hansen and Kawaler (1994) and Kawaler *et. al.* (1997).

2. *Section 5.3:* For more detailed discussion see, Meszaros (1992); Michel (1991), Lyne Graham-Smith (1998) and Blandford *et al.* (1993).

3. *Section 5.4:* Black hole physics is covered in all good texts in general relativity, like, for example, Misner Thorne and Wheeler (1973).

4. *Section 5.5:* The evolution of binary stars is a major branch of study all by itself. Some text books which deal with it are Padmanabhan (2001), de Loore and Doom (1992), Frank King and Raine (1992). A nice discussion of PSR 1913+16 is available in the article by J.H. Taylor in Bondi and Weston-Smith (1998).

# Chapter 6

# Cosmology and the Early Universe

Having studied the physics of stars and stellar remnants, one might be tempted to proceed directly to the next level in astrophysical structures, viz., galaxies. But the formation (and evolution) of galaxies is so closely tied up with cosmological considerations that it is necessary to grapple with aspects of cosmology to understand galaxies in their totality. Keeping this in mind, we will devote this chapter to the physics of the evolving universe and take up the study of galaxies in the coming chapters.

## 6.1 Evolution of the universe

Observations suggest that the universe at large scales is homogeneous ('same at all points') and isotropic ('same in all directions'). One way to quantify the idea of homogeneity is the following: Suppose you place a sphere of radius $R$ at different locations in the universe and compute the mass (or, what is more relevant, the energy) contained within it. In a strictly uniform universe with density $\bar{\rho}$, you will always get the mass to be $\bar{M}(R) = (4\pi/3)\bar{\rho}R^3$. In a real universe, you will get different values when the sphere is centered at different points so that there will be a fractional rms fluctuation $\sigma(R) = \langle (M - \bar{M})^2 \rangle^{1/2}/\bar{M} \equiv (\delta M/M)_R$ around the mean mass. If this quantity decreases with $R$ and becomes sufficiently small for large $R$, then it makes sense to talk about a universe which is smooth at large scales with a mean density $\bar{\rho}$. In that case, if we average the inhomogeneities over a sufficiently large scale $R$, the fractional fluctuations will be small, making the average density a useful quantity. In our universe, $\sigma(R) \approx 0.1$ at $R \approx 50h^{-1}$ Mpc (where 1 Mpc $\approx 3 \times 10^{24}$ cm is a convenient unit for cosmological distances) and $\sigma(R)$ decreases roughly as $R^{-2}$ at larger scales.

You can, of course, do the same kind of observations with the density (rather than mass) and you will find that the fractional fluctuations $(\delta\rho/\rho)_R$ in the mass density (and energy density) $\rho$, within a randomly placed sphere of radius $R$, also decreases with $R$ as a power law. We will use the fractional contrast in the fluctuation of either mass or density interchangeably.

This suggests that we can model the universe as made up of a smooth background with an average density $\rho(t)$ — which is, at best, only a function of time and we have omitted the overbar on $\bar{\rho}$ to simplify the notation — superposed with fluctuations in the density $\delta\rho(t, \mathbf{x})$, which are large at small scales but decreases with scale. At sufficiently large scales, the universe may be treated as being homogeneous and isotropic with a uniform density.

Any large scale motion of matter in such a universe must also be consistent with homogeneity and isotropy. That is, all observers, located at different spatial points, must agree on the nature of the large scale motion. Of course, this condition is trivially satisfied if there is no large scale motion at all and all observers see the universe as static. But it is possible to come up with a more general form of the motion which has this feature. Suppose the velocity of an object located at a point $\mathbf{r}$ (with respect to some origin) is given by $\dot{\mathbf{r}}(t) = \mathbf{v}(t) = f(t)\mathbf{r}$. An observer at the origin will see all particles moving with a speed proportional to the distance from him/her. Consider now another observer at a different location $\mathbf{r}_1$. In the Newtonian picture, (s)he will see the object at $\mathbf{r}$ to move with the velocity $\mathbf{v}' = \mathbf{v} - \mathbf{v}_1 = f(t)(\mathbf{r} - \mathbf{r}_1)$ so that (s)he too will conclude that the particles are moving away with a speed proportional to the distance from him/her! So, at least in the Newtonian context of velocity addition, one can have a motion of the form $\dot{\mathbf{r}}(t) = \mathbf{v}(t) = f(t)\mathbf{r}$ that is consistent with homogeneity.

Integrating this relation, $\dot{\mathbf{r}} = f(t)\mathbf{r}$, we can describe the position $\mathbf{r}$ of any material body in the universe in the form $\mathbf{r} = a(t)\mathbf{x}$ where $a(t)$ is another arbitrary function related to $f(t)$ by $f(t) = (\dot{a}/a)$ and $\mathbf{x}$ is a constant for any given material body in the universe. It is conventional to call $\mathbf{x}$ and $\mathbf{r}$ the *comoving* and *proper* coordinates of the body and $a(t)$ the *expansion factor*. The dynamics of the universe is entirely determined by the function $a(t)$. The simplest choice will be $a(t) =$ constant, in which case there will be no motion in the universe and all matter will be distributed uniformly in a static configuration. It is, however, clear that such a configuration will be violently unstable when the mutual gravitational forces of the bodies are taken into account. Any such instability will eventually lead to random motion of particles in localized regions thereby destroying the initial homogeneity. Observations, however, indicate that this is not true and that the relation $\mathbf{v} = (\dot{a}/a)\mathbf{r}$ does hold in the observed universe with $\dot{a} > 0$.

In that case, the dynamics of $a(t)$ can be qualitatively understood along the following lines. Consider a particle of *unit* mass at the location $r$, with respect to some coordinate system. Equating the sum of its kinetic energy $v^2/2$ and gravitational potential energy $(-GM(r)/r)$ due to the attraction of matter inside a sphere of radius $r$, to a constant, one finds that $a(t)$ should satisfy the condition

$$\frac{1}{2}\dot{a}^2 - \frac{4\pi G\rho(t)}{3}\, a^2 = \text{ constant} \tag{6.1}$$

where $\rho$ is the mean density of the universe; that is,

$$\frac{\dot{a}^2}{a^2} + \frac{k}{a^2} = \frac{8\pi G}{3}\, \rho(t) \tag{6.2}$$

where $k$ is a constant.

While the arguments given above to determine this equation (as well as the relation $\mathbf{v} = f\mathbf{r})$) are fallacious, the results and — in particular, Eq. (6.2) — happens to be exact and arises from the application of general theory of relativity to a homogeneous and isotropic distribution of matter with $\rho$ interpreted as the energy density. We shall now briefly describe how these results arise in general relativity.

We saw in chapter 1 that general relativity describes gravity as a manifestation of spacetime curvature. In the case of a homogeneous and isotropic spacetime, the spacetime interval has a particularly simple form:

$$ds^2 = dt^2 - a^2(t)\left[\frac{dr^2}{1 - kr^2} + r^2(d\theta^2 + \sin^2\theta d\phi^2)\right] \tag{6.3}$$

where the parameter $k$ modifies the metric of 3-space and the function $a(t)$ scales the distances between any two points in the 3-dimensional space. One of the Einstein's equations reduces to Eq. (6.2) while the rest of the equations are satisfied if the equations of motion for matter are satisfied. When $k = 0$, life becomes particularly simple and the line interval reduces to

$$ds^2 = dt^2 - a^2(t)\left[dx^2 + dy^2 + dz^2\right]. \tag{6.4}$$

This is the form we will use quite often.

Having settled that Eq. (6.2) describes the dynamics of the universe, let us look at it more closely. Observations suggest that our universe today (at

$t = t_0$, say) is governed by Eq. (6.2) with

$$\left(\frac{\dot{a}}{a}\right)_0 \equiv H_0 = 0.3 \times 10^{-17} h \text{ s}^{-1} = 100 \, h \text{ km s}^{-1} \text{ Mpc}^{-1} \quad (6.5)$$

where $h \approx (0.6 - 0.8)$. We will always refer to quantities evaluated at the present epoch $(t = t_0)$ with a subscript zero. The relation $\mathbf{v} = f(t)\mathbf{r} = (\dot{a}/a)\mathbf{r} = H_0\mathbf{r}$ is known as *Hubble's law* and $H_0$ is called *Hubble's constant*. From $H_0$ we can form the time scale $t_{\text{univ}} \equiv H_0^{-1} \approx 10^{10} h^{-1}$ yr and the length scale $cH_0^{-1} \approx 3000 h^{-1}$ Mpc; $t_{\text{univ}}$ characterizes the evolutionary time scale of the universe and $cH_0^{-1}$ is of the order of the largest length scales currently accessible in cosmological observations. More importantly, one can also construct from $H_0$ a quantity with the dimensions of density called *critical density*:

$$\rho_c = \frac{3H_0^2}{8\pi G} = 1.88 h^2 \times 10^{-29} \text{ gm cm}^{-3} = 2.8 \times 10^{11} h^2 M_\odot \text{ Mpc}^{-3}$$
$$= 1.1 \times 10^4 h^2 \text{ eV cm}^{-3} = 1.1 \times 10^{-5} h^2 \text{ protons cm}^{-3}. \quad (6.6)$$

The last two "equalities" should be interpreted in terms of conversion of mass into energy by a factor $c^2$ and the conversion of mass into number of baryons by a factor $m_p^{-1}$ where $m_p$ is the proton mass. It is useful to measure all other mass and energy densities in the universe in terms of the critical density. If $\rho_i$ is the mass or energy density associated with a particular species of particles (say, photons, baryons *etc*....), then we define a density parameter $\Omega_i$ through the ratio $\Omega_i \equiv (\rho_i/\rho_c)$. More generally, $H(t) = (\dot{a}/a), \rho_i(t)$ and hence $\Omega_i(t)$ can be a function of time.

It is obvious from Eq. (6.2) that the numerical value of $k$ can be absorbed into the definition of $a(t)$ by rescaling it so that we can treat $k$ as having one of the three values $(0, -1, +1)$. The choice among these three values for $k$ is decided by Eq. (6.2) depending on the value of the total $\Omega \equiv \sum \Omega_i$; we see that $k = 1, 0$ or $-1$ depending on whether $\Omega$ greater than, equal or less than unity. The fact that $k$ is proportional to the total energy of the dynamical system described by Eq. (6.2) shows that $a(t)$ will have a maximum value followed by a contracting phase to the universe if $k = 1, \Omega > 1$.

The light emitted at an earlier epoch by an object will reach us today with the wavelength stretched due to the expansion. If the light was emitted at $a = a_e$ and received today (when $a = a_0$), the wavelength will change by the factor $(1 + z_e) = (a_0/a_e)$ where $z_e$ is called the *redshift* corresponding to the epoch of emission, $a_e$. Correspondingly, the frequencies will decrease with expansion as $\nu \propto (1+z)^{-1}$. Thus we associate a redshift $z(t)$ to any epoch $a(t)$ by $(1+z)^{-1} = a(t)/a_0$. It is very common in cosmology to use $z$ as a time coordinate to charatecterise an epoch in the past. Since the observed luminosity $L$ of a source is proportional to $(|\mathbf{p}_\gamma|c)d^3\mathbf{p}_\gamma \propto \nu^3 d\nu \propto (1+z)^{-4}$, where $p_\gamma = (\epsilon/c) = (h\nu/c)$ is the photon momentum, $L$ will decrease as $(1+z)^{-4}$ in an expanding universe.

To determine the nature of the cosmological model we need to determine the value of $\Omega$ for the universe, taking into account all forms of energy densities which exist at present. Further, to determine the form of $a(t)$ we need to determine how the energy density of any given species varies with time. Let us turn to this task.

Suppose a particular species contributes an energy density $\rho$ and pressure $p$. Then we need to integrate the equation of motion $d(\rho a^3) = -pd(a^3)$ (which is essentially $dE = -pdV$ with $V \propto a^3$) to determine how $\rho$ varies with $a$. The simplest type of equation of state which is adequate for describing the large scale dynamics of the universe is $p = w\rho$ with $w =$ constant. Then the equation $d(\rho a^3) = -w\rho d(a^3)$ can be immediately integrated to give $\rho \propto a^{-3(1+w)}$. Doing this for each component of energy density, Eq.(6.2) can be now written in the form

$$\frac{\dot{a}^2}{a^2} = H_0^2 \sum_i \Omega_i \left(\frac{a_0}{a}\right)^{3(1+w_i)} - \frac{k}{a^2} \qquad (6.7)$$

where each of these species is identified by density parameter $\Omega_i$ and the equation of state characterized by $w_i$.

The most familiar forms of energy densities are those due to *pressureless matter* or *radiation*. For pressureless matter — usually called 'dust' — we can take $w_i = 0$ (this represents non relativistic matter with the rest mass energy density $\rho c^2$ dominating over the kinetic energy density $(\rho v^2/2)$). On the other hand, for radiation or extreme relativistic particles, we have (see Sec. 3.1) $w_i = (1/3)$. Observational situation regarding the composition of our universe, however, suggests a more complex situation and can be summarised as follows:

- Our universe has $\Omega_{tot} \approx 1$ or, more precisely, $0.98 \lesssim \Omega_{tot} \lesssim 1.08$. The value of $\Omega_{tot}$ can be determined from the angular anisotropy spectrum of the Cosmic Microwave Background Radiation (CMBR), with the reasonable assumption that $h > 0.5$, and these observations (which we will discuss in Sec. 6.6) now show that we live in a universe with density close to the critical density.

- Observations of primordial deuterium produced in big bang nucleosynthesis (which took place when the universe was about 1 minute in age; see Sec. 6.2) as well as the CMBR observations show that the *total* amount of baryons in the universe contributes about $\Omega_B = (0.024 \pm 0.0012)h^{-2}$. Given the independent observations on the Hubble constant which fix $h = 0.72 \pm 0.07$, we conclude that $\Omega_B \cong 0.04 - 0.06$. These observations take into account *all* baryons which exist in the universe today irrespective of whether they are luminous or not. *When combined with the previous item we conclude that most of the universe is non-baryonic.*

- Host of observations related to large scale structure and dynamics (such as rotation curves of galaxies, estimate of cluster masses, gravitational lensing, galaxy surveys) all suggest that the universe is populated by a non-luminous component of matter (dark matter; DM hereafter) made of weakly interacting massive particles which *does* cluster at galactic scales. This component contributes about $\Omega_{DM} \cong 0.20 - 0.35$.

- Combining the last observation with the first we conclude that there must be (at least) one more component to the energy density of the universe contributing about 70% of critical density. Early analysis of several observations indicated that this component is unclustered and has *negative* pressure. This is confirmed dramatically by more recent supernova observations and analysis of CMBR data. The observations suggest that the missing component (called *dark energy*) has $w = p/\rho \lesssim -0.78$ and contributes $\Omega_{DE} \cong 0.60 - 0.75$.

  The simplest choice for dark energy is a 'fluid' with $p = -\rho$ such that $w = -1$. In this case, the equation $d(\rho a^3) = -pda^3$ is identically satisfied with $\rho = -p = $ constant! That is, a fluid with $w = -1$ will have the same energy density and pressure at all times as the universe expands. This is to be contrasted with the energy density of normal matter which decreases as the universe expands — because of the work done by the $pdV$ term with $p > 0$. When the pressure is negative, the decrease in the energy density due to expansion can be compensated by the negative work done thereby maintaining constant energy density. It is indeed possible to mimic such a fluid by adding a term to Einstein's equations called the *cosmological constant*. It is also possible that the energy of the vacuum state of the universe is non zero and exerts a gravitational influence. You will expect the vacuum to have an energy density and pressure which is constant in space and time; the equation $d(\rho a^3) = -p\,da^3$ then *demands* that it should have an equation of state $p = -\rho$. (Of course, one way of satisfying this is to have $\rho = p = 0$ but it also admits $\rho = -p=$ constant with any constant value.) Hence one often calls the dark energy arising from a fluid with equation of state $p = -\rho$ either as *cosmological constant* or *vacuum energy*. We shall use these expressions interchangeably.

- The universe also contains radiation (with $p = \rho/3$) contributing an energy density $\Omega_R h^2 = 2.56 \times 10^{-5}$ today most of which is due to photons in the CMBR. Assuming that most of the energy density is at temperature $T = 2.73$ K today, we get $\rho_R = (\pi^2/15)(k_B^4 T^4/c^3 \hbar^3)$. Dividing this by $\rho_c \simeq 1.88 \times 10^{-29} h^2 \text{g cm}^{-3}$ we find that $\Omega_R h^2 = 2.56 \times 10^{-5}$.

Taken together we conclude that our universe has (approximately) $\Omega_{DE} \simeq 0.7, \Omega_{DM} \simeq 0.26, \Omega_B \simeq 0.04, \Omega_R \simeq 5 \times 10^{-5}$. All known observations are consistent with such an — admittedly weird — composition for the universe.

As far as dynamics is concerned, we therefore need to deal with three different forms of energy densities in the universe: $\rho_{NR}, \rho_R$ and $\rho_V$. If neither particles nor photons are created or destroyed during the expansion, then the number density $n$ of particles or photons will decrease as $n \propto a^{-3}$ as $a$ increases. In the case of photons, the wavelength will also get stretched during expansion with $\lambda \propto a$; since the energy density of material particles is $nmc^2$ while that of photons of frequency $\nu$ will be $nh\nu = (nhc/\lambda)$, it follows that the energy densities of radiation and non relativistic matter vary as $\rho_R \propto a^{-4}$ and $\rho_{NR} \propto a^{-3}$. Since the vacuum energy density should be Lorentz invariant, it has to be a constant independent of space or time giving $\rho_V =$ constant. Of course, you get the same result by integrating $d(\rho a^3) = -p\,da^3 = -w\rho\,da^3$ to give $\rho \propto a^{-3(1+w)}$ and using $w = (0, 1/3, -1)$ for non relativistic matter, radiation and dark energy respectively.

Note that as we go to smaller $a(t)$ in the past, radiation energy density grows faster (as $\Omega_R a^{-4}$) compared to matter energy density (which grows as $\Omega_{NR} a^{-3}$). So even though radiation is dynamically irrelevant today, it would have been the dominant component in the universe at sufficiently small $a$: when $a < a_{eq} = a_0(\Omega_{NR}/\Omega_R)$. That is, at redshifts larger that $z_{eq} \simeq \Omega_{NR}/\Omega_R \simeq 4 \times 10^4 \Omega_{NR} h^2$. Further, combining $\rho_R \propto a^{-4}$ with the result $\rho_R \propto T^4$ for thermal radiation, it follows that any thermal spectrum of photons in the universe will have its temperature varying as $T \propto a^{-1}$.

In the past, when the universe was smaller, it would also have been: (i) denser (ii) hotter and — at sufficiently early epochs — (iii) dominated by radiation energy density since $(\rho_R/\rho_{NR}) \propto (1/a)$.

Given all these, the total energy density in the universe at any epoch can be expressed as

$$\rho_{\text{total}}(a) = \rho_R(a) + \rho_{NR}(a) + \rho_V(a)$$
$$= \rho_c \left[ \Omega_R \left(\frac{a_0}{a}\right)^4 + (\Omega_B + \Omega_{DM}) \left(\frac{a_0}{a}\right)^3 + \Omega_V \right] \qquad (6.8)$$

where $\rho_c$ and various $\Omega$'s refer to their values at $a = a_0$; $\Omega_B$ refers to the density parameter of baryons. Substituting this into Eq. (6.7) we get

$$\frac{\dot{a}^2}{a^2} + \frac{k}{a^2} = H_0^2 \left[ \Omega_R \left(\frac{a_0}{a}\right)^4 + \Omega_{NR} \left(\frac{a_0}{a}\right)^3 + \Omega_V \right] \qquad (6.9)$$

with $\Omega_{NR} = \Omega_B + \Omega_{DM}$. This equation can be cast in a more suggestive form. We can write $(k/a^2)$ as $(\Omega_{\text{tot}} - 1)H_0^2(a_0/a)^2$ and move it to the right hand side. Introducing a dimensionless time coordinate $\tau = H_0 t$ and writing $a = a_0 q(\tau)$ our equation becomes

$$\frac{1}{2} \left(\frac{dq}{d\tau}\right)^2 + V(q) = E \qquad (6.10)$$

where

$$V(q) = -\frac{1}{2} \left[ \frac{\Omega_R}{q^2} + \frac{\Omega_{NR}}{q} + \Omega_V q^2 \right]; \quad E = \frac{1}{2}(1 - \Omega_{\text{tot}}). \qquad (6.11)$$

Equation (6.10) has the structure of the first integral for motion of a particle with energy $E$ in a potential $V(q)$. For models with $\Omega_{\text{tot}} = \Omega_{NR} + \Omega_V + \Omega_R \cong \Omega_{NR} + \Omega_V = 1$, we can take $E = 0$ so that $(dq/d\tau) = \sqrt{-2V(q)}$. Figure 6.1 shows the velocity $(dq/d\tau)$ as a function of the position $q = (1 + z)^{-1}$ for such models. Several features are clear from this figure: (i) At high redshift (small q) the universe is radiation dominated (with $V(q) \simeq -(1/2)\Omega_R q^{-2}$) and $\dot{q}$ is independent of the other cosmological parameters; hence all the curves asymptotically approach each other at the left end of the figure. (ii) At lower redshifts $[\mathcal{O}(1) < z < 10^4]$ the universe is matter dominated and $V(q) \simeq -(1/2)\Omega_{NR} q^{-1}$. (iii) At still lower redshifts $[0 < z < \mathcal{O}(1)]$ the presence of cosmological constant makes a difference and – in fact – the velocity $\dot{q}$ changes from being a decreasing function to an increasing function. In other words, the presence of a cosmological constant leads to an *accelerating* universe at low redshifts.

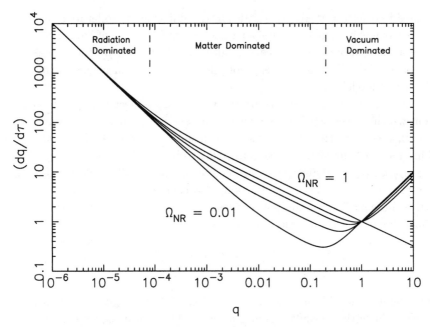

Fig. 6.1   The dynamical evolution of the universe depicted as a phase space trajectory. The vertical axis gives the rate of expansion of the universe (in rescaled units) and the horizontal axis gives the expansion factor of the universe (in rescaled units). The different curves are for different cosmological models with $\Omega_R = 2.56 \times 10^{-5} h^{-2}, h = 0.5, \Omega_{\mathrm{NR}} + \Omega_V = 1$. Curves are parameterized by the value of $\Omega_{\mathrm{NR}} = 0.01, 0.1, 0.3, 0.5, 1.0$ going from bottom to top as indicated.

Based on the above considerations, we can identify three distinct phases in the evolution of the universe depending on which form of the energy density dominates the expansion. At very early epochs, the radiation will dominate over other forms of energy density and Eq. (6.10) can be easily integrated to give $a(t) \propto t^{1/2}$. As the universe expands, a time will come when (at $t = t_{\mathrm{eq}}$ with $a = a_{\mathrm{eq}}$ and $z = z_{\mathrm{eq}}$, say) the matter energy density will be comparable to radiation energy density. From the known value of $\Omega_R$, it follows that $(1 + z_{\mathrm{eq}}) = (\Omega_{\mathrm{NR}}/\Omega_R) \simeq 3.9 \times 10^4 (\Omega_{\mathrm{NR}} h^2)$. Similarly, at very late epochs, the vacuum energy density (cosmological constant) will dominate over non relativistic matter and the universe will become "vacuum dominated". This occurs at a redshift of $z_v$ where $(1 + z_v) = (\Omega_v/\Omega_{\mathrm{NR}})^{1/3}$. For $\Omega_v \approx 0.7, \Omega_{\mathrm{NR}} \approx 0.3$, this occurs at $z_v \approx 0.33$. During $z_v \ll z \ll z_{\mathrm{eq}}$, the universe is matter dominated with $a \propto t^{2/3}$.

The discussion so far did not really use the fact that the geometry of the universe is curved. This fact becomes relevant when one tries to find expressions for observable quantities in the universe at moderate redshifts. Since the line interval of the universe is different from that of flat spacetime, several observable (geometrical) quantities are affected by the spacetime curvature. We shall now briefly consider a few of them concentrating on the $k = 0$ model.

Cosmological observations are mostly based on electromagnetic radiation which is received from far away sources. Let an observer, located at $r = 0$, receive at time $t = t_0$ radiation from a source located at $r = r_{em}$. This radiation must have been emitted at some earlier time $t_e$ such that the events $(t_e, r_{em})$ and $(t_0, 0)$ are connected by a null geodesic. Taking the propagation of the ray to be along $\theta =$ constant, $\phi =$ constant, we can write the equation for the null geodesic to be $dt^2 = a^2(t)dr^2$. Integrating this, we can find the relation between $r_{em}$ and $t_e$:

$$r_{em} = \int_{t_e}^{t_0} \frac{dt}{a(t)}. \qquad (6.12)$$

The right hand side is a definite function of $t_e$ in a given cosmological model. Since the redshift $z$ is also a unique function of $t_e$, we can express the right hand side as a function of $z$ and hence $r_{em}$ can be expressed as a function of $z$. This function $r_{em}(z)$ is of considerable use in observational astronomy, since it relates the radial distance to an object (from which the light is received) to the redshift at which the light is emitted. We see from Eq. (6.2) that Einstein's equations allow us to determine $(\dot{a}/a)$ more easily than $a(t)$ itself. It is, therefore, convenient to define a quantity (called *Hubble radius*) by

$$d_H(t) = d_H(z) \equiv \left(\frac{\dot{a}}{a}\right)^{-1} \qquad (6.13)$$

where the first equality reminds you that $d_H$ can also be thought of as a function of the redshift $z$. This quantity allows us to convert integrals over $dt$ to integrals over $dz$ by using

$$dt = \left(\frac{dt}{da}\right)\left(\frac{da}{dz}\right) dz = -d_H(z)\left(\frac{dz}{1+z}\right). \qquad (6.14)$$

It is now possible to write Eq. (6.12) as

$$r_{em} = \int_0^z d_H(z)dz \qquad (6.15)$$

if we normalize $a(t)$ by $a(t_0) = 1$. As an example, consider the matter dominated case in which $(\dot{a}/a) = H_0\sqrt{\Omega_{\text{NR}}}a^{-3/2}$. Then $d_H = H_0^{-1}\Omega_{\text{NR}}^{-1/2}(1 + z)^{-3/2}$ and

$$r_{\text{em}}(z) = 2H_0^{-1}\Omega_{\text{NR}}^{-1/2}\left[1 - \frac{1}{\sqrt{1+z}}\right]. \qquad (6.16)$$

Two important observable quantities in which $r_{\text{em}}(z)$ plays a role are in: (i) relating the luminosity of distant objects with the observed flux and (ii) in the measurement of angular sizes of distant objects which we shall now discuss. Let $\mathcal{F}$ be the flux received from a source of luminosity $\mathcal{L}$ when the photons from the source reach us with a redshift $z$. The flux can be expressed as

$$\mathcal{F} = \frac{1}{(\text{Area})}\left(\frac{dE_{\text{rec}}}{dt_{\text{rec}}}\right) = \frac{1}{(\text{Area})}\frac{1}{(1+z)^2}\left(\frac{dE_{\text{em}}}{dt_{\text{em}}}\right) \qquad (6.17)$$

since $dE_{\text{rec}} = (1+z)^{-1}dE_{\text{em}}$ and $dt_{\text{rec}} = (1+z)dt_{\text{em}}$ due to redshift. The proper area at $a = a_0 = 1$ of a sphere of comoving radius $r_{\text{em}}(z)$, is $4\pi r_{\text{em}}^2$, giving

$$\mathcal{F} = \frac{1}{4\pi r_{\text{em}}^2}\frac{1}{(1+z)^2}\left(\frac{dE_{\text{em}}}{dt_{\text{em}}}\right) = \frac{1}{4\pi r_{\text{em}}^2}\frac{1}{(1+z)^2}\mathcal{L} \qquad (6.18)$$

where $\mathcal{L} = (dE_{\text{em}}/dt_{\text{em}})$ is the luminosity of the source. Since the distance to an object of luminosity $\mathcal{L}$ and flux $\mathcal{F}$ can be expressed as $(\mathcal{L}/4\pi\mathcal{F})^{1/2}$ in flat Euclidean geometry, it is convenient to define distance $d_L(z)$ (called *luminosity distance*) by

$$d_L(z) \equiv \left(\frac{\mathcal{L}}{4\pi\mathcal{F}}\right)^{1/2} = r_{\text{em}}(z)(1+z) = (1+z)\int_0^z d_H(z)dz. \qquad (6.19)$$

From Eq. (6.15) and Eq. (6.19) we have the inverse relation

$$d_H(z) = \frac{d}{dz}\left[\frac{d_L(z)}{1+z}\right]. \qquad (6.20)$$

Thus if $d_L(z)$ can be determined from observations, one can determine $d_H(z)$ and thus $(\dot{a}/a)$. This is the key to supernova observations which determined that the universe is accelerating in its expansion in the recent past.

Another observable parameter for distant sources is the angular diameter. If $D$ is the physical size of the object which subtends an angle $\delta$ to the observer, then, for small $\delta$, we have $D = r_{\text{em}}a(t_e)\delta$. The "angular diameter distance" $d_A(z)$ for the source is defined via the relation $\delta = (D/d_A)$; so we have:

$$d_A(z) = r_{\rm em} a(t_e) = \frac{1}{1+z} \int_0^z d_H(z) dz. \qquad (6.21)$$

Comparing Eq. (6.19) and Eq. (6.21) we see that $d_L = (1+z)^2 d_A$. For example, in the matter dominated case,

$$d_A = \frac{2H_0^{-1}\Omega_{\rm NR}^{-1/2}}{(1+z)} \left[ 1 - \frac{1}{\sqrt{1+z}} \right] \simeq 2H_0^{-1}\Omega_{\rm NR}^{-1/2} z^{-1} \qquad (6.22)$$

with second equality being valid at large $z$. We will need this result in Sec. 6.6. Also note that $d_A(z)$ is not monotonic in $z$. It increases as $d_A \propto z$ for small $z$, reaches a maximum and decreases as $z^{-1}$ at large $z$.

Finally we note that the proper volume in the universe spanned by the region between comoving radii $r_{\rm em}$ and $r_{\rm em} + dr_{\rm em}$ can also be expressed in terms of $d_H(z)$ and $r_{\rm em}(z)$. This infinitesimal proper volume element $dV = \sqrt{^3g}\, d^3\mathbf{x}$ is given by

$$dV = a^3\, r_{\rm em}^2\, dr_{\rm em}\, \sin\theta\, d\theta\, d\phi. \qquad (6.23)$$

If we interpret this volume as the volume of the universe spanned along the backward light cone by the photons which are receive today with redshifts between $z$ and $z + dz$, then we can express $dr_{\rm em}$ in terms of $dt$ using Eq. (6.12) and express $dt$ in terms of $dz$ using Eq. (6.14) and obtain

$$dV = a^2 dt \left( r_{\rm em}^2 \sin\theta d\theta d\phi \right) = \frac{r_{\rm em}^2(z) d_H(z)}{(1+z)^3} dz \sin\theta d\theta d\phi. \qquad (6.24)$$

This is used extensively in observational cosmology.

## 6.2    Primordial nucleosynthesis

Let us now try to figure out the key events in the life of such a universe. When the temperature of the universe was higher than the temperature corresponding to the atomic ionisation energy, the matter content in the universe will be a high temperature plasma. For $t < t_{\rm eq}$, in the radiation dominated phase, we can also ignore the contribution of non-relativistic particles to $\rho$. We have seen earlier that, during the radiation dominated phase, $a(t) \propto t^{1/2}$; therefore

$$\left(\frac{\dot{a}}{a}\right)^2 = H^2(t) = \frac{1}{4t^2} = \frac{8\pi G}{3}\rho = \frac{8\pi G}{3}g\left(\frac{\pi^2}{30}\right)T^4. \tag{6.25}$$

The last equality follows from the expression for energy density for radiation (see Eq. (2.20)) in units with $\hbar = c = k_B = 1$. The factor $g$ in these expressions counts the degrees of freedom of those particles which are *still* relativistic at this temperature $T$. As the temperature decreases, more and more particles will become non-relativistic and $g$ will decrease; thus $g = g(T)$ is a slowly decreasing function of $T$. Numerically,

$$t \cong 0.3g^{-\frac{1}{2}}\left(\frac{m_{\text{Pl}}}{T^2}\right) \simeq 1s\left(\frac{T}{1\text{MeV}}\right)^{-2}g^{-\frac{1}{2}}. \tag{6.26}$$

In the standard models for particle interactions, $g \simeq 10^2$ at $T \gtrsim 1\text{GeV}$; $g \simeq 10$ for $T \simeq (100 - 1)\text{MeV}$ and $g \simeq 3$ for $T < 0.1\text{MeV}$. These relations give the temperature of the universe as a function of time.

When the temperature of the universe is higher than the binding energy of the nuclei ($\sim$ MeV), none of the heavy elements (helium and the metals) could have existed in the universe. The binding energies of the first four light nuclei, $^2H$, $^3H$, $^3He$ and $^4He$ are 2.22 MeV, 6.92 MeV, 7.72 MeV and 28.3 MeV respectively. This would suggest that these nuclei could be formed when the temperature of the universe is in the range of $(1 - 30)\text{MeV}$. The actual synthesis takes place only at a much lower temperature, $T_{\text{nuc}} = T_n \simeq 0.1\text{MeV}$. The main reason for this delay is the 'high entropy' of our universe, i.e., the high value for the photon-to-baryon ratio, $\eta^{-1}$. Numerically,

$$\eta = \frac{n_B}{n_\gamma} = 5.5 \times 10^{-10}\left(\frac{\Omega_B h^2}{0.02}\right). \tag{6.27}$$

We have already seen a similar effect in chapter 3 Sec. 3.4.1 in the case of ionisation equilibrium, in which this effect was described by Eq. (3.75). Let us assume, for a moment, that the nuclear (and other) reactions are fast enough to maintain thermal equilibrium between various species of particles and nuclei. In thermal equilibrium, the number density of a nuclear species $^AN_Z$ with atomic mass $A$ and charge $Z$ will be

$$n_A = g_A\left(\frac{m_A T}{2\pi}\right)^{3/2}\exp\left[-\left(\frac{m_A - \mu_A}{T}\right)\right]. \tag{6.28}$$

By repeating exactly the same kind of analysis we did for ionisation equilibrium for this case, one can obtain the equation for the temperature $T_A$

at which the mass fraction of a particular species-A will be of order unity $(X_A \simeq 1)$

$$T_A \simeq \frac{B_A/(A-1)}{\ln(\eta^{-1}) + 1.5\ln(m_B/T)} \qquad (6.29)$$

where $B_A$ is the binding energy of the species. (This is analogous to Eq. (3.75).) This temperature will be fairly lower than $B_A$ because of the large value of $\eta^{-1}$. For $^2H$, $^3He$ and $^4He$ the value of $T_A$ is 0.07MeV, 0.11MeV and 0.28MeV respectively. Comparison with the binding energy of these nuclei shows that these values are lower than the corresponding binding energies $B_A$ by a factor of about 10, at least.

Thus, even when the thermal equilibrium is maintained, significant synthesis of nuclei can occur only at $T \lesssim 0.3$MeV and not at higher temperatures. If such is the case, then we would expect significant production $(X_A \lesssim 1)$ of nuclear species-A at temperatures $T \lesssim T_A$. It turns out, however, that the rate of nuclear reactions is *not* high enough to maintain thermal equilibrium between various species. We have to determine the temperatures up to which thermal equilibrium can be maintained and redo the calculations to find non-equilibrium mass fractions.

The general procedure for studying non equilibrium abundances in an expanding universe is based on *rate equations*. Since we will require this formalism again in Section 6.3 (for the study of recombination), let us spend a little time and develop it in a somewhat general context.

Consider a reaction in the early universe in which two particles 1 and 2 interact to form two other particles 3 and 4. For example, $n + \nu_e \rightleftharpoons p + e$ constitutes one such reaction which converts neutrons into protons in the forward direction and protons into neutrons in the reverse direction; another example you will come across in the next section is $p + e \rightleftharpoons H + \gamma$ where the forward reaction describes recombination of electron and proton forming a neutral hydrogen atom (with the emission of a photon), while the reverse reaction is the photoionisation of a hydrogen atom. In general, we are interested in how the number density $n_1$ of particle species 1, say, changes due to a reaction of the form $1 + 2 \rightleftharpoons 3 + 4$.

We first note that even if there is no reaction, the number density will change as $n_1 \propto a^{-3}$ due to the expansion of the universe; so what we are really after is the change in $n_1 a^3$. Further, the forward reaction will be proportional to the product of the number densities $n_1 n_2$ while the reverse reaction will be proportional to $n_3 n_4$. Hence we can write an equation for the rate of change of particle species $n_1$ as

$$\frac{1}{a^3}\frac{d(n_1 a^3)}{dt} = \mu(An_3 n_4 - n_1 n_2). \tag{6.30}$$

The left hand side is the relevant rate of change over and above that due to the expansion of the universe; on the right hand side, the two proportionality constants have been written as $\mu$ and $(A\mu)$, both of which, of course, will be functions of time. The quantity $\mu$ has the dimensions of $cm^3 s^{-1}$, so that $n\mu$ has the dimensions of $s^{-1}$; usually $\mu \simeq \sigma v$ where $\sigma$ is the cross-section for the relevant process and $v$ is the relative velocity. The left hand side has to vanish when the system is in thermal equilibrium with $n_i = n_i^{eq}$, where the superscript 'eq' denotes the equilibrium densities for the different species labeled by $i = 1 - 4$. This condition allows us to rewrite $A$ as $A = n_1^{eq} n_2^{eq}/(n_3^{eq} n_4^{eq})$. Hence the rate equation becomes

$$\frac{1}{a^3}\frac{d(n_1 a^3)}{dt} = \mu n_1^{eq} n_2^{eq}\left(\frac{n_3 n_4}{n_3^{eq} n_4^{eq}} - \frac{n_1 n_2}{n_1^{eq} n_2^{eq}}\right). \tag{6.31}$$

In the left hand side, one can write $(d/dt) = Ha(d/da)$ which shows that the relevant time scale governing the process is $H^{-1}$. Clearly, when $H/n\mu \gg 1$ the right hand side becomes ineffective because of the $(\mu/H)$ factor and the number of particles of species 1 does not change. You see that when the expansion rate of the universe is large compared to the reaction rate, the given reaction is ineffective in changing the number of particles — which is what you will expect. This certainly does *not* mean that the reactions have reached thermal equilibrium and $n_i = n_i^{eq}$; in fact, it means exactly the opposite: The reactions are not fast enough to drive the number densities towards equilibrium densities and the number densities "freeze out" at non equilibrium values. Of course, the right hand side will also vanish when $n_i = n_i^{eq}$ which is the other extreme limit of thermal equilibrium.

Having taken care of the general formalism, let us now apply it to the process of nucleosynthesis which requires protons and neutrons combining together to form bound nuclei of heavier elements like deuterium, helium *etc.*. The abundance of these elements are going to be determined by the relative abundance of neutrons and protons in the universe. Therefore, we need to first worry about the maintenance of thermal equilibrium between protons and the neutrons in the early universe. As long as the inter-conversion between $n$ and $p$ through the weak interaction processes $(\nu + n \leftrightarrow p + e)$, $(\bar{e} + n \leftrightarrow p + \bar{\nu})$ and the 'decay' $(n \leftrightarrow p + e + \bar{\nu})$, is rapid (compared to the expansion rate of the universe), thermal equilibrium will be maintained. Then the equilibrium $(n/p)$ ratio will be

$$\left(\frac{n_n}{n_p}\right) = \frac{X_n}{X_p} = \exp(-Q/T), \tag{6.32}$$

where $Q = m_n - m_p = 1.293$ MeV. Therefore, at high $(T \gg Q)$ temperatures, there will be equal number of neutrons and protons but as the temperature drops below about 1.3 MeV, the neutron fraction will start dropping exponentially provided thermal equilibrium is still maintained. To check whether thermal equilibrium is indeed maintained, we need to compare the expansion rate with the reaction rate. The expansion rate is given by $H = (8\pi G\rho/3)^{1/2}$ where $\rho = (\pi^2/30)gT^4$ with $g \approx 10.75$ representing the effective relativistic degrees of freedom present at these temperatures. At $T = Q$, this gives $H \approx 1.1$ s$^{-1}$. The reaction rate needs to be computed from weak interaction theory. The neutron to proton conversion rate, for example, is well approximated by

$$\lambda_{np} \approx 0.29 \text{ s}^{-1} \left(\frac{T}{Q}\right)^5 \left[\left(\frac{Q}{T}\right)^2 + 6\left(\frac{Q}{T}\right) + 12\right]. \tag{6.33}$$

At $T = Q$, this gives $\lambda \approx 5$ s$^{-1}$, slightly more rapid than the expansion rate. But as $T$ drops below $Q$, this decreases rapidly and the reaction ceases to be fast enough to maintain thermal equilibrium. Hence we need to work out the neutron abundance by using Eq. (6.31).

Using $n_1 = n_n, n_3 = n_p$ and $n_2, n_4 = n_l$ where the subscript $l$ stands for the leptons, Eq. (6.31) becomes

$$\frac{1}{a^3}\frac{d(n_n a^3)}{dt} = \mu n_l^{\text{eq}}\left(\frac{n_p n_n^{\text{eq}}}{n_p^{\text{eq}}} - n_n\right). \tag{6.34}$$

We now use Eq. (6.32), write $(n_l^{\text{eq}}\mu) = \lambda_{np}$ which is the rate for neutron to proton conversion and introduce the fractional abundance $X_n = n_n/(n_n + n_p)$. Simple manipulation then leads to the equation

$$\frac{dX_n}{dt} = \lambda_{np}\left((1 - X_n)e^{-Q/T} - X_n\right). \tag{6.35}$$

Converting from the variable $t$ to the variable $s = (Q/T)$ and using $(d/dt) =$

$-HT(d/dT)$, the equations we need to solve reduce to

$$-Hs\frac{dX_n}{ds} = \lambda_{np}\left((1 - X_n)e^{-s} - X_n\right); \quad H = (1.1\ \text{sec}^{-1})\ s^{-4};$$

$$\lambda_{np} = \frac{0.29\ \text{s}^{-1}}{s^5}\left[s^2 + 6s + 12\right]. \tag{6.36}$$

It is now straightforward to integrate these equations numerically and determine how the neutron abundance changes with time. The results (of a more exact treatment) are shown in Fig. 6.2. The neutron fraction falls out of equilibrium when temperatures drop below 1 MeV and it freezes to about 0.15 at temperatures below 0.5 MeV.

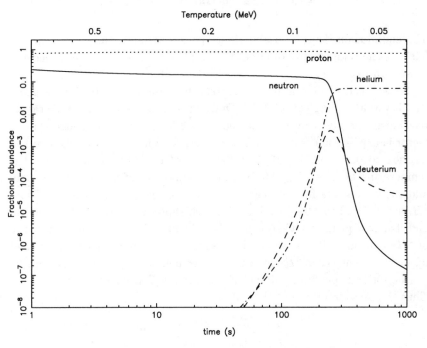

Fig. 6.2   Primordial abundance of elements as a function of time. (Data for the figure: Courtesy Jim Kneller.)

As the temperature decreases further, the neutron decay with a half life of $\tau_n \approx 886.7$ sec (which is *not* included in the above analysis) becomes important and starts depleting the neutron number density. The only way neutrons can survive is through the synthesis of light elements. As the temperature falls further to $T = T_{\text{He}} \simeq 0.28\text{MeV}$, significant amount of He could have been produced if the nuclear reaction rates were high enough. The possible reactions which produces $^4He$ are $[D(D, n)\ ^3He(D, p)\ ^4He,$ $D(D, p)\ ^3H(D, n)\ ^4He, D(D, \gamma)\ ^4He]$. These are all based on $D$, $^3He$ and $^3H$ and do not occur rapidly enough because the mass fraction of $D$, $^3He$ and $^3H$ are still quite small $[10^{-12}, 10^{-19}$ and $5 \times 10^{-19}$ respectively] at $T \simeq 0.3$MeV. The reactions $n + p \rightleftharpoons d + \gamma$ will lead to an equilibrium abundance ratio of deuterium given by

$$\frac{n_p n_n}{n_d n} = \frac{4}{3} \left( \frac{m_p m_n}{m_d} \right)^{3/2} \frac{(2\pi k_B T)^{3/2}}{(2\pi\hbar)^3 n} e^{-B/k_B T}$$

$$= \exp \left[ 25.82 - \ln \Omega_B h^2 T_{10}^{3/2} - \left( \frac{2.58}{T_{10}} \right) \right]. \qquad (6.37)$$

The equilibrium deuterium abundance passes through unity (for $\Omega_B h^2 = 0.02$) at the temperature of about 0.07 MeV which is when the nucleosynthesis can really begin.

So we need to determine the neutron fraction at $T = 0.07$ MeV given that it was about 0.15 at 0.5 MeV. During this epoch, the time-temperature relationship is given by $t = 130$ sec $(T/0.1 \text{ MeV})^{-2}$. The neutron decay factor is $\exp(-t/\tau_n) \approx 0.74$ for $T = 0.07$ MeV. This decreases the neutron fraction to $0.15 \times 0.74 = 0.11$ at the time of nucleosynthesis. When the temperature becomes $T \lesssim 0.07$MeV, the abundance of $D$ and $^3H$ builds up and these elements further react to form $^4He$. A good fraction of $D$ and $^3H$ is converted to $^4He$ (See Fig.6.2 which shows the growth of deuterium and its subsequent fall when helium is built up). The resultant abundance of $^4He$ can be easily calculated by assuming that almost all neutrons end up in $^4He$. Since each $^4He$ nucleus has two neutrons, $(n_n/2)$ helium nuclei can be formed (per unit volume) if the number density of neutrons is $n_n$. Thus the mass fraction of $^4He$ will be

$$Y = \frac{4(n_n/2)}{n_n + n_p} = \frac{2(n/p)}{1 + (n/p)} = 2x_c \qquad (6.38)$$

where $x_c = n/(n+p)$ is the neutron abundance at the time of production of deuterium. For $\Omega_B h^2 = 0.02$, $x_c \approx 0.11$ giving $Y \approx 0.22$. Increasing baryon density to $\Omega_B h^2 = 1$ will make $Y \approx 0.25$. An accurate fitting formula for the dependence of helium abundance on various parameters is given by

$$Y = 0.226 + 0.025 \log \eta_{10} + 0.0075(g_* - 10.75) + 0.014(\tau_{1/2}(n) - 10.3 \text{ min})$$
$$(6.39)$$

where $\eta_{10}$ measures the baryon-photon ratio today via the relation

$$\Omega_B h^2 = 3.65 \times 10^{-3} \left( \frac{T_0}{2.73 \text{ K}} \right)^3 \eta_{10} \tag{6.40}$$

and $g_*$ is the effective number of relativistic degrees of freedom contributing to the energy density and $\tau_{1/2}(n)$ is the neutron half life.

As the reactions converting $D$ and $^3H$ to $^4He$ proceed, the number density of $D$ and $^3H$ is depleted and the reaction rates - which are proportional to $\Gamma \propto X_A(\eta n_\gamma) < \sigma v >$ - become small. These reactions soon freeze-out leaving a residual fraction of $D$ and $^3H$ (a fraction of about $10^{-5}$ to $10^{-4}$). Since $\Gamma \propto \eta$ it is clear that the fraction of $(D,^3H)$ left unreacted will decrease with $\eta$. In contrast, the $^4He$ synthesis - which is not limited by any reaction rate - is fairly independent of $\eta$ and depends only on the $(n/p)$ ratio at $T \simeq 0.1\text{MeV}$. The best fits, with typical errors, to deuterium abundance calculated from the theory, for the range $\eta = (10^{-10} - 10^{-9})$ is given by

$$Y_2 \equiv \left( \frac{D}{H} \right)_p = 3.6 \times 10^{-5 \pm 0.06} \left( \frac{\eta}{5 \times 10^{-10}} \right)^{-1.6}. \tag{6.41}$$

The production of still heavier elements - even those like $^{16}C$, $^{16}O$ which have *higher* binding energies than $^4He$ - is suppressed in the early universe. Two factors are responsible for this suppression: (1) For nuclear reactions to proceed, the participating nuclei must overcome their Coulomb repulsion. The probability to tunnel through the Coulomb barrier is governed by the factor $F = \exp[-2A^{1/3} (Z_1 Z_2)^{2/3}(T/1\text{MeV})^{-1/3}]$ where $A^{-1} = A_1^{-1} + A_2^{-1}$. For heavier nuclei (with larger $Z$), this factor suppresses the reaction rate. (2) Reaction between helium and proton would have led to an element with atomic mass 5 while the reaction of two helium nuclei would have led to an element with atomic mass 8. However, there are no stable elements in the periodic table with the atomic mass of 5 or 8! The $^8Be$, for example, has a half life of only $10^{-16}$ seconds. One can combine $^4He$ with $^8Be$ to produce $^{12}C$ but this can occur at significant rate only if it is a resonance reaction. That is, there should exist an excited state $^{12}C$ nuclei which has an energy close to the interaction energy of $^4He + {}^8Be$. Stars, incidentally, use this route to synthesize heavier elements. It is this triple-alpha reaction which allows the synthesis of heavier elements in stars but it is not fast enough in the early universe. (You must thank your stars that there is no such resonance in $^{16}O$ or in $^{20}Ne$ — which is equally important for the survival of carbon and oxygen.)

The current observations indicate, with reasonable certainty that: (i) $(D/H) \gtrsim 1 \times 10^{-5}$. (ii) $[(D +^3 He)/H] \simeq (1 - 8) \times 10^{-5}$ and (iii) $0.236 < (^4He/H) < 0.254$. These observations are consistent with the predictions if $10.3 \, \text{min} \lesssim \tau \lesssim 10.7 \, \text{min}$, and $\eta = (3 - 10) \times 10^{-10}$. Using $\eta = 2.68 \times 10^{-8}\Omega_B h^2$, this leads to the important conclusion: $0.011 \leq \Omega_B h^2 \leq 0.037$. When combined with the broad bounds on $h$, $0.6 \lesssim h \lesssim 0.8$, say, we can constrain the baryonic density of the universe to be: $0.01 \lesssim \Omega_B \lesssim 0.06$. These are the typical bounds on $\Omega_B$ available today. It shows that, if $\Omega_{\text{total}} \simeq 1$ then most of the matter in the universe must be non baryonic.

Since the $^4He$ production depends on $g$, the observed value of $^4He$ restricts the total energy density present at the time of nucleosynthesis. In particular, it constrains the number $(N_\nu)$ of light neutrinos (that is, neutrinos with $m_\nu \lesssim 1\text{MeV}$ which would have been relativistic at $T \simeq 1\text{MeV}$). The observed abundance is best explained by $N_\nu = 3$, is barely consistent with $N_\nu = 4$ and rules out $N_\nu > 4$. The laboratory bound on the total number of particles including neutrinos, which couples to the $Z^0$ boson is determined by measuring the decay width of the particle $Z^0$; each particle with mass less than $(m_z/2) \simeq 46$ GeV contributes about 180 MeV to this decay width. This bound is $N_\nu = 2.79 \pm 0.63$ which is consistent with the cosmological observations.

## 6.3 Decoupling of matter and radiation

In the early hot phase, the radiation will be in thermal equilibrium with matter; as the universe cools below $k_B T \simeq (\epsilon_a/10)$ where $\epsilon_a$ is the binding energy of atoms, the electrons and ions will combine to form neutral atoms and radiation will decouple from matter (see Eq. (3.75)). This occurs at $T_{\text{dec}} \simeq 3 \times 10^3$ K. As the universe expands further, these photons will continue to exist without any further interaction. It will retain thermal spectrum since the redshift of the frequency $\nu \propto a^{-1}$ is equivalent to changing the temperature in the spectrum by the scaling $T \propto (1/a)$. It turns out that the major component of the extragalactic background light (EBL) which exists today is in the microwave band and can be fitted very accurately by a thermal spectrum at a temperature of about 2.73 K. It seems reasonable to interpret this radiation as a relic arising from the early, hot, phase of the evolving universe. This relic radiation, called *cosmic microwave background radiation*, turns out to be a gold mine of cosmological information and is extensively investigated in recent times. We shall now discuss some details

related to the formation of neutral atoms and the decoupling of photons.
The relevant reaction is, of course, $e + p \rightleftharpoons H + \gamma$ and if the rate of this reaction is faster than the expansion rate, then one can calculate the neutral fraction using Saha's equation developed in chapter 3, Sec. 3.4.1. Introducing the fractional ionisation, $X_i$, for each of the particle species and using the facts $n_p = n_e$ and $n_p + n_H = n_B$, it follows that $X_p = X_e$ and $X_H = (n_H/n_B) = 1 - X_e$. Saha's equation Eq. (3.73) now gives

$$\frac{1 - X_e}{X_e^2} \cong 3.84\eta(T/m_e)^{3/2}\exp(B/T) \tag{6.42}$$

where $\eta = 2.68 \times 10^{-8}(\Omega_B h^2)$ is the baryon-to-photon ratio. We may define $T_{\text{atom}}$ as the temperature at which 90 percent of the electrons, say, have combined with protons: i.e. when $X_e = 0.1$. This leads to the condition:

$$(\Omega_B h^2)^{-1}\tau^{-\frac{3}{2}}\exp\left[-13.6\tau^{-1}\right] = 3.13 \times 10^{-18} \tag{6.43}$$

where $\tau = (T/1\text{eV})$. For a given value of $(\Omega_B h^2)$, this equation can be easily solved by iteration. Taking logarithms and iterating once we find $\tau^{-1} \cong 3.084 - 0.0735\ln(\Omega_B h^2)$ with the corresponding redshift $(1 + z) = (T/T_0)$ given by

$$(1 + z) = 1367[1 - 0.024\ln(\Omega_B h^2)]^{-1}. \tag{6.44}$$

For $\Omega_B h^2 = 1, 0.1, 0.01$ we get $T_{\text{atom}} \cong 0.324\text{eV}, 0.307\text{eV}, 0.292\text{eV}$ respectively. These values correspond to the redshifts of $1367, 1296$ and $1232$.

Because the preceding analysis was based on equilibrium densities, it is important to check that the rate of the reactions $p + e \leftrightarrow H + \gamma$ is fast enough to maintain equilibrium. For $\Omega_B h^2 \approx 0.02$, the equilibrium condition is only marginally satisfied, making this analysis suspect. More importantly, the direct recombination to the ground state of the hydrogen atom — which was used in deriving the Saha's equation — is not very effective in producing neutral hydrogen in the early universe. The problem is that each such recombination releases a photon of energy 13.6 eV which will end up ionizing another neutral hydrogen atom which has been formed earlier. As a result, the direct recombination to the ground state does not change the neutral hydrogen fraction at the lowest order. Recombination through the excited states of hydrogen is more effective since such a recombination ends up emitting more than one photon each of which has an energy less than 13.6 eV. Given these facts, it is necessary to once again use the rate equation developed in the previous section to track the evolution of ionisation fraction. A simple procedure for doing this which captures the essential physics is as follows:

We again begin with Eq. (6.31) and repeating the analysis done in the last section, now with $n_1 = n_e, n_2 = n_p, n_3 = n_H$ and $n_4 = n_\gamma$, and defining $X_e = n_e/(n_e + n_H) = n_p/n_H$ one can easily derive the rate equation for this case:

$$\frac{dX_e}{dt} = [\beta(1 - X_e) - \alpha n_b X_e^2] = \alpha\left(\frac{\beta}{\alpha}(1 - X_e) - n_b X_e^2\right). \qquad (6.45)$$

This equation is analogous to Eq. (6.35); the first term gives the photoionisation rate which produces the free electrons and the second term is the recombination rate which converts free electrons into hydrogen atom and we have used the fact $n_e = n_b X_e$ etc.. Since we know that direct recombination to the ground state is not effective, the recombination rate $\alpha$ is the rate for capture of electron by a proton forming an excited state of hydrogen. To a good approximation, this rate is given by

$$\alpha = 9.78 r_0^2 c \left(\frac{B}{T}\right)^{1/2} \ln\left(\frac{B}{T}\right) \qquad (6.46)$$

where $r_0 = e^2/m_e c^2$ is the classical electron radius. To integrate Eq. (6.45) we also need to know $\beta/\alpha$. This is easy because in thermal equilibrium the right hand side of Eq. (6.45) should vanish and Saha's equation tells us the value of $X_e$ in thermal equilibrium. On using Eq. (6.42), this gives

$$\frac{\beta}{\alpha} = \left(\frac{m_e T}{2\pi}\right)^{3/2} \exp[-(B/T)]. \qquad (6.47)$$

We can now integrate Eq. (6.45) using the variable $B/T$ just as we used the variable $Q/T$ in solving Eq. (6.35). The result shows that the actual recombination proceeds more slowly compared to that predicted by the Saha's equation. The actual fractional ionisation is higher than the value predicted by Saha's equation at temperatures below about 1300. For example, at $z = 1300$, these values differ by a factor 3; at $z \simeq 900$, they differ by a factor of 200. The value of $T_{atom}$, however, does not change significantly. A more rigorous analysis shows that, in the redshift range of $800 < z < 1200$, the fractional ionisation varies rapidly and is given (approximately) by the formula,

$$X_e = 2.4 \times 10^{-3} \frac{(\Omega_{NR} h^2)^{1/2}}{(\Omega_B h^2)} \left(\frac{z}{1000}\right)^{12.75}. \qquad (6.48)$$

This is obtained by fitting a curve to the numerical solution.

The formation of neutral atoms makes the photons decouple from the matter. The redshift for decoupling can be determined as the epoch at which the optical depth for photons is unity. Using Eq. (6.48), we can compute the optical depth for photons to be

$$\tau = \int_0^t n(t) X_e(t) \sigma_T dt = \int_o^z n(z) X_e(z) \sigma_T \left(\frac{dt}{dz}\right) dz \simeq 0.37 \left(\frac{z}{1000}\right)^{14.25}$$

(6.49)

where we have used the relation $H_0 dt \cong -\Omega_{NR}^{-1/2} z^{-5/2} dz$ which is valid for $z \gg 1$ (see Eq. (6.14)). This optical depth is unity at $z_{\text{dec}} = 1072$. (Alternatively, one can compare the scattering rate of photons $n_e \sigma_T = X_e n_b \sigma_T$ with the expansion rate $H$. This calculation shows that $n_e \sigma_T / H$ is of order unity when $X_e \approx 10^{-2}$ if $\Omega_B h^2 = 0.02$ and $\Omega_{NR} = 0.3$ leading to substantially the same conclusion.) From the optical depth, we can also compute the probability that the photon was last scattered in the interval $(z, z + dz)$. This is given by $(\exp -\tau) (d\tau/dz)$ which can be expressed as

$$P(z) = e^{-\tau} \frac{d\tau}{dz} = 5.26 \times 10^{-3} \left(\frac{z}{1000}\right)^{13.25} \exp\left[-0.37 \left(\frac{z}{1000}\right)^{14.25}\right].$$

(6.50)

This $P(z)$ has a sharp maximum at $z \simeq 1067$ and a width of about $\Delta z \cong 80$. It is therefore reasonable to assume that decoupling occurred at $z \simeq 1070$ in an interval of about $\Delta z \simeq 80$. We shall see later that the finite thickness of the surface of last scattering has important observational consequences.

## 6.4 Formation of dark matter halos

Having discussed the life history of the homogeneous universe, we shall now turn to the issue of formation of different structures like galaxies, clusters etc in the universe.

A totally uniform universe, of course, will never lead to any of the inhomogeneous structures seen today. However, if the universe had even slightest inhomogeneity, then gravitational instability can amplify the density perturbations. To see how this comes about in the simplest context, let us consider a universe with a single component of matter with the equation of state $p = w\rho$. The $w = 0$ case is the most important but we shall have occasion to use others as well; so we will keep a general $w$. Integrating $d(\rho a^3) = -p da^3$, we find that the background density evolves as $\rho \propto a^{-3(1+w)}$, Further, differentiating Eq. (6.2), we get $\ddot{a} = -(4\pi G/3)(1 + 3w)\rho a$ which has the solution $a = (t/t_0)^{2/(3+3w)}$ describing the background universe. We now perturb $a(t)$ slightly to $a(t) + \delta a(t)$ so that the corresponding fractional density perturbation is $\delta \equiv (\delta\rho/\rho) \propto (\delta a/a)$. Using the form of the background solution, it is easy to show that the perturbation $\delta a$ satisfies the equation

$$t^2 \frac{d^2\delta a}{dt^2} = m\delta a; \qquad m = \frac{2}{9}\frac{(1 + 3w)(2 + 3w)}{(1 + w)^2}. \tag{6.51}$$

This equation has power law solutions $\delta a \propto t^p$ with $p(p - 1) = m$. Solving the quadratic equation and choosing the growing mode, we find that the density contrast grows as

$$\delta \propto \frac{\delta a}{a} \propto t^\alpha; \qquad \alpha = \frac{2}{3}\frac{(1 + 3w)}{(1 + w)}. \tag{6.52}$$

This result can be expressed more concisely. It can be directly verified that this expression scales exactly as $(\rho a^2)^{-1}$ where $\rho$ is the background density. Thus we get the result that

$$\delta \propto (\rho a^2)^{-1}. \tag{6.53}$$

In the matter dominated universe $\rho \propto a^{-3}$ and the density perturbation grows as $\delta \propto a$.

When the perturbations have grown sufficiently, their self gravity will start dominating and the matter can collapse to form a gravitationally bound system. The dark matter will form virialised, gravitationally bound, structures with different masses and radii. The physics of such dark matter halos is somewhat easier to understand since they are not affected by gas dynamical processes. In contrast, the baryonic matter will cool by radiating energy, sink to the centers of the dark matter halos and will form galaxies. The formation of galaxies, which is more complex, will be discussed in the next chapter. But to discuss properly the features of a scenario for the formation of dark matter halos, we need a model that will describe the *nonlinear* epochs of the collapse. We will develop that model first.

Let us consider a spherically symmetric region of radius, $R(t)$, which has a density $\rho$ that is higher than the background density $\rho_b$. In a spherically symmetric context, the matter outside this region does not influence the evolution of this over dense region. The radius $R(t)$ will grow at a rate slightly slower than the background expansion (because of the over-density inside), reach a maximum value and contract back again. This model is particularly simple for the $\Omega = 1$, matter dominated universe, in which we have to solve the equation

$$\frac{d^2 R}{dt^2} = -\frac{GM}{R^2}. \tag{6.54}$$

The density contrast of non relativistic matter in this region will be given by:

$$1 + \delta = \frac{\rho}{\rho_b} = \frac{3M}{4\pi R^3(t)} \frac{1}{\rho_b(t)} = \frac{2GM}{H_0^2} \left[\frac{a(t)}{R(t)}\right]^3. \tag{6.55}$$

Solving Eq. (6.54), the evolution of a spherical over-dense region can now be summarized by the following relations:

$$R(t) = A(1 - \cos\theta), \tag{6.56}$$

$$t = B(\theta - \sin\theta); \qquad A^3 = GMB^2. \tag{6.57}$$

Substituting into Eq. (6.55) and using $H_0^2 t_0^2 = 4/9$ we find that the density and density contrast evolve as:

$$\rho(t) = \rho_b(t) \frac{9(\theta - \sin\theta)^2}{2(1 - \cos\theta)^3}, \tag{6.58}$$

$$\delta = \frac{9}{2} \frac{(\theta - \sin\theta)^2}{(1 - \cos\theta)^3} - 1. \tag{6.59}$$

Given the solution in Eq. (6.56), Eq. (6.57) we can equate the energy of the motion to the initial kinetic energy of expansion, $K_i = (1/2)H_i^2 r_i^2$ and thus fix $A$ and $B$. We get, for $\delta_i \ll 1$, the result $A \simeq r_i/2\delta_i$, $B = 3t_i/4\delta_i^{3/2}$. For small $t$, this leads to $\delta = (3/5)\delta_i(t/t_i)^{2/3}$ which — in turn — fixes the overall normalisation. We can now express $\delta$ and $\theta$ in terms of the redshift by using the relation $(t/t_i)^{2/3} = (1 + z_i)(1 + z)^{-1}$. This gives

$$(1 + z) = \left(\frac{4}{3}\right)^{2/3} \frac{\delta_i(1 + z_i)}{(\theta - \sin\theta)^{2/3}} \equiv \left(\frac{5}{3}\right) \left(\frac{4}{3}\right)^{2/3} \frac{\delta_0}{(\theta - \sin\theta)^{2/3}}, \tag{6.60}$$

where $\delta_0 \equiv (3/5)\delta_i(1 + z_i)$ is the linearly extrapolated density contrast today, starting from $\delta = \delta_i$ at $z = z_i$. Given an initial density contrast $\delta_i$ at redshift $z_i$, these equations define (implicitly) the function $\delta(z)$ for $z > z_i$. Equation Eq. (6.60) defines $\theta$ in terms of $z$ (implicitly); Eq. (6.59) gives the density contrast at that $\theta(z)$.

For comparison, note that linear evolution (with $\delta_L \propto a \propto t^{2/3}$) obtained earlier (from Eq. (6.53) for $\rho \propto a^{-3}$) leads to a density contrast $\delta_L$ where

$$\delta_L = \frac{\bar{\rho}_L}{\rho_b} - 1 = \frac{3}{5}\frac{\delta_i(1+z_i)}{1+z} = \frac{3}{5}\left(\frac{3}{4}\right)^{2/3}(\theta - \sin\theta)^{2/3}. \qquad (6.61)$$

We can estimate the accuracy of the linear theory by comparing $\delta(z)$ and $\delta_L(z)$. To begin with, for $z \gg 1$, we have $\theta \ll 1$ and we get $\delta(z) \simeq \delta_L(z)$. When $\theta = (\pi/2)$, $\delta_L = (3/5)(3/4)^{2/3}(\pi/2 - 1)^{2/3} = 0.341$ while $\delta = (9/2)(\pi/2 - 1)^2 - 1 = 0.466$; thus the actual density contrast is about 40 percent higher. When $\theta = (2\pi/3)$, $\delta_L = 0.568$ and $\delta = 1.01 \simeq 1$. If we interpret $\delta = 1$ as the transition point to nonlinearity, then such a transition occurs at $\theta = (2\pi/3)$, $\delta_L \simeq 0.57$. From Eq. (6.60), we see that this occurs at the redshift $(1 + z_{nl}) = 1.06\delta_i(1 + z_i) = (\delta_0/0.57)$.

The spherical region reaches the maximum radius of expansion at $\theta = \pi$. From our equations, we find that the redshift $z_m$, the proper radius of the shell $r_m$ and the average density contrast $\delta_m$ at 'turn-around' are:

$$(1 + z_m) = \frac{\delta_i(1 + z_i)}{\pi^{2/3}(3/4)^{2/3}} = 0.57(1 + z_i)\delta_i$$

$$= \frac{5}{3}\frac{\delta_0}{(3\pi/4)^{2/3}} \simeq \frac{\delta_0}{1.062},$$

$$r_m = \frac{3x}{5\delta_0}, \quad \left(\frac{\bar{\rho}}{\rho_b}\right)_m = 1 + \bar{\delta}_m = \frac{9\pi^2}{16} \approx 5.6. \qquad (6.62)$$

The first equation gives the redshift at turn-around for a region, parametrized by the (hypothetical) linear density contrast $\delta_0$ extrapolated to the present epoch. If, for example, $\delta_i \simeq 10^{-3}$ at $z_i \simeq 10^4$, such a perturbation would have turned around at $(1 + z_m) \simeq 5.7$ or when $z_m \simeq 4.7$. The second equation gives the maximum radius reached by the perturbation. The third equation shows that the region under consideration is nearly six times denser than the background universe, at turn-around. This corresponds to a density contrast of $\delta_m \approx 4.6$ which is definitely in the nonlinear regime. The linear evolution gives $\delta_L = 1.063$ at $\theta = \pi$.

After the spherical overdense region turns around it will continue to contract. Equation (6.58) suggests that at $\theta = 2\pi$ all the mass will collapse to a point. However, long before this happens, the approximation that matter is distributed in spherical shells and that random velocities of the particles are small, (implicit in the assumption of homogeneous over dense region) will break down. The collisionless (dark matter) component will relax to a configuration with radius $r_{\text{vir}}$, velocity dispersion $v$ and density $\rho_{\text{coll}}$. After virialization of the collapsed shell, the potential energy $U$ and the kinetic energy $K$ will be related by $|U| = 2K$ so that the total energy $\mathcal{E} = U + K = -K$. At $t = t_m$ all the energy was in the form of potential energy. For a spherically symmetric system with constant density, $\mathcal{E} \approx -3GM^2/5r_m$. The 'virial velocity' $v$ and the 'virial radius' $r_{\text{vir}}$ for the collapsing mass can thus be estimated by the equations:

$$K \equiv \frac{Mv^2}{2} = -\mathcal{E} = \frac{3GM^2}{5r_m}; \quad |U| = \frac{3GM^2}{5r_{\text{vir}}} = 2K = Mv^2. \qquad (6.63)$$

We get:

$$v = (6GM/5r_m)^{1/2}; \quad r_{\text{vir}} = r_m/2. \qquad (6.64)$$

The time taken for the fluctuation to reach virial equilibrium, $t_{\text{coll}}$, is essentially the time corresponding to $\theta = 2\pi$. From equation Eq. (6.60), we find that the redshift at collapse, $z_{\text{coll}}$, is

$$(1 + z_{\text{coll}}) = \frac{\delta_i(1 + z_i)}{(2\pi)^{2/3}(3/4)^{2/3}} = 0.36\delta_i(1 + z_i) = 0.63(1 + z_m) = \frac{\delta_0}{1.686}. \qquad (6.65)$$

The density of the collapsed object can also be determined fairly easily. Since $r_{\text{vir}} = (r_m/2)$, the mean density of the collapsed object is $\rho_{\text{coll}} = 8\rho_m$ where $\rho_m$ is the density of the object at turn-around. We have, $\rho_m \cong 5.6\rho_b(t_m)$ and $\rho_b(t_m) = (1 + z_m)^3 \, (1 + z_{\text{coll}})^{-3}\rho_b(t_{\text{coll}})$. Combining these relations, we get

$$\rho_{\text{coll}} \simeq 2^3\rho_m \simeq 44.8\rho_b(t_m) \simeq 170\rho_b(t_{\text{coll}}) \simeq 170\rho_0(1 + z_{\text{coll}})^3 \qquad (6.66)$$

where $\rho_0$ is the present cosmological density. This result determines $\rho_{\text{coll}}$ in terms of the redshift of formation of a bound object. Once the system has virialised, its density and size does not change. Since $\rho_b \propto a^{-3}$, the density *contrast* $\delta$ increases as $a^3$ for $t > t_{\text{coll}}$. The evolution is described schematically in Fig. 6.3.

Such virialised, gravitationally bound, structures — once formed — will remain frozen at a mean density $\bar{\rho}$ which is about $f_c \simeq 200$ times the background density of the universe at the redshift of formation, $z$. Taking the background density of the universe at redshift $z$ to be $\rho_{\text{bg}}(z) = \rho_c\Omega_{\text{NR}}(1 + z)^3$, the mean density $\bar{\rho}$ of an object which would have collapsed at redshift $z$ is given by $\bar{\rho} \simeq \Omega_{\text{NR}}\rho_c f_c(1+z)^3$. We define the *circular velocity*

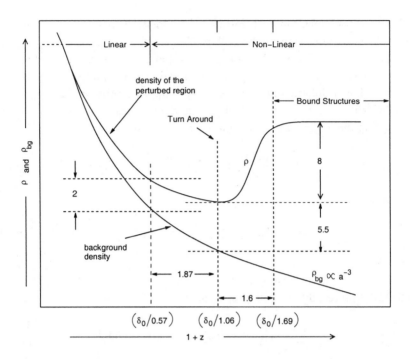

Fig. 6.3  Modelling the nonlinear growth of the density contrast using the dynamics of a spherical overdense region. The two curves indicate the density evolution of a spherical overdense region and the background universe.

$v_c$ for such a collapsed body to be,

$$v_c^2 \equiv \frac{GM}{r} \equiv \frac{4\pi G}{3}\,\bar{\rho}r^2. \tag{6.67}$$

Eliminating $\bar{\rho}$ in terms of $v_c$, the redshift of formation of an object can be expressed in the form

$$(1+z) \cong 5.8 \left(\frac{200}{\Omega_{\mathrm{NR}}\,f_c}\right)^{1/3} \frac{\left(v_c/200 \text{ km s}^{-1}\right)^{2/3}}{\left(r/h^{-1} \text{ Mpc}\right)^{2/3}}. \tag{6.68}$$

It is interesting that such a fairly elementary calculation leads to an acceptable result regarding the redshift for the formation of first structures. If one considers small scale halos (about a few kpc), the formation redshift can go up to, say, 20. This calculation also introduces the notion of *hierarchical clustering* in which smaller scales go non-linear and virialise earlier on and the merging of these smaller structures leads to hierarchically bigger and bigger structures. Of course, the process is supplemented by the larger scales going non-linear by themselves but the importance of mergers cannot be ignored in galaxy formation scenarios.

The description so far concentrated on a single spherical patch in the universe which was overdense. In practice, we are more interested in describing the *statistical* properties of the structures in the universe rather in the dynamics of individual patches. At any given moment of time, the actual density of matter $\rho(t, \mathbf{x})$ will differ from the averaged out density $\rho_{bg}(t)$ and the density contrast $\delta(t, \mathbf{x}) \equiv (\rho(t, \mathbf{x}) - \rho_{bg})/\rho_{bg}$ provides a description of the inhomogeneous universe. To switch from this to a statistical description, it is better to concentrate on the spatial Fourier transform $\delta_{\mathbf{k}}(t)$ of the density contrast $\delta(t, \mathbf{x})$. We will assume that the density fluctuation $\delta_{\mathbf{k}}(t)$ is just one particular realization of a random processes. Then, one can define the power spectrum of fluctuations at a given wavenumber $k$ by $P(k, t) = < |\delta_{\mathbf{k}}(t)|^2 >$ where the averaging symbol denotes that we are treating $P(k, t)$ as a statistical quantity averaged over an ensemble of possibilities; statistical isotropy of the universe implies that this power spectrum can only depend on the magnitude $|\mathbf{k}|$ of the wave number. This quantity provides a suitable statistical measure of the fluctuations in the universe which we shall use extensively. The power per logarithmic band in $k$ will be given by

$$\Delta_k^2(t) = \frac{k^3 \langle |\delta_k(t)|^2 \rangle}{2\pi^2} = \frac{k^3 P(k, t)}{2\pi^2}. \tag{6.69}$$

(Quite often we will just use $P(k, t) = |\delta_k(t)|^2$ without bothering to show the averaging symbol $\langle \cdots \rangle$ explicitly.) For smoothly varying power spectrum, this quantity is related to the mean square fluctuation in density (or mass) at the scale $R \approx k^{-1}$ in the universe by

$$\Delta_k^2 = \left( \frac{\delta\rho}{\rho} \right)_{R \simeq k^{-1}}^2 = \left( \frac{\delta M}{M} \right)_{R \simeq k^{-1}}^2 \cong \sigma^2(R, t). \tag{6.70}$$

If we approximate the power spectrum of fluctuations in the universe by a power law in $k$ locally then $P(k) \propto k^n$ at any given time. When the perturbation is small, ($\delta \ll 1$) we can use the result derived above, $\delta \propto a$, to obtain

$$\Delta_k^2(t) \propto a^2 k^{n+3}; \qquad \sigma^2(R, t) \propto a^2 R^{-(n+3)} \tag{6.71}$$

as long as $\sigma \ll 1$ with $n$ being a slowly varying function of scale $k$ or $R$.

When $\sigma(R,t) \to 1$ that particular scale characterized by $R$ will go non linear and matter at that scale will collapse and form a bound structure. Since this occurs when the density contrast $\sigma$ reaches some critical value $\sigma_c \approx 1$, it follows from Eq. (6.71) that the scale which goes non linear at any given time $t$ in the past (corresponding to a redshift $z$) obeys the relation

$$R_{\mathrm{NL}}(t) \propto a(t)^{2/(n+3)} = R_{\mathrm{NL}}(t_0)(1+z)^{-2/(n+3)}. \qquad (6.72)$$

Equivalently, structures with mass $M \propto R_{\mathrm{NL}}^3$ will form at a redshift $z$ where

$$M_{\mathrm{NL}}(z) = M_{\mathrm{NL}}(t_0)(1+z)^{6/(n+3)}. \qquad (6.73)$$

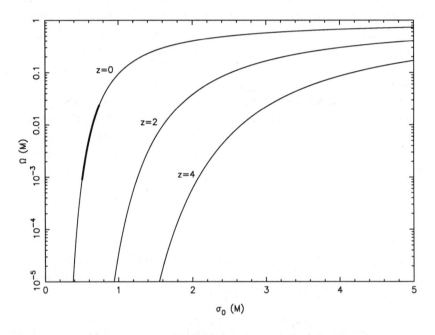

Fig. 6.4   The fraction of the critical density $\Omega(M)$ contributed by collapsed halos with mass $M$, plotted against the density contrast at that mass scale, at different redshifts. Given the abundance of structures at different redshifts, one can constrain the $\sigma$ at that mass scale. The abundance of massive clusters is over plotted on the $z = 0$ curve to illustrate this procedure.

It is, in fact, possible to work out many properties of collapsed structures from the above formalism. To do this, let us consider a spherical region of radius $R$ which had very small mean fluctuation $\delta_i$ at some initial time $t = t_i$. The $\sigma(M, t_i)$ in Eq. (6.71) denotes the *typical* value of density fluctuations which exists in the universe at a time $t_i$. In a statistical description, we could think of regions having a density contrast, say, $\nu$ times larger than the typical value $\sigma(M, t)$ occurring with a probability $\mathcal{P}(\nu)$ which — in turn — could be approximated as a Gaussian of unit variance in most models at early times. Later on, the fluctuation grows, turns around, virialises and forms a bound structure. We know that this happens when the density contrast, *computed using linear theory*, becomes $\delta_c \approx 1.68$. Let $\delta_i(t, t_i)$ be the density contrast needed at time $t_i$ so that $\delta = \delta_c = 1.68$ by the time $t$. For the $\Omega = 1$, matter dominated universe, $\delta_i(t, t_i) = \delta_c(t_i/t)^{2/3}$. As a first approximation, we may assume that a region with $\delta > \delta_i$ $(t, t_i)$ (when smoothed on the scale $R$ at time $t_i$) will form a gravitationally bound object with mass $M \propto \bar{\rho} R^3$ by the time $t$. Therefore, the fraction of bound objects with mass greater than $M$ will be

$$f_{\text{coll}}(> M, t) \propto \int_{\delta_i(t, t_i)}^{\infty} P(\delta, R, t_i) d\delta \qquad (6.74)$$

$$\propto \frac{1}{\sqrt{2\pi}} \frac{1}{\sigma(R, t_i)} \int_{\delta_i(t, t_i)}^{\infty} \exp\left(-\frac{\delta^2}{2\sigma^2(R, t_i)}\right) d\delta \propto \text{erfc}\left(\frac{\delta_i(t, t_i)}{\sqrt{2}\sigma(R, t_i)}\right)$$

where $\text{erfc}(x)$ is the complementary error function. This expression can be written in a more convenient form using

$$\frac{\delta_i(t, t_i)}{\sigma(R, t_i)} = \frac{\delta_c(t_i/t)^{2/3}}{\sigma(R, t_i)} = \frac{\delta_c}{\sigma_L(R, t)} = \frac{\delta_c(1 + z)}{\sigma_0(M)} \qquad (6.75)$$

where $\sigma_L(R, t) \equiv \sigma(R, t_i)(t/t_i)^{2/3}$ is the *linear* density contrast at $t$ and $\sigma_0(M)$ is its value at present $t = t_0$. We can express it either in terms of the mass $M = (4\pi/3)R^3\bar{\rho}$ or in terms of $R$ depending on the convenience. The overall normalization of $f_{coll}$ needs to be done by hand, so that the integral of $f_{\text{coll}}(M)$ over all $M$ is unity. (There are some subtleties about normalizing this expression, which we shall not go into.) Then we get, in terms of redshift,

$$f_{\text{coll}}(> M, z) = \text{erfc}\left[\frac{\delta_c(1 + z)}{\sqrt{2}\sigma_0(M)}\right]. \qquad (6.76)$$

This magical formula (called the *Press-Schechter* formula) gives you nonlinear results for the price of linear effort! Differentiating Eq. (6.76), one can compute the number of halos with mass between $M$ and $M + dM$ as well as the fraction of the critical density $\Omega(M)$ contributed by halos in this mass range. You can easily convince yourself that $\Omega(M)$ has a fairly

simple expression (since you are now differentiating what you originally got
by integrating a Gaussian!):

$$\Omega(M) = \left| \frac{d \ln \sigma}{d \ln M} \right| \sqrt{\frac{2}{\pi}} \, \nu \, \exp\left( -\frac{\nu^2}{2} \right) \qquad (6.77)$$

where $\nu = \delta_c(1+z)/\sigma(M)$. If you want the baryon density in the collapsed
halos, you only need to multiply this expression by $\Omega_B/\Omega_{\rm NR}$. Figure 6.4
gives $\Omega(M)$ for a more realistic universe with $\Omega_{\rm NR} = 0.35, \Omega_V = 0.65$ with
a more precise linear evolution numerically coded in.

An interesting application of this result is to constrain cosmological
models by the known abundance of classes of objects like e.g. galaxy clusters. From the fact that typical cluster has a mass of about $5 \times 10^{14} M_\odot$
and an abundance of about $4 \times 10^{-6} h^3$ Mpc$^{-3}$, one can work out the $\Omega$
contributed by the clusters to be about $8 \times 10^{-3}$. Because of various uncertainties in these measurements, this could vary anywhere between, say
$8 \times 10^{-4}$ to $2.4 \times 10^{-2}$, that is, by a factor 30. This range of $\Omega$ at $z = 0$ is
also shown in the Fig. 6.4 by a thick line superposed on the $z = 0$ curve.
You will notice that, because of the steep rise of the $z = 0$ curve, the observed range of $\Omega$ maps to a very tiny range in $\sigma \approx (0.5 - 0.8)$ ! In other
words, at the scale corresponding to the mass of the clusters, the density
contrast must be in this range for consistency with observations. Since a
mass of $10^{14} M_\odot$ corresponds to a radius of about $8 \, h^{-1}$ Mpc with the current cosmological density, one often states this result by saying that at the
length scale of $8 \, h^{-1}$ Mpc the density contrast should be in the range 0.5 to
0.8 — or, more simply, that $\sigma_8 = (0.5 - 0.8)$. You will hear a lot about $\sigma_8$
in the study of structure formation since it is a fairly powerful constraint.

## 6.5   Generation of initial perturbations

For the cosmological paradigm described in the previous sections to work,
we need to have a mechanism to generate small perturbations in the early
universe. This turns out to be more non trivial than one would have at
first imagined, because of the following reason.

Consider a perturbation at a proper wavelength $\lambda$ which will expand
with the universe as $\lambda(t) \propto a(t) = \lambda_0 a(t)$ where $\lambda_0$ is the wavelength at the
present epoch at which $a = a_0 = 1$. We need to compare this length scale
— which characterizes the inhomogeneity — with the characteristic length
scale for the expansion of the universe. The latter, of course, is the Hubble

radius, given by $d_H(t) = (\dot{a}/a)^{-1}$. If the universe is matter or radiation dominated, then $a(t) \propto t^n$ with $n < 1$. In this case, $\lambda(t) \propto t^n$ will be *bigger* than $d_H(t) \propto t$ at sufficiently small $t$.

This result leads to a major difficulty in conventional cosmology. Normal physical processes can act coherently only over length scales smaller than the Hubble radius. Thus any physical process leading to density perturbations at some early epoch, $t = t_i$, could only have operated at scales smaller than $d_H(t_i)$. But most of the relevant astrophysical scales (corresponding to clusters, groups, galaxies, *etc.*) were much bigger than $d_H(t)$ at sufficiently early epochs. Thus, if we want the seed perturbations to have originated in the early universe, then it is difficult to understand how any physical process could have generated them. To tackle this difficulty, we must arrange matters such that at sufficiently small $t$, we have $\lambda(t) < d_H(t)$. If this can be done, then physical processes can lead to an initial density perturbation.

We will first show that $d_H(t)$ will rise faster than $a(t)$ as $t \to 0$ if the source of expansion of the universe has positive energy density and positive pressure. Thus the above problem *cannot* be tackled in any Friedmann model with $\rho > 0$ and $p > 0$ at all times. To see this, we begin by noting that as $t \to 0$, we can ignore the $k/a^2$ term in Eq. (6.2) and write

$$d_H^2 = \frac{a^2}{\dot{a}^2} = \frac{3}{8\pi G\rho}. \tag{6.78}$$

So

$$\frac{d\ln d_H}{dt} = -\frac{1}{2}\frac{d\ln\rho}{dt} = \frac{3}{2}\left(1 + \frac{p}{\rho}\right)\frac{d\ln a}{dt} \tag{6.79}$$

where we have used the relation $d(\rho a^3) = -p\,da^3$. The proper wavelengths scale as $\lambda \propto a(t)$ giving $d(\ln\lambda) = d(\ln a)$. Combining with Eq. (6.79), we get

$$\frac{d\ln d_H}{d\ln\lambda} = \frac{3}{2}\left(1 + \frac{p}{\rho}\right). \tag{6.80}$$

For $(p/\rho) > 0$, the right hand side is greater than unity. So $d_H$ grows as $\lambda^n$ with $n > 1$ during the early epochs of evolution. Of course, $n$ itself could vary with time if $(p/\rho)$ is not a constant; but we will always have the condition $n > 1$. It follows that $(d_H/\lambda) < 1$ for sufficiently small $a$ proving that the Hubble radius will be smaller than the wavelength of the perturbation at *sufficiently* early epochs (as $a \to 0$), provided $(p/\rho) > 0$.

Hence, in order to generate density perturbations from a region which was smaller than Hubble radius, it is necessary to have models with either $\rho$ or $p$ negative. But to satisfy the Friedmann equation with $k = 0$ we need $\rho > 0$; thus the only models in which we can have $d_H > \lambda$ for small $t$ are the ones for which $p < 0$.

One simple way of achieving this is to make $a(t)$ increase rapidly with $t$ (for example, exponentially or as $a \propto t^n$ with $n \gg 1$) for a brief period of time. Such a rapid growth is called *inflation*. In conventional models of inflation, the energy density during the inflationary phase is provided by a scalar field with a potential $V(\phi)$. If the potential energy dominates over the kinetic energy, such a scalar field can act like an ideal fluid with the equation of state $p = -\rho$. Since the field is assumed to be inherently quantum mechanical, it will also have characteristic quantum fluctuations. It is possible for these quantum fluctuations to eventually manifest as classical density perturbations. The situation is illustrated schematically in Fig. 6.5 which shows the Hubble radius $d_H = (\dot{a}/a)^{-1}$ by a thick line and the wavelength of a given perturbation $\lambda(t) \propto a(t)$ by a thin line. If the inflation is exponential with $a(t) \propto \exp(Ht)$, then the wavelength of the perturbation increases exponentially during the inflationary phase while the Hubble radius remains constant. This allows the wavelength of the perturbation to start with a size which is smaller than the Hubble radius and grow to a value that is bigger than the Hubble radius during the inflationary phase (the wavelength will be equal to the Hubble radius at the point marked A in the figure; one usually says that the perturbation 'leaves' the horizon at A). After the inflation ends — at the epoch marked by a broken vertical line — the Hubble radius $d_H(t) \propto t \propto a^2$ grows faster than the wavelength $\lambda \propto a$ and catches up with it. That is, the perturbation again 'enters' the Hubble radius at B. The idea now is that quantum fluctuations of a field at scales smaller than the Hubble radius at A can lead to classical energy density fluctuations at B.

To do this, we have to discuss little bit of field theory in the Friedman universe. We will do this in two easy steps, in case you are not familiar with it. First, recall that in classical mechanics, the dynamical equation for a particle moving in a potential is given by $\ddot{q}(t) = -\partial V/\partial q$. When we proceed from a point particle (for which the dynamical variable $q(t)$ depends only on time $t$) to a field $\phi(t, \mathbf{x})$ (which depends on both space and time), the time derivatives will be supplemented by spatial derivatives, with the usual minus sign in special relativity. Thus, in special relativity, we will have the

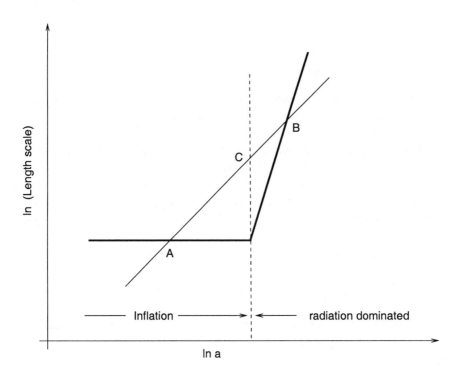

Fig. 6.5   The behaviour of Hubble radius (thick lines) and the wavelength of a perturbation (thin line) in an inflationary universe.

field equation for a scalar field:

$$\Box\phi \equiv \frac{\partial^2\phi}{\partial t^2} - \nabla^2\phi = -\frac{\partial V}{\partial\phi}. \qquad (6.81)$$

(We are using units with $c = 1$). Next we need to generalise this equation
from flat spacetime to FRW background. This is easy if we remember that,
when we proceed from Cartesian coordinates to spherical coordinates, we
just use the replacement

$$\frac{\partial^2\phi}{\partial x^2} + \frac{\partial^2\phi}{\partial y^2} + \frac{\partial^2\phi}{\partial z^2} \rightarrow \frac{1}{\sqrt{g}}\frac{\partial}{\partial x^\alpha}(\sqrt{g}g^{\alpha\beta}\frac{\partial\phi}{\partial x^\beta}) \qquad (6.82)$$

where $g_{\alpha\beta}$ is the three dimensional metric with $\alpha, \beta = 1, 2, 3$ and $g$ is the
determinant of $g_{\alpha\beta}$. (We are also using the convention that any index
that is repeated in a term is summed over its range of values.) The same
applies in four dimensions (with the indices now running over 0, 1, 2, 3)

and Eq. (6.81) becomes, in the $k = 0$ FRW universe:

$$\frac{1}{a^3}\frac{\partial}{\partial t}(a^3\frac{\partial\phi}{\partial t}) - \frac{1}{a^2}\nabla^2\phi = -\frac{dV}{d\phi}. \tag{6.83}$$

The $a^3$ factors in the first term comes from $\sqrt{-g} = a^3$; the $\nabla^2$ gets changed to $a^{-2}\nabla^2$ because of the expansion of spatial scales. This is what we will work with.

Consider a scalar field $\phi(t, \mathbf{x})$ which is nearly homogeneous in the sense that we can write $\phi(t, \mathbf{x}) = \phi(t) + \delta\phi(t, \mathbf{x})$ with $\delta\phi \ll \phi$. We will think of $\phi(t)$ as the mean, homogeneous, part of the scalar field and $\delta\phi$ as due to quantum fluctuations around the mean value. Let us first ignore the fluctuations and consider how one can use the mean value to drive a rapid expansion of the universe. Since $\phi$ is independent of spatial coordinates, the the equation of motion Eq. (6.83) reduces to

$$\frac{1}{a^3}\frac{d}{dt}\left(a^3\frac{d\phi}{dt}\right) = \ddot{\phi} + 3H\dot{\phi} = -\frac{dV}{d\phi}. \tag{6.84}$$

It is easy to verify that this equation is the same as $d(\rho_\phi a^3) = -p_\phi da^3$, if we attribute to the scalar field the energy density $\rho_\phi$ and pressure $p_\phi$ given by

$$\rho_\phi = \frac{1}{2}\dot{\phi}^2 + V(\phi); \qquad p_\phi = \frac{1}{2}\dot{\phi}^2 - V(\phi) \tag{6.85}$$

where $V(\phi)$ is the potential for the scalar field (There are more formal ways of getting this result from the energy momentum tensor but this will do for our purpose). The Einstein's equation (for the $k = 0$ Friedmann model) with the above source can be written in the form

$$\frac{\dot{a}^2}{a^2} = H^2(t) = \frac{8\pi G}{3}\left[V(\phi) + \frac{1}{2}\dot{\phi}^2\right]. \tag{6.86}$$

In principle, Eq. (6.86) and Eq. (6.84), can be integrated to give $a(t)$ and $\phi(t)$ once the form of $V(\phi)$ is given. The actual nature of solutions, of course, will depend critically on the form of $V(\phi)$ as well as the initial conditions. Among these solutions, there exists a subset in which $a(t)$ is a rapidly growing function of $t$, either exponentially or as a power law $a(t) \propto t^\alpha$ with an arbitrarily large value of $\alpha$. We shall now provide a general procedure for determining a form of $V(\phi)$ such that any reasonable evolution characterized by a function $a(t)$ can be reproduced using a scalar field.

It is fairly easy to verify that the solutions to Eqs. (6.86), (6.84) can be expressed in the parametric form

$$V(t) = \frac{3H^2}{8\pi G}\left[1 + \frac{\dot{H}}{3H^2}\right]; \quad \phi(t) = \int dt \left[-\frac{\dot{H}}{4\pi G}\right]^{1/2}. \tag{6.87}$$

Given an $a(t)$ these equations give $V(t)$ and $\phi(t)$ and thus determine $V(\phi)$ implicitly. Thus Eqs. (6.87) completely solves the problem of finding a potential $V(\phi)$ which will lead to a given $a(t)$. As an example of using Eq. (6.87), let us design a universe in which $a(t) = a_0 t^\alpha$. Elementary algebra using Eq. (6.87) now gives the potential to be of the form

$$V(\phi) = V_0 \exp\left(-\sqrt{\frac{16\pi G}{\alpha}}\,\phi\right) \tag{6.88}$$

where $V_0$ and $\alpha$ are constants. The corresponding evolution of $\phi(t)$ is given by

$$\phi(t) = \left(\frac{\alpha}{4\pi G}\right)^{1/2} \ln\left(\sqrt{\frac{8\pi G V_0}{\alpha(3\alpha - 1)}}\,t\right). \tag{6.89}$$

A second example, consider an evolution of the form

$$a(t) \propto \exp(At^f), \quad f = \frac{\beta}{4 + \beta}, \quad 0 < f < 1, \quad A > 0. \tag{6.90}$$

In this case, we can determine the potential to be

$$V(\phi) \propto \phi^{-\beta}\left(1 - \frac{\beta^2}{48\pi G}\frac{1}{\phi^2}\right) \tag{6.91}$$

where $\beta$ is a constant. Both these examples show that it is indeed possible to have rapid growth of $a(t)$ for a suitable choice for the potential. We shall next enquire about the general conditions $V(\phi)$ should satisfy in order to lead to this behaviour.

One possible way of achieving this is through potentials which allow what is known as *slow roll-over*. Such potentials have a gently decreasing form for $V(\phi)$ for a range of values for $\phi$ during which $\phi(t)$ evolves very slowly. Assuming a sufficiently slow evolution of $\phi(t)$ we can ignore the $\ddot{\phi}$ term in Eq. (6.84). Similarly, we can ignore the kinetic energy term $\dot{\phi}^2$ in comparison with the potential energy $V(\phi)$ in Eq. (6.86). In this limit,

Eq. (6.86) and Eq. (6.84) become

$$H^2 \simeq \frac{8\pi G V(\phi)}{3}; \qquad 3H\dot{\phi} \simeq -V'(\phi) \equiv \frac{dV}{d\phi}. \tag{6.92}$$

Let us assume that the slow roll-over approximation is valid until a time $t = t_{\text{end}}$ when $\phi = \phi_{\text{end}}$. The amount of inflation can be characterized by the ratio $a(t_{\text{end}})/a(t)$; a large value for this number indicates that the universe expands by a large factor from the time $t$ till the time $t_{\text{end}}$. Since this is a large number in most models, it is conventional to work with the quantity $N(t) \equiv \ln[a(t_{\text{end}}/a(t)]$. From Eq. (6.92) it is easy to obtain an expression for $N$ in terms of the potential $V(\phi)$;

$$N \equiv \ln \frac{a(t_{\text{end}})}{a(t)} = \int_t^{t_{\text{end}}} H \, dt \simeq 8\pi G \int_{\phi_{\text{end}}}^{\phi} \frac{V}{V'} \, d\phi. \tag{6.93}$$

This provides a general procedure for generating and quantifying the rapid growth of $a(t)$ arising from a given potential. The validity of slow roll over, in turn, requires the following two parameters to be sufficiently small:

$$\epsilon(\phi) = \frac{1}{16\pi G} \left(\frac{V'}{V}\right)^2; \qquad \eta(\phi) = \frac{1}{8\pi G} \frac{V''}{V}. \tag{6.94}$$

Of these, $\epsilon$ has to be small in order to ignore the kinetic energy in comparison with the potential energy, while $\eta$ has to be small in order to ignore the acceleration of the scalar field. The end point for inflation can be taken to be the epoch at which $\epsilon$ becomes comparable to unity.

We now turn to the inhomogeneous part $\delta\phi(t, x)$ of the scalar field and consider the spectrum of density perturbations which are generated from the quantum fluctuations of the scalar field. This requires the study of quantum field theory in a time dependent background which is quite nontrivial. We shall, therefore, adopt a heuristic approach, as follows:

Let us assume that, during the inflationary phase, the universe expands exponentially as $a(t) \propto \exp(Ht)$. Then all the frequencies will be redshifted exponentially, $\omega(t) \propto \exp(-Ht)$. We saw in Sec. 5.4 that such a situation is equivalent to the existence of a temperature [see Eq. (5.54)]. This shows that quantum fluctuations in an exponentially expanding universe behave like thermal fluctuations at temperature $T = (H/2\pi)$ — just as in the case of black holes. (The analysis is identical to that which led to Eq. (5.54) from Eq. (5.51) with $H$ replacing $\kappa$.) Hence the scalar field will have an intrinsic rms fluctuation $\delta\phi \approx T = (H/2\pi)$ in the inflationary phase at the scale of the Hubble radius. This will cause a time shift $\delta t \approx \delta\phi/\dot{\phi}$ in the evolution of the field between patches of the universe of size about $H^{-1}$. This, in turn, will lead to an rms fluctuation $\Delta = (k^3 P)^{1/2}$ of amplitude

$$\Delta = \frac{\delta a}{a} = \left(\frac{\dot{a}}{a}\right) \delta t \approx \frac{H^2}{2\pi \dot{\phi}} \qquad (6.95)$$

at the Hubble scale. Since the wavelength of the perturbation is equal to Hubble radius at A (see Fig. 6.5), we conclude that the rms amplitude of the perturbation when it leaves the Hubble radius is: $\Delta_A \approx H^2/(2\pi \dot{\phi})$.

Between A and B (in Fig. 6.5) the wavelength of the perturbation is bigger than the Hubble radius and one can show that $\Delta_A = \Delta(\text{at } A) \approx \Delta(\text{at } B)$. To see this, let us consider the density fluctuations at $A, B$ and at $C$ which is the epoch at which the universe makes the transition from the inflationary phase to the radiation dominated phase. Using Eq. (6.53), which gives $\delta \propto (\rho a^2)^{-1}$, we know that $\rho \approx$ constant and $\delta \propto a^{-2}$ during the inflationary phase, while $\rho \propto a^{-4}$ and $\delta \propto a^2$ during the radiation dominated phase. This gives

$$\frac{\delta_C}{\delta_A} = \frac{a_A^2}{a_C^2}; \qquad \frac{\delta_B}{\delta_C} = \frac{a_B^2}{a_C^2}. \qquad (6.96)$$

Further, we know that $(a_C/a_B) = (t_C/t_B)^{1/2} = (H_B/H_C)^{1/2} = (H_B/H_A)^{1/2}$. The first equality follows from $a \propto t^{1/2}$ in radiation dominated phase and the last equality follows from the fact that $H_C = H_A$. Therefore, we get

$$a_C^2 = a_B^2 \left(\frac{H_B}{H_A}\right). \qquad (6.97)$$

Combining the two relations in Eq. (6.96) and using Eq. (6.97) we get

$$\left(\frac{\delta_B}{\delta_A}\right)^{1/2} = \frac{a_A a_B}{a_C^2} = \frac{a_A a_B}{a_B^2} \frac{H_A}{H_B} = \frac{a_A H_A}{a_B H_B} = 1. \qquad (6.98)$$

The last result follows from the fact that the points $A$ and $B$ are defined as the exit and entry at which $(2\pi/k)a = H^{-1}$ making $aH = (k/2\pi) =$ constant at these two epochs, implying $a_A H_A = a_B H_B$. We thus conclude that

$$\Delta(\text{at } B) = \Delta(\text{at } A) = \frac{H^2}{2\pi \dot{\phi}}. \qquad (6.99)$$

Since this expression is independent of $k$, all perturbations will enter the Hubble radius with constant power per decade. That is $\Delta^2(k, a) \propto$

$k^3 P(k, a)$ is independent of $k$ when evaluated at $a = a_{enter}(k)$ for the relevant mode. Observations suggest (see Sec. 6.6) that $\Delta_B \simeq 10^{-5}$.

From this, one can show that $P(k, a) \propto k$ at constant $a$. To see this, note that if $P \propto k^n$, then the power per logarithmic band of wave numbers is $\Delta^2 \propto k^3 P(k) \propto k^{n+3}$. Further, when the wavelength of the mode is larger than the Hubble radius, during the radiation dominated phase, the perturbation grows as $\delta \propto a^2$ making $\Delta^2 \propto a^4 k^{(n+3)}$. The epoch $a_{enter}$ at which a mode enters the Hubble radius is determined by the relation $2\pi a_{enter}/k = d_H$. Using $d_H \propto t \propto a^2$ in the radiation dominated phase, we get $a_{enter} \propto k^{-1}$ so that

$$\Delta^2(k, a_{enter}) \propto a_{enter}^4 k^{(n+3)} \propto k^{(n-1)}. \qquad (6.100)$$

So if the power $\Delta^2(k, a_{\text{enter}})$ per octave in $k$ is independent of scale $k$, at the time of entering the Hubble radius, then $n = 1$. In fact, a *prediction* that the initial fluctuation spectrum will have a power spectrum $P = Ak^n$ with $n = 1$ was made by Harrison and Zeldovich, years before inflationary paradigm, based on general arguments of scale invariance. Inflation is one possible mechanism for realising the scale invariance.

As a specific example, consider the case of $V(\phi) = \lambda\phi^4$, for which Eq.(6.93) gives $N = H^2/2\lambda\phi^2$ and the amplitude of the perturbations at Hubble scale is:

$$\Delta \simeq \frac{H^2}{\dot\phi} \simeq \frac{3H^3}{V'} \simeq \lambda^{1/2} N^{3/2}. \qquad (6.101)$$

If we want inflation to last for reasonable amount of time ($N \gtrsim 60$, say) and demand that $\Delta \approx 10^{-5}$ (as determined from CMBR temperature anisotropies; see Sec. 6.6) then we require $\lambda \lesssim 10^{-15}$. This has been a serious problem in virtually any reasonable model of inflation: *The parameters in the potential need to be extremely fine tuned to match observations.*

The qualitative reason for this result is the following: The fluctuations arises from $(\delta a/a) = H\delta t \simeq H(\delta\phi/\dot\phi)$. To obtain *slow* roll-over and sufficient inflation we need to keep $\dot\phi$ small; this leads to an increase in the value of $\delta$. The difficulty could have been avoided if it were possible to keep $\delta\phi$ arbitrarily small; unfortunately, the inflationary phase induces a fluctuation of about $(H/2\pi)$ on any quantum field due to the non zero temperature. This lower bound prevents us from getting acceptable values for $\delta$ unless we fine-tune the dimensionless parameters of $V(\phi)$.

At the end of inflation, one expects the energy in the scalar field transfered to other matter fields, though the details of this process are somewhat model dependent. By and large, this leads to the same relative changes in the number density of all species leaving the relative abundances invariant. In particular, the photon-to-baryon ratio $(n_\gamma/n_B)$ does not change. Since $n_\gamma \propto T^3 \propto$ entropy of radiation, the condition $\delta(n_\gamma/n_B) = 0$ leads to the entropy per baryon remaining the same. Such fluctuations are called *isoentropic* or adiabatic.

## 6.6    Temperature anisotropies in the CMBR

We have seen in Sec. 6.3 that the photons in the universe decoupled from matter at a redshift of about $10^3$. These photons have been propagating freely in spacetime since then and can be detected today. In an *ideal* Friedmann universe, a *comoving* observer will see these photons as a blackbody spectrum at some temperature $T_0$. The deviations in the metric from that of the Friedmann universe, motion of the observer with respect to the comoving frame, the astrophysical processes which take place along the trajectory of the photon, can all lead to potentially observable effects in this radiation. These effects can be of two kinds: (i) The spectrum may not be strictly blackbody; that is, the radiation intensity $I(\omega)$ at different frequencies, coming to us from a *particular* direction in the sky, may not correspond to a Planck spectrum with a single temperature $T_0$. (ii) The spectrum may be Planckian in any given direction; but the temperature of the radiation may be different in *different* directions; that is $T_0 = T_0(\theta, \phi)$ where $(\theta, \phi)$ are the angular coordinates on the sky.

The first kind of deviation is not detected in the CMBR and there exist tight observational bounds on the spectral deviations of CMBR from the ideal blackbody form. This (negative) result is useful in constraining the models describing structure formation, especially those aspects which are related to the intergalactic medium. On the other hand, CMBR does show the distortions of the second kind at different angular scales. Direct measurements of these distortions, which are currently available at different angular scales, turn out to be very restrictive. Because of this reason, we shall concentrate on the angular anisotropies.

A physical process characterized by a length scale $L$ at $z = z_{\text{dec}}$, will subtend an angle $\theta(L) = [L/d_A(z_{\text{dec}})]$ in the sky today where $d_A(z)$ is the angular diameter distance. We saw in Sec. 6.1 that $d_A = r_{\text{em}}(z)(1+z)^{-1}$ where $r_{\text{em}}(z)$ is given by Eq. (6.16). For large $z$, this reduces to $r_{\text{em}}(z) \approx 2H_0^{-1}\Omega_{\text{NR}}^{-1/2}$, so that, for $z \gg 1$, the angular diameter distance becomes

$d_A(z) \cong 2H_0^{-1}(\sqrt{\Omega_{\rm NR}}\, z)^{-1}$, giving

$$\theta(L) \cong \left(\frac{\sqrt{\Omega_{\rm NR}}}{2}\right)\left(\frac{Lz}{H_0^{-1}}\right) = 34.4''\,(\Omega_{\rm NR}h^2)^{1/2}\left(\frac{\lambda_0}{1{\rm Mpc}}\right). \qquad (6.102)$$

(It is conventional to quote the numerical values with the length scales extrapolated to the present epoch. Thus $\lambda_0 = L(1 + z_{\rm dec}) \simeq Lz_{\rm dec}$ is the proper length *today* which would have been $L$ at the redshift of $z_{\rm dec}$. We will not bother to indicate this fact with a subscript '0' when no confusion is likely to arise.) In particular, consider the angle subtended by the region which has the size equal to that of the Hubble radius at $z_{\rm dec}$; that is, we take $L$ to be

$$d_H(z_{\rm dec}) = H^{-1}[z = z_{\rm dec}] = H_0^{-1}(\Omega_{\rm NR}z_{\rm dec})^{-1/2}z_{\rm dec}^{-1} \qquad (6.103)$$

so that

$$\lambda = d_H(z_{\rm dec})[1 + z_{\rm dec}] \cong d_H(z_{\rm dec})z_{\rm dec} \cong H_0^{-1}(\Omega_{\rm NR}z_{\rm dec})^{-1/2}. \qquad (6.104)$$

Then Eq. (6.102) gives:

$$\theta_H \equiv \theta(d_H) \cong 0.87°\left(\frac{z_{\rm dec}}{1100}\right)^{-\frac{1}{2}} \simeq 1°. \qquad (6.105)$$

Therefore, angular separation of more than one degree in the sky would correspond to regions which were bigger than the Hubble radius at the time of decoupling.

Let us now estimate the magnitude of the temperature anisotropy $(\Delta T/T)$ that is caused by different processes. We begin with the intrinsic anisotropy in the radiation field at large angular scales, $\theta > \theta_H$.

If baryons and photons are tightly coupled, then we expect the anisotropy in the energy density of the radiation, $\delta_R$, to be comparable to $\delta_B$. The exact relation between these two depends on the initial process which generated the fluctuations. As we mentioned at the end of Sec. 6.5, inflationary models generate fluctuations which are isoentropic, in the sense that, the photon to baryon ratio (which is proportional to the entropy per baryon) is unaffected by the fluctuation. These isoentropic fluctuations are characterized by

$$0 = \delta\left[\frac{s_R}{n_B}\right] = \frac{s_R}{n_B}\left[\frac{\delta s_R}{s_R} - \frac{\delta\rho_B}{\rho_B}\right]. \qquad (6.106)$$

Using $s_R \propto T^3$ and $\rho_R \propto T^4$, it follows that $(\delta s_R/s_R) = (3/4)\delta_R$. Therefore, for isoentropic fluctuations, we will have $(3/4)\delta_R = \delta_B$. In general, we expect $\delta_B \propto \delta_{\mathrm{DM}}$ at large scales in which gravity dominates over pressure forces. Hence,

$$\left(\frac{\Delta T}{T}\right)_{\mathrm{int}} \propto \delta_{\mathrm{DM}}. \tag{6.107}$$

To determine the angular scale over which this effect is significant, we proceed as follows: Each Fourier mode of $\delta_{\mathrm{DM}}(k)$, labelled by a wave vector $k$, will correspond to a wavelength $\lambda \propto k^{-1}$ and will contribute at an angle $\theta$ given by Eq. (6.102). The mean square fluctuation in temperature will therefore scale with the angle $\theta$ as

$$\left(\frac{\Delta T}{T}\right)_{\mathrm{int}}^2 \propto d^3 k P(k) \propto \frac{dk}{k} P(k) k^3 \propto \lambda^{-(3+n)} \propto \theta^{-(3+n)} \tag{6.108}$$

for $\theta > \theta_H$. If $n = 1$, which is the preferred choice, then $(\Delta T/T)_{\mathrm{in}} \propto \theta^{-2}$.

Consider next the contribution to $(\Delta T/T)$ from the velocity of the scatterers in the last scattering surface (LSS). If a particular photon which we receive was last scattered by a charged particle moving with a velocity $v$, then the photon will suffer a Doppler shift of the order $(v/c)$. This will contribute a $(\Delta T/T)^2 \simeq (v^2/c^2)$. At large scales, (for $\lambda \gtrsim d_H$), the velocity of the baryons is essentially due to the gravitational force exerted by the dark matter potential wells and $v = g t_{\mathrm{dec}} \propto t_{\mathrm{dec}} \nabla \phi$ captures the spatial variation of $\mathbf{v}$ on the LSS. By an analysis similar to the above, we conclude that

$$\left(\frac{\Delta T}{T}\right)_{\mathrm{Dopp}}^2 \propto d^3 k\, k^2 \frac{P}{k^4} \propto \frac{dk}{k} k P(k) \propto \lambda^{-(1+n)} \propto \theta^{-(1+n)} \tag{6.109}$$

for $\theta > \theta_H$. If $n = 1$, then $(\Delta T/T)_{\mathrm{Dopp}} \propto \theta^{-1}$.

Finally, let us consider the contribution to $(\Delta T/T)$ from the gravitational potential along the path of the photons. When the photons travel through a region of gravitational potential $\phi$, they undergo a redshift of the order of $(\phi/c^2)$. The gravitational potential is related to the perturbed density contrast by $\nabla_x^2 \phi = 4\pi G \rho a^2 \delta$. (Note that $\rho \delta = (\delta \rho)$ is the perturbed density; the $a^2$ factor arises in transforming from $\nabla_r^2$ to $\nabla_x^2$ using $\mathbf{r} = a\mathbf{x}$.) In the $\Omega = 1$, matter dominated universe, $\rho \propto a^{-3}$ and $\delta \propto a$ so that $4\pi G \rho a^2 \delta$ is independent of time making the perturbed gravitational potential independent of time, i.e., $\phi(t, \mathbf{x}) = \phi(\mathbf{x})$. We may therefore estimate the temperature anisotropies due to gravitational potential as $(\Delta T/T) \simeq (\phi/c^2)$ where $\phi$ is the gravitational potential *on* the LSS. This contribution to temperature anisotropy is usually called *Sachs-Wolfe* effect. Its angular dependence will be

$$\left(\frac{\Delta T}{T}\right)^2_{\text{SW}} \propto d^3k\frac{P(k)}{k^4} \propto \frac{dk}{k^2}P(k) \propto \theta^{1-n} \tag{6.110}$$

for $\theta > \theta_H$. If $n = 1$, this contribution is a constant.

Let us next consider the anisotropies at small angular scales $\theta < \theta_H$. It can be shown by a more detailed analysis that the contribution $(\Delta T/T)^2_{\text{SW}}$ varies as $\theta^4$ and hence is not very significant at small $\theta$. As regards $(\Delta T/T)^2_{\text{int}}$ and $(\Delta T/T)^2_{\text{Dopp}}$, we should note the following: At small scales, the photon-baryon fluid will have strong pressure support and will oscillate as an acoustic wave. To study the properties of this acoustic wave and its effect on $(\Delta T/T)$, we need to obtain some preliminary results. First of all, note that acoustic disturbances in a photon-baryon fluid propagates with the speed $c_s = (\partial p/\partial \rho)^{1/2} = (\dot{p}/\dot{\rho})^{1/2}$. Since $\dot{p} = \dot{p}_R = (1/3)\dot{\rho}_R$, $\dot{\rho} = \dot{\rho}_R + \dot{\rho}_B$, it follows that

$$c_s^2 = \frac{1}{3}\frac{1}{1 + (\dot{\rho}_B/\dot{\rho}_R)} = \frac{1}{3}\frac{1}{1 + (3\rho_B/4\rho_R)} \approx \frac{1}{3} \tag{6.111}$$

if we ignore the contribution of baryons. Second, we note that the phase of the wave will vary as

$$\omega \int \frac{dt}{a(t)} \approx c_s k\eta \approx \frac{1}{\sqrt{3}}k\eta \tag{6.112}$$

where we have defined a new time coordinate $\eta$ by $d\eta = dt/a$. This follows from the fact that the equation to the light ray is given by $ds = 0$. Given all these, you will expect the photon-baryon fluid to oscillate in time as $\cos[(k\eta/\sqrt{3})]$, say. But the CMBR is observed from a fixed surface $\eta = \eta_{\text{rec}}$. On this surface, the density fluctuations will vary in space in the same manner. The crusts and troughs of the wave will now be separated by $\Delta k \approx (\sqrt{3}\pi/\eta_{\text{rec}})$. The corresponding length scale is $\Delta\lambda \approx (\eta_{\text{rec}}/\sqrt{3}\pi)$. This length scale will subtend an angle $\theta = (\Delta\lambda/D)$ where $D \approx \eta_0 - \eta_{\text{rec}} \approx \eta_0$ is the distance up to the recombination epoch. Using $\eta \propto a^{1/2}$ in the matter dominated phase, we get

$$\theta = \frac{\Delta\lambda}{D} = \frac{1}{\sqrt{3}\,\pi}\left(\frac{a_{\text{rec}}}{a_0}\right)^{1/2} = \frac{1}{\sqrt{3}\,\pi}(1 + z_{\text{rec}})^{-1/2} \approx 5.8 \times 10^{-3} \approx 0.33°. \tag{6.113}$$

A more exact calculation gives a slightly larger value. This shows that we will expect alternating maxima and minima in $\Delta T/T$ in the sky at small scales separated by about a degree or so. There is, however, a minor subtlety. Both the density fluctuations and velocity fluctuations will lead to these oscillations but they will be opposite in phase in a pure radiation fluid. We get a net effect only because of the existence of baryons which prevents them from canceling each other completely. The angular scale we have computed, however, is unaffected by this effect.

There is, however, another effect which needs to be taken into account at small angular scales. The above analysis assumes that decoupling took place at a sharp value of redshift $z_{\text{dec}}$. In reality, the LSS has a finite thickness $\Delta z \approx 80$ (see Sec. 6.3, Eq. (6.50)). Any photon detected today has a probability $\mathcal{P}(z)dz$ to have been last scattered in the redshift interval $(z, z + dz)$. The observed $(\Delta T/T)$ has to be computed as

$$\left(\frac{\Delta T}{T}\right)_{\text{obs}} = \int dz \left\{ \begin{array}{l} (\Delta T/T) \text{ if the last} \\ \text{scattering was at } z \end{array} \right\} \times \mathcal{P}(z). \qquad (6.114)$$

We saw in Sec. 6.3 that $\mathcal{P}(z)$ is very well approximated by a Gaussian peaked at $z = z_{\text{dec}}$ with a width $\Delta z \approx 80$. This width corresponds to a line of sight (comoving) distance of

$$\Delta l = c\left(\frac{dt}{dz}\right)\Delta z \cdot (1 + z_{\text{dec}}) \approx H_0^{-1} \frac{\Delta z}{\Omega_{\text{NR}}^{1/2} z_{\text{dec}}^{3/2}} \approx 8 \left(\Omega_{\text{NR}} h^2\right)^{-1/2} \text{ Mpc}. \qquad (6.115)$$

Hence the anisotropies at length scales smaller than $\Delta l$ will be suppressed due to the finite thickness of LSS. This length scale corresponds to the angular scale of about $\theta_{\Delta z} \simeq 3.8' \Omega_{\text{NR}}^{1/2}$.

Another effect which operates at small scales is process called *photon diffusion*. As recombination proceeds, the diffusion length approaches infinity and hence can be important. The actual length scales below which this effect can wipe out fluctuations is somewhat model dependent but could be about $l_{\text{diff}} \simeq 35$ Mpc, corresponding to $\theta_{\text{diff}} \simeq 24'$. This result can be understood as follows: Consider a time interval $\Delta t$ in which a photon suffers $N = (\Delta t/l(t))$ collisions. Between successive collisions it travels a proper distance $l(t)$, or - equivalently - a coordinate distance $[l(t)/a(t)]$. Because of this random walk, it acquires a mean-square coordinate displacement:

$$(\Delta x)^2 = N\left(\frac{l}{a}\right)^2 = \frac{\Delta t}{l(t)}\frac{l^2}{a^2} = \frac{\Delta t}{a^2}l(t). \qquad (6.116)$$

The total mean-square coordinate distance traveled by a typical photon till the time of decoupling is

$$x^2 \equiv \int_0^{t_{\rm dec}} \frac{dt}{a^2(t)} l(t) = \frac{3}{5} \frac{t_{\rm dec} l(t_{\rm dec})}{a^2(t_{\rm dec})} \qquad (6.117)$$

where we have used the fact that $l(t) = [\sigma n_e(t)]^{-1} \simeq 10^{29}$ cm $(1 + z)^{-3}(\Omega_B h^2)^{-1}$ and that $a \propto t^{2/3}$. This corresponds to the proper distance

$$l_{\rm diff} = a(t_{\rm dec})x = \left[\frac{3}{5} t_{\rm dec} l(t_{\rm dec})\right]^{1/2} \simeq 35 {\rm Mpc} \left(\frac{\Omega_B h^2}{0.02}\right)^{-1/2} (\Omega h_{50}^2)^{-1/4}. \qquad (6.118)$$

The effect of this diffusion is equivalent to multiplying each mode $(\Delta T/T)_k$ in the Fourier space by the Gaussian $\exp[-k^2(\Delta l)^2/2] = \exp[-(k/k_T)^2]$ with $k_T \simeq 0.02 h {\rm Mpc}^{-1}$.

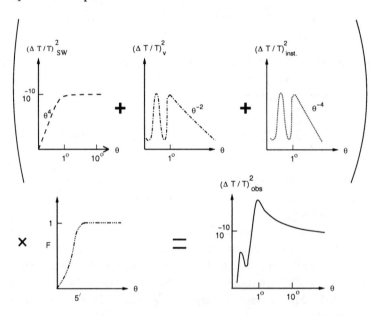

Fig. 6.6 The angular pattern of CMBR temperature anisotropies due to different physical processes are indicated schematically. See text for discussion.

Taking all these effects into account, we arrive at the final pattern of anisotropies which will look like the one in Fig. 6.6. The contribution at large angular scales due to intrinsic anisotropy, Doppler effect and Sachs-Wolfe effect are shown by the dotted line, dot-dash line and the dashed line respectively. The dot-triple dash line indicates the Gaussian cutoff due to the photon diffusion. The final thick line clearly shows the peaks (called *acoustic peaks*) and the suppression at small scale. At large scales, the anisotropies are dominated by the Sachs-Wolfe effect and are independent of the angular scale. As we move to smaller scales, other contributions add up making $(\Delta T/T)$ increase. At scales $\theta \lesssim \theta_{\text{eq}}$, the oscillations in $\delta_B$ and $v$ make their presence felt and we see characteristic acoustic peaks. At still smaller scales, the effects due to finite thickness of LSS drastically reduce the amplitude of fluctuations.

The fact that several different processes contribute to the structure of angular anisotropies makes CMBR a valuable tool for extracting cosmological information. To begin with, the anisotropy at very large scales directly probe modes which are bigger than the Hubble radius at the time of decoupling and allows us to directly determine the primordial spectrum. As we have seen, the effect of initial fluctuations in the scalar gravitational potential leads to a temperature anisotropy with the angular dependence, $\Delta T/T \propto \lambda^{1-n} \propto \theta^{1-n}$. In particular, if $\Delta T/T$ is independent of the angular scale at large angles, then the spectrum is scale invariant with $n = 1$ in a $\Omega = \Omega_{\text{NR}} = 1$ universe. The behaviour is different if there is a cosmological constant (or if $\Omega < 1$), but in a predictable way. Thus, in general, if the angular dependence of the spectrum at very large scales is known, one can work backwards and determine the initial power spectrum.

The angles subtended by the acoustic peaks will depend on the geometry of the universe and provides a reliable procedure for estimating the cosmological parameters. Detailed computations show that: (i) The multipole index $l$ corresponding to the first acoustic peak has a strong, easily observable, dependence on $\Omega_{\text{tot}}$ and scales as $l_p \approx 220\Omega_{\text{tot}}^{-1/2}$ if there is *no* dark energy and $\Omega_{tot} = \Omega_{NR}$. (ii) But if both non-relativistic matter and dark energy is present, with $\Omega_{\text{NR}} + \Omega_{DE} = 1$ and $0.1 \lesssim \Omega_{\text{NR}} \lesssim 1$, then the peak has only a very weak dependence on $\Omega_{NR}$ and $l_p \approx 220\Omega_{NR}^{0.1}$. Thus the observed location of the peak (which is around $l \sim 220$) can be used to infer that $\Omega_{\text{tot}} \simeq 1$. More precisely, the current observations show that $0.98 \lesssim \Omega_{\text{tot}} \lesssim 1.08$; combining with $h > 0.5$, this result implies the existence of dark energy.

The heights of acoustic peaks also contain important information. In particular, the height of the first acoustic peak relative to the second one depends sensitively on $\Omega_B$ and the current results are consistent with that obtained from big bang nucleosynthesis.

## Exercises

**Exercise 6.1**   *Spheres and planes:*

The purpose of this exercise is to motivate the form of 3-dimensional space in Eq. (6.3) as one which is homogeneous and isotropic. An obvious example of space which is homogeneous and isotropic is a plane which has the metric $ds^2 = dr^2 + r^2 d\theta^2$ and a second example will be the surface of a sphere with metric $ds^2 = d\alpha^2 + \sin^2 \alpha \, d\beta^2$. Show that one can write these metrics together in the form $ds^2 = dx^2(1 - kx^2)^{-1} + x^2 d\theta^2$ with $k = (0, 1)$. Further argue that this metric represents a homogeneous and isotropic space even when $k = -1$. Finally generalize the result from two to three dimension and obtain Eq. (6.3).

**Exercise 6.2**   *Age of the Universe:*

Another quantity of interest in Friedmann universe, which we did not discuss, is its age. Since the age of the universe should be larger than the age of any object in it, this can be used to constrain cosmological models. Consider the age of a universe with $\Omega_V + \Omega_{NR} = 1$, ignoring all other components. It is convenient to use the variable $y = (a/a_V)$ where $\Omega_V a_V^3 = \Omega_{NR}$. (So that $y = 1$ when the matter and cosmological constant contributes equal amounts to energy density.) Show that $y(t) = \sinh^{2/3}(3H_V t/2)$ where $H_V^2 = (8\pi G \rho_V/3)$. Also show that the product $tH(t)$ is given by

$$tH(t) = \frac{2}{3(1-f)^{1/2}} \sinh^{-1}\left[\left(\frac{1-f}{f}\right)^{1/2}\right] \tag{6.119}$$

where $f = (1 + y^3)^{-1}$. Use this to estimate the current age of the universe for the two models with $(\Omega_{NR} = 1, \Omega_V = 0)$ and $(\Omega_{NR} = 0.3, \Omega_V = 0.7)$.

**Exercise 6.3**   *Non standard nucleosynthesis:*

Determine whether the primordial nucleosynthesis will produce more (or less) helium and deuterium compared to the standard model if: (i) there are many more neutrinos than anti neutrinos (or photons) in the universe, (ii) there are many more anti neutrinos than neutrinos (or photons) in the universe, (iii) there is a significant contribution from gravitational radiation to the energy density of the early universe.

**Exercise 6.4**   *Approximating a sphere:*

Equations (6.59), (6.61) give $\delta$ in terms $\delta_L$ only in parametric form through the angle $\theta$. Try an approximate fit of the form $1 + \delta = [1 - (\delta_L/n)]^{-n}$ with $n$ as a free parameter. Determine $n$ so that this relation fits the spherical evolution as accurately as possible. Comment on the result.

**Exercise 6.5** *Putting asunder what gravity has put together:*

We saw in chapter 4 that a supernova explosion releases about $10^{51}$ erg of energy. Show that this hydrodynamic energy can, in principle, unbind the baryonic gas in the dark matter halos at $z > 5$. Take the binding energy of baryons to be about a factor $(\Omega_B/\Omega_{\mathrm{DM}}) \approx 0.15$ less than the binding energy $U$ computed above.

**Exercise 6.6** *Inflation with $V(\phi) = \lambda\phi^4$:*

It is possible to satisfy the conditions of slow roll over even in fairly simple potentials. As an example, let us consider the case of $V(\phi) = \lambda\phi^4$ with $\phi(t)$ evolving from some initial value $\phi_i$ at $a = a_i$ towards $\phi = 0$. Show that the slow roll over solutions are given by

$$\phi = \phi_i \exp\left[-\sqrt{\frac{32\lambda M_{\mathrm{Pl}}^2}{6}}\,(t - t_i)\right], \tag{6.120}$$

$$a = a_i \exp\left(\frac{\phi_i^2}{8M_{\mathrm{Pl}}^2}\left\{1 - \exp\left[-\sqrt{\frac{64\lambda M_{\mathrm{Pl}}^2}{3}}\,(t - t_i)\right]\right\}\right). \tag{6.121}$$

Determine when the inflation ends and the value of $N$ in such a model.

**Exercise 6.7** *Angular diameter distance:*

Show that the angular diameter distance in a universe with $\Omega_{\mathrm{NR}} + \Omega_V = 1$ is

$$d_A = \frac{c}{H_0(1 + z)} \int_1^{(1+z)} \frac{dx}{[\Omega_{\mathrm{NR}}x^3 + (1 - \Omega_{\mathrm{NR}})]^{1/2}}. \tag{6.122}$$

Plot this as a function of redshift for different values of $\Omega_{\mathrm{NR}}$. Use this to estimate the angle subtended by the comoving scales $1h^{-1}$ Mpc, $10h^{-1}$ Mpc, $100h^{-1}$ Mpc and $3000h^{-1}$ Mpc at $z = 10^3$.

**Exercise 6.8** *Ruling models out:*

Comparing the spectrum of a galaxy at the redshift $z = 1.5$ with computer simulation suggests that it is about 3 Gyr old. Check whether this is consistent with a $\Omega_{\mathrm{NR}} = 0.3, \Omega_V = 0.7, h = 0.7$ universe assuming that stars can form only after the protogalactic material has turned around, collapsed and virialized.

**Exercise 6.9** *Lensing distance:*

While discussing gravitational lenses in chapter 1, we needed the expressions for distances between source and the lens *etc.*. We now know that one can define different kinds of distances (angular diameter distance, luminosity distance, ...) so one needs to be careful in using the lensing formula in curved spacetime. (a) Argue that the correct distance scale to use for gravitational lensing is the angular diameter distance. (b) Show that in the case of a flat universe with $\Omega_{NR} = 1$ the relevant distance between redshifts $z_i$ and $z_j$ is given by

$$D_{ij} = \frac{2c}{H_0} \frac{(1+z_i)^{1/2}(1+z_j) - (1+z_i)(1+z_j)^{1/2}}{(1+z_i)(1+z_j)^2}. \tag{6.123}$$

**Exercise 6.10** *Real life astronomy: CMBR anisotropy:*

A favourite tool of theoreticians is the CMBFAST program which allows the computation of temperature anisotropies for different cosmological models. This is also an excellent tool to learn how varying different parameters changes the pattern of anisotropies. You can download the program from http://www.cmbfast.org. Get it and install it properly in your computer. Run it and find the temperature anisotropy pattern for the following models: (a) $\Omega_V = 0$ with $\Omega_{NR}$ varying from 0.2 to 1.0; (b) $\Omega_{tot} = 1$ with $\Omega_V$ varying from 0.1 to 0.9. (c) $\Omega_V = 0.7, \Omega_{tot} = 1, \Omega_B h^2 = 0.02$ with $\Omega_B$ varying from 0.01 to 0.05; of course, you need to vary $h$ to maintain the last constraint. (d) Same as in (c) but keep $h = 0.72$ rather than fixing $\Omega_B h^2$. See whether you can understand the varying patterns.

**Notes and References**

Modern cosmology is discussed in the textbooks Padmanabhan (2002); Peacock (1999) and Dodelson (2003). Figure 6.3 and Fig. 6.6 are adapted from chapters 5, 6 of Padmanabhan (2002). For a radically different approach to cosmology, see Hoyle, Narlikar and Burbridge (2000).

1. *Section 6.1:* The composition of the universe described here was obtained from several pieces of work. Some of the original references can be found in the review articles Padmanabhan (2003) and Padmanabhan (2005).

2. *Section 6.2:* For a simple analytical model describing primordial nucleosynthesis, see Esmailzadeh et al. (1991).

3. *Section 6.6:* There exists an excellent code called CMBFAST which can compute the temperature anisotropies in the microwave radiation for a wide class of models and parameters; see Seljak and Zaldarriaga (1996)

# Chapter 7

# Universe at $z < 20$

Let us now turn our attention to the formation of galaxies which arises from the non linear growth of the density perturbations in normal gaseous matter, made of baryons. This is, of course, *the* key problem in the study of structure formation since all structures which we observe directly are baryonic and most of them are in the non linear phase of gravitational collapse. For the same reason, this is also the area in which progress has been slow because of the inherent complexity of gas dynamics and radiative processes. We shall, therefore, rely heavily on simple minded approaches to understand this phase of the universe.

## 7.1 Galaxy formation

The description of gravitational collapse in the last chapter applies only to dark matter which is not directly coupled to radiation. Since there is at least six times more dark matter than baryons in the universe, the former defines the pattern of gravitational potential wells in which structure formation occurs. You saw in Sec. 6.4 that when the dark matter fluctuation at a given mass scale $M$ becomes non-linear, collapses and virialises, it leads to a formation of a halo of mass $M$. The resulting dark matter halo could be described reasonably accurately as an isothermal sphere (see Sec. 3.2) with a given velocity dispersion $v^2$ and an asymptotic density distribution $\rho(r)$ given by

$$\rho(r) = \frac{v^2}{2\pi G r^2}. \tag{7.1}$$

Based on the discussion in the previous chapter (see Eq. (6.76)), we can also work out the abundance of dark matter halos at different redshifts.

Let us now consider the baryons located in these dark matter halos. The baryonic gas can radiate energy, cool and sink to the centers of the dark matter halos. When the dark matter halos merge, the baryonic structures can survive provided they have condensed to sufficiently small size within these halos. This scenario suggests that it may be possible to understand the basic features of galaxies, at a broad level, by studying the interplay between cooling and gravitational collapse. As you shall see, it is indeed possible, among other things, to understand the characteristic sizes and masses of the galaxies by such simple arguments.

To do this, let us consider in some detail how the baryons cool. Collisional excitation followed by radiative de-excitation allows the gas to lose its energy in random motions and thus cool. At the time of recombination, the temperature of the universe is $k_B T \approx 0.3$ eV and as the universe expands, the temperature decreases further. Since the excitation from $n = 1$ to $n = 2$ state of hydrogen requires about 9 eV, the gas is too cold to be excited either by collisions or by radiation. But later on, when the baryons undergo nonlinear collapse in the potential wells of dark matter, they acquire velocities of the order of 100 km s$^{-1}$. A proton with this velocity has an energy of about 50 eV which is more than enough to ionize the hydrogen atom in a collision. Once heated to the virial temperature of the dark matter potential well, the gas gets ionized and the resulting plasma can cool, mainly due to two different processes. radiation emitted during recombination of electrons and ions which, if it escapes the plasma, can be a source of recombination cooling. From the discussion in chapter 2, Sec. 2.5.3 we know that this rate varies as $n^2 T^{-1/2}$. The second is the bremsstrahlung cooling with an energy loss rate proportional to $n^2 T^{1/2}$ [see chapter 2, Sec. 2.4.1]. For baryons, virialized inside a dark matter halo and heated to the temperature $k_B T \simeq (GMm_p/R)$ which is much higher than the ionisation potential, $\alpha^2 m_e c^2$, bremsstrahlung cooling dominates over recombination cooling and so we will concentrate on bremsstrahlung. The cooling time for this process is (see Eq. (2.44)):

$$t_{\rm cool} \simeq \frac{nk_B T}{(d\mathcal{E}/dtdV)} = \left(\frac{\hbar}{m_e c^2}\right)\left(\frac{1}{n\lambda_e^3}\right)\left(\frac{k_B T}{m_e c^2}\right)^{1/2}\frac{1}{\alpha^3} \qquad (7.2)$$

which should be compared to the time-scale for gravitational collapse:

$$t_{\rm grav} \simeq \left(\frac{GM}{R^3}\right)^{-1/2}. \qquad (7.3)$$

The condition for efficient cooling $t_{\rm cool} < t_{\rm grav}$, coupled to $k_B T \simeq$

$GMm_p/R$ now leads to the constraint $R < R_g$ with

$$R_g \simeq \alpha^3 \alpha_G^{-1} \lambda_e \left(\frac{m_p}{m_e}\right)^{1/2} \simeq 74 \text{ kpc} \qquad (7.4)$$

where $\alpha_G \equiv (Gm_p^2/\hbar c) \approx 6 \times 10^{-39}$ is the gravitational (equivalent of) fine structure constant which you first met in chapter 4 in connection with the stars. The analysis assumed that $k_B T > \alpha^2 m_e c^2$; for $R \simeq R_g$ this constraint is equivalent to the condition $M > M_g$ with

$$M_g \simeq \alpha_G^{-2} \alpha^5 \left(\frac{m_p}{m_e}\right)^{1/2} m_p \simeq 3 \times 10^{44} \text{gm}. \qquad (7.5)$$

This result suggests that systems having a mass of about $3 \times 10^{44}$ gm and radius of about 70 kpc could rapidly cool, fragment and form gravitationally bound structures. Most galaxies have masses around this region and this is one possible scenario for forming galaxies. It is remarkable that the mass and length scales in Eqs. (7.5) and (7.4) describing the galaxies arise entirely from the fundamental physical constants.

Comparison with Eq. (4.7) shows that the number of stars $N_{\text{star}} \simeq (M_g/M_*)$ in a typical galaxy is also be given by the combination of fundamental constants $N_{\text{star}} = \alpha^{7/2} \alpha_G^{-1/2} (m_e/m_p)^{1/4} \simeq 10^{12}$. Typical galaxies indeed have about $(10^{11} - 10^{12})$ stars though there is a fair amount of spread in this number. The mean distance between the stars in a galaxy will be $d_{\text{star}} \approx (R_{\text{gal}}/N_{\text{star}}^{1/3}) \approx 1$ pc.

You can also estimate a few more numbers fairly easily. The original (maximum) radius of the cooling plasma estimated above in Eq. (7.4) is about 50 kpc. After the matter has cooled and contracted, the final radius of a galaxy is more like (10 – 20) kpc. For $M_g \simeq 3 \times 10^{44}$ gm and $R_g \simeq 20$ kpc, the density is $\rho_{\text{gal}} \simeq 10^{-25}$ gm cm$^{-3}$ which is about $10^5$ times larger than the current mean density, $\rho_0 \simeq 10^{-30}$ gm cm$^{-3}$, of the universe. If we assume (based on the discussion of spherical top-hat model in chapter 6, Sec. 6.4) that high density regions with $\bar{\rho} \gtrsim 100\bar{\rho}_{\text{univ}}$ would have collapsed to form the galaxies, then the galaxy formation must have taken place when the density of the universe was about $(10^5/100) = 1000$ times larger; the value of $a(t)$ would have been 10 times smaller and the redshift of the galaxy formation should have been $z_{\text{gal}} \lesssim 9$. That appears to be reasonable.

Further, if the protogalactic plasma condensations were almost touching each other at the time of formation, these centers (which would have been at a separation of about $2 \times 74 \simeq 150$ kpc) would have now moved apart due to cosmic expansion to a distance of $150(1 + z_{\mathrm{gal}})$ kpc $\approx 1500$ kpc $= 1.5$ Mpc. This is indeed the mean separation between the large galaxies today. The nearest galaxy of radius about 10 kpc, at a distance of 1 Mpc, would subtend an angle of $\theta_{\mathrm{gal}} \approx 10^{-2}$ rad $\approx 30'$. A galaxy at the Hubble distance of about 4000 Mpc will subtend $0.5''$.

One can even venture to do a somewhat more detailed modeling based on these ideas. Suppose $\bar{\rho}$ is the density, averaged over a radius $r$, within a protogalaxy forming at a redshift $z_f$. Based on the spherical top-hat model discussed in the last chapter, we can assume that such a structure will form if $\bar{\rho} \gtrsim f_c \rho_{\mathrm{bg}}(z_f)$ where $f_c \approx 200$ and $\rho_{\mathrm{bg}}$ is the mean density of the background universe. This condition translates to

$$\frac{\bar{\rho}}{\rho_b} = \frac{3}{4\pi r^3} \frac{v_c^2 r}{G} \frac{8\pi G}{3H_0^2 \Omega_{\mathrm{NR}}(1+z_f)^3} = \frac{2v_c^2}{\Omega_{\mathrm{NR}}(H_0 r)^2 (1+z_f)^3} \gtrsim f_c \quad (7.6)$$

where $v_c$ is the circular velocity of the isothermal halo. You can now determine the redshift at which the mass inside the radius of, say, $r = (10, 30, 100)h^{-1}$ kpc could have been assembled. Assuming that the circular velocity of the protogalaxy is $v_c \approx 200$ km s$^{-1}$, we find that $(1 + z_f) = (7.6, 3.0, 1.5)\Omega^{-1/3}$ for $r = (10, 30, 100)h^{-1}$ kpc. Among other things, this simple calculation shows that the galaxy formation is a fairly extended process in time.

Incidentally, we note that still larger structures than galaxies, called *galaxy clusters*, with masses of about $10^{47}$ gm, radius of about 3 Mpc and mean density of $10^{-27}$ gm cm$^{-3}$ exist in the universe as gravitationally bound systems. Our argument given above shows that the gas in these structures could *not* have yet cooled and will have a virial temperature of about $T \approx (GMm_p/Rk_B) \approx 4 \times 10^7$ K. This is also seen more directly by comparing the age of the universe with the free fall collapse time. The cosmic time $t$ and the redshift $z$ can be related to each other once the form of $a(t)$ is known. In the case of a matter dominated universe with $a(t) = (t/t_0)^{2/3} = (1 + z)^{-1}$, we have

$$t(z) = t_0(1+z)^{-3/2} = (2/3)H_0^{-1}(1+z)^{-3/2} \approx 6.5h^{-1}(1+z)^{-3/2} \text{ Gyr} \quad (7.7)$$

while other cosmological models can change this value by a factor of few. On the other hand, the free fall time scale under gravity for a system of mass $M$ and initial radius $R_{\text{init}}$ is given by

$$t_{\text{ff}} = \left(\frac{\pi^2}{8G}\right)^{1/2} R_{\text{init}}^{3/2} M^{-1/2} \approx 0.5 \text{ Gyr} \left(\frac{R_{\text{init}}}{100 \text{ kpc}}\right)^{3/2} \left(\frac{M}{10^{12} M_\odot}\right)^{-1/2}$$

$$\approx 1.6 \text{ Gyr} \left(\frac{R_{\text{init}}}{10 \text{ Mpc}}\right)^{3/2} \left(\frac{M}{10^{15} M_\odot}\right)^{-1/2}. \tag{7.8}$$

The two lines give scalings appropriate for galaxies and clusters. Comparing Eq. (7.7) with the two lines of Eq. (7.8) we see that the peak epoch of galaxy formation should be in the range $z \approx (2-10)$ depending on the cosmological models while cluster formation is probably a still ongoing process.

Since the process of gravitational instability, leading to the condensation of galaxy like objects cannot be 100 per cent efficient, it would leave some amount of matter distributed (almost) uniformly in between the galaxies. The light from distant galaxies will have to pass through this matter and will contain signature of the state of such an *intergalactic medium* (IGM). The photons (with $\nu > \nu_I$ where $h\nu_I = 13.6$ eV) produced in the first generation objects could cause significant amount of ionisation of the IGM, especially the low density regions. When a flux of photons (with $\nu > \nu_I$) impinge on a gas of neutral hydrogen with number density $n_H$, it will have an ionisation optical depth of $\tau = n_H \sigma_{\text{bf}} R$. Setting $\tau = 1$ gives a critical *column* density for ionisation to be $N_c \equiv n_H R = \sigma_{\text{bf}}^{-1} \simeq 10^{17}$ cm$^{-2}$ (see the discussion after Eq. (2.108)). Regions with hydrogen column density greater than $10^{17}$ cm$^{-2}$ will appear as patches of neutral regions in the ionized plasma of IGM. Such regions can be studied by absorption of light from more distant sources, (especially through Lyman alpha absorption corresponding to the transition between $n = 1$ and $n = 2$ levels) and are called *Lyman alpha clouds*. We will discuss them in detail in Sec. 9.4.

Let us next consider some aspects of the processes which actually generate the luminosity of the galaxies. If a fraction (say, half) of the typical binding energy $(GM^2/R)$ of an average galaxy is radiated over the free fall time, $(GM/R^3)^{-1/2}$, then a simple calculation shows that the luminosity is about $L \simeq 5 \times 10^9 L_\odot$ — which is not much. But then, galaxies mostly shine through starlight and not by radiating the gravitational potential energy. Higher luminosities can be easily achieved by nucleosynthesis which produces the stellar light. About 30 MeV of energy is released for every baryon which is converted to metals in stellar nucleosynthesis. If the mass of metals produced is $M_Z \equiv Z M_*$, where $M_*$ is the total mass burnt in stars during the protogalactic star bursts, the energy released is

$$E_{\mathrm{nucl}} = 30 \text{ MeV} \left( \frac{M_Z}{m_p} \right) = 3 \times 10^{-2} Z M_* c^2. \qquad (7.9)$$

Taking $Z \simeq Z_\odot \simeq 10^{-2}$ (which is the typical metallicity of a sun-like star) we find that $(E_{\mathrm{nucl}}/M_* c^2) \equiv \Delta X \approx 3 \times 10^{-4}$. The energy released by this process will be

$$E_{\mathrm{nucl}} \simeq M_* c^2 \Delta X \simeq 7.2 \times 10^{61} \text{ erg} \left( \frac{M_*}{10^{11} M_\odot} \right) \left( \frac{\Delta X}{4 \times 10^{-4}} \right). \qquad (7.10)$$

If *this* energy is released in a time scale $t_{\mathrm{ff}} \simeq 10^8$ yr, the resulting luminosity is about $5 \times 10^{12} L_\odot$ and the mean star formation rate would be several hundred solar mass per year. Even if the process proceeded on a more extended, hierarchical, time scale of $10^9$ yr, the resulting luminosity will be comparable to that of a bright galaxy. This suggests that a generic protogalaxy will have a luminosity of $L_{\mathrm{gal}} \simeq (10^{10} - 10^{12}) L_\odot$. Taking a luminosity distance of about $d_L \simeq H_0^{-1} c \simeq 2 \times 10^{28}$ cm, we find that the expected bolometric flux will be about $(10^{-14} - 10^{-16})$ erg cm$^{-2}$ s$^{-1}$. Equating this bolometric luminosity to $\nu F_\nu$ we get the R band magnitude to be about 25, which — in more sensible units — is $F_\nu \approx 0.1$ mJy at $\lambda \approx 0.3$ mm.

The above discussion also illustrates the important link between star formation, production of metals and production of starlight. These are related by:

$$\frac{dM_Z}{dV dt} \approx Z \frac{dM_*}{dV dt}; \qquad \frac{dE}{dV dt} \approx 0.03 Z \frac{dM_*}{dV dt}; \qquad \frac{dE}{dV dt} \approx 0.03 \frac{dM_Z}{dV dt}. \qquad (7.11)$$

Thus the density of starlight produced per second and the density of metals produced per second are proportional to the star formation rate. These are of use in different contexts in the study of structure formation.

These processes eventually lead to a background of photons in different wave bands, usually called *Extragalactic Background Light* (called EBL) which we will discuss briefly. The unit for measuring the $\nu I_\nu$ of EBL can be conveniently taken to be nW m$^{-2}$ ster$^{-1}$ = $10^{-6}$ ergs s$^{-1}$ cm$^{-2}$ ster$^{-1}$; multiplying this quantity by $(4\pi/c)$ will give the spectral energy density $U_\nu = (4\pi/c)(\nu I_\nu)$ with the dimensions of ergs cm$^{-3}$. Dividing the energy density by $\rho_c c^2$ will give the density parameter $\Omega$ contributed by the radiation background; roughly 1 nW m$^{-2}$ ster$^{-1}$ corresponds to $\Omega h_{50}^2 = 10^{-7}$. The majority of this EBL at UV to IR wavelengths originates from the

starlight in the universe. In particular, the far UV (1000 − 1600) Å light
from the galaxies is produced by the same hot massive stars which also
generate most of the metals. Thus the EBL in this band is a good tracer
of star formation rate and also traces the enhancement of the metallicity
in the universe. Observations have converged to a EBL flux in the range
of (2 − 5) nW m$^{-2}$ sr$^{-1}$ in this band. (Of this, quasars and active galactic
nuclei — which we will discuss in chapter 9 — probably contribute less than
0.2 nW m$^{-2}$ sr$^{-1}$ with the remaining amount arising from the galaxies.)
Comparing the cumulative flux of detected sources in the optical with the
EBL flux suggests that nearly 50 per cent of the flux arises from unresolved
sources. These observations also suggest that the energy density does not
vary too much over the spectral range $(0.1 − 10^3)\mu$.

The sources of EBL at optical (3000 Å - 1$\mu$) and IR $(1 − 1000)\mu$ are more
complex. The spectral energy density of a conventional stellar population
peaks around $(1 − 1.5)\mu$ during most of the stellar ages; hence the optical
EBL will include older populations at low redshifts and younger population
at high redshifts. More importantly even the normal stellar emission at UV
- optical bands will indirectly contribute to IR band along the following
route: During the period of active star formation, most of the light in a
galaxy will be emitted in the optical and UV bands and the spectrum (if
not obscured by dust) will be almost flat ($I_\nu \approx$ constant) up to the Lyman
continuum cut-off at 912 Å. This light is associated with the most luminous
stars which have the shortest life time. The star formation rate of $1M_\odot$
yr$^{-1}$ leads to a luminosity of about $2.2 \times 10^9 L_\odot$. The presence of dust in
galaxies lead to re-processing of the optical-UV light into FIR. This will
lead to correspondingly large amount of FIR radiation from the primeval
galaxies.

If the total bolometric intensity of star light is taken to be about 50 nW
m$^{-2}$ ster$^{-1}$, for the sake of estimate, then the corresponding energy density
is

$$U = 2 \times 10^{-14}(1 + \bar{z}) \text{ erg cm}^{-3} = 1 \times 10^{-8}(1 + \bar{z}) \text{ MeV cm}^{-3} \quad (7.12)$$

where $(1 + \bar{z})$ factor corrects for the cosmological redshift with $\bar{z}$ denoting
the mean epoch of emission. Each baryon that is converted to helium
releases about 25 MeV of energy; this figure rises to a total of about 30
MeV per baryon when we take into account the conversion from hydrogen
to heavy elements. Dividing $U$ by 30 MeV gives the mean baryon number
density in metals that is involved in the production of background light to

be $n_Z \approx 4 \times 10^{-10}(1 + \bar{z})$ cm$^{-3}$. If we take the baryon number density to be $n_B = 1.1 \times 10^{-7}$ cm$^{-3}$, then the ratio of mass fraction in heavy elements which can produce the observed EBL is $Z \approx 4 \times 10^{-3}(1 + \bar{z})$. If the bulk of radiation was produced at $\bar{z} = 2$ then the corresponding metallicity is $Z \approx 0.01$. This gives the ball park figure for consistency between production of metals and EBL.

## 7.2　Star formation history of the universe

The previous discussion highlights the importance of understanding the star formation history of different types of galaxies in order to make useful predictions about them. We shall now discuss some aspects of this.

Observationally the star formation in galaxies is studied by using different diagnostics all of which have different levels of reliability. One procedure is to study the UV emission from the galaxies — which — is directly connected to the formation of metals since the same massive stars which produced the bulk of the UV photons also synthesize the metals and disperse them through supernova. In contrast, the relation between UV emission and *total* star formation rate is lot more problematic since the low mass stars of any standard IMF contains most of the mass while it is the high mass stars which produces most of the UV emission and the metals. Nevertheless, there have been several attempts to connect the luminosity density in the high redshift galaxies with the results of low redshift surveys and obtain an overall picture of the star formation rate (SFR) as a function of redshift. These approaches estimate the amount of stars produced at any given epoch from the abundance of UV luminosity of the star forming galaxies. A simple phenomenological relationship between star formation rate and far UV luminosity, which seems to be consistent with observations, is

$$\text{SFR}(M_\odot yr^{-1}) = 1.4 \times 10^{-28} L_{\text{FUV}}(\text{erg s}^{-1} \text{ Hz}^{-1}). \qquad (7.13)$$

This relation applies to continuous star formation with a Salpeter initial mass function between $(0.1 - 100)M_\odot$.

The key difficulty in using such a relation arises from unknown amount of dust obscuration in the star forming galaxies. If the ISM of primeval galaxies contains large amount of dust which absorbs UV radiation and converts it into IR radiation, then the estimated star formation rate will be lower than the actual rate. In fact, the presence of dust in galaxies can lead to anything between 50 to 80 per cent of UV light being absorbed. There is also some uncertainty arising from the fact that stellar IMF is poorly known at the low mass end where most of the mass is contained.

If we assume that the properties of the dust and intrinsic spectra of galaxies at high redshifts are similar to those of star burst galaxies in the local universe, then the correction is about factor 5 at $z = 3$. Different models and assumptions could vary this correction factor anywhere in the range of $(2 - 5)$. This, in turn, suggests that the star formation rate observed in the universe is consistent with the assumption of near constancy at about $10^{-0.8} M_\odot$ yr$^{-1}$ Mpc$^{-3}$ in a redshift range of $(1 \lesssim z \lesssim 4.5)$. One convenient way of parameterizing the star formation rate as a function of redshift is through a fit

$$\dot{\rho}_{\rm sfr} = \frac{Ae^{az}}{e^{bz} + B} \tag{7.14}$$

with some constants $A, B, a, b$. Direct translation of UV luminosities suggests the parameters $a = 3.4, b = 3.8, B = 44.7, A = 0.11 M_\odot$ Mpc$^{-3}$ yr$^{-1}$. Correcting for the dust could change the parameters of the fit to $a = 2.2, b = 2.2, B = 6, A = 0.13 M_\odot$ Mpc$^{-3}$. In the first case, the star formation decreases exponentially at high redshifts, while in the second case it stays nearly constant with redshift. (Of course, these fits are valid only in the range $z \lesssim 5$.) Dividing $\dot{\rho}_{\rm sfr}$ by the critical density $\rho_c = 2.8 \times 10^{11} h^2 M_\odot$ Mpc$^{-3}$, we get the rate of change of $\Omega_{\rm sfr}$ with respect to time. Given the SFR as a function of $z$, one can also obtain the total amount of gas which has been processed into stars by some epoch $t$ from the integral

$$\rho_*(t) = \int_0^t dz \left(\frac{dt}{dz}\right) \dot{\rho}_*(z). \tag{7.15}$$

A somewhat more relevant quantity is the amount of gas that has been converted into stars per Hubble time which is given by $(a/\dot{a})\rho_*$. This quantity — which is not constrained to be monotonic, unlike $\rho_*(t)$ — is shown in Fig. 7.1 for the two sets of fitting functions, covering an extreme range of assumptions regarding the dust. It is likely that the actual value is somewhere between these two curves at any given $z$.

These curves illustrate several interesting points. First of all, note that most of the star formation activity is in the recent past; the peak activity over a Hubble time is around $z \approx 1.2$ (which corresponds to a cosmic time of about $3h_{50}^{-1}$ Gyr). This feature is more emphatically brought out by the time scales shown in the frames. The universe spends less time at high

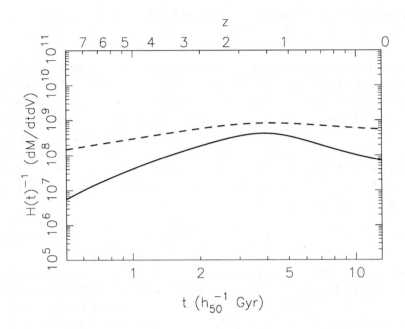

Fig. 7.1    Rate of production of stars in the universe. The curves are fits to data based
on the UV radiation background in the universe. The broken and unbroken lines give the
amount of star formation within a Hubble time in units of $M_\odot$ Mpc$^{-3}$ for two different
parameterization discussed in the text. (Figure based on data provided by C. C. Steidel.)

redshifts and treated as a function of *time* rather than redshift, much of
the star formation occurs recently.

Second, one can estimate the total star formation up to any given time
by integrating the star formation rate. The results could vary by a factor of
few between the two curves but the typical value at $z = 0$ is $\Omega_* \approx 5 \times 10^{-3}$
which is adequate to account for the entire stellar content in disks and
spheroids of the present day universe. Half of this value is reached at
redshift of about $z \approx 2$ in these models.

Further, the stars will also produce the metals in the universe and one
could ask for the amount of metal enrichment at any redshift $z$ due to
a given star formation rate. From standard stellar evolution theory, one
could make a rough estimate that the rate of production of metals which
are ejected to enrich the surrounding medium to be $\dot{\rho}_{metals} \approx 2.4 \times 10^{-2}\dot{\rho}_*$.
For the SFR given above, the amount of metals produced by $z = 2.5$ is about
$0.04\ Z_\odot$ which is in rough agreement with the measurement of metallicity

in quasar absorption systems at high redshifts.

This scenario could also account for, quite naturally, the existence of the cosmic infrared background radiation (CIBR) mentioned in Sec. 7.1. Observations suggest that this background has an intensity of about $\nu I_\nu \approx (24 \pm 5)$ nW m$^{-2}$ sr$^{-1}$ at $\lambda = 140\mu$m and $\nu I_\nu \approx (15 \pm 5)$ nW m$^{-2}$ sr$^{-1}$ at $\lambda = 240\mu$m. The intensity is probably about $\nu I_\nu \approx (40 \pm 5)$ nW m$^{-2}$ sr$^{-1}$ in the wavelength band $\lambda = (100 - 1000)\mu$m. This flux is larger than the "optical" background of about $\nu I_\nu \approx 17$ nW m$^{-2}$ sr$^{-1}$ obtained by counting all the galaxies detected by HST in the band between 0.3 and 3 $\mu$m. To see what these numbers imply, let us make some simple estimates based on a fiducial star formation scenario. The mass density in baryons at a redshift $z$ is:

$$\rho_B = \frac{3H_0^2(1+z)^3}{8\pi G}\,\Omega_B \simeq 7 \times 10^{10}(1+z)^3\,h_{50}^2\,\Omega_B M_\odot\ \text{Mpc}^{-3}. \quad (7.16)$$

If a fraction $f_*$ of the baryons undergoes processing, either through star formation or by gravitational accretion, with a radiative efficiency $\epsilon$, then the locally observed energy density of photons will be

$$\rho_\gamma \cong \rho_B \frac{c^2 \epsilon f_*}{(1+z)^4} \quad (7.17)$$

$$\simeq 5 \times 10^{-30} \left(\frac{\Omega_B h_{50}^2}{0.05}\right) \left(\frac{f_*}{0.1}\right) \left(\frac{2.5}{(1+z)}\right)^4 \left(\frac{\epsilon}{0.001}\right)\ \text{gm cm}^{-3}.$$

(For a salpeter IMF with lower cut-off of $0.1M_\odot$, one can use the standard stellar evolution models to computer $\epsilon$; one gets the efficiency to be about $\epsilon \approx 10^{-3}$ which is the value used in the above scaling). The intensity corresponding to this photon density will be $I = (c/4\pi)(\rho_\gamma c^2)$; equating this total energy to $\nu I_\nu$ we get:

$$\nu I_\nu \simeq 20\ \left(\frac{\Omega_B h_{50}^2}{0.05}\right) \left(\frac{f_*}{0.1}\right) \left(\frac{2.5}{(1+z)}\right) \left(\frac{\epsilon}{0.001}\right)\ \text{nW m}^{-2}\ \text{sr}^{-1} \quad (7.18)$$

which may be compared with the optical-UV background with a bolometric intensity of about

$$\nu I_\nu \bigg|_{\text{opt}} \approx (17 \pm 3)\ \text{nW m}^{-2}\ \text{sr}^{-1} \quad (7.19)$$

in the range $\lambda = (0.1 - 7)\mu$m due to distant galaxies. We are not too widely off.

In this picture, the optical emission arises from quiescent star formation in galaxies by intermediate and low mass stars. This is consistent with the assumption that 10 percent all baryons are processed in the low mass stars at $z \simeq 1.5$. The local density in low mass stars will then be

$$\rho_B(\text{stars}) \simeq 7 \times 10^{10} f_* \, \Omega_B \simeq 3.4 \times 10^8 M_\odot \ \text{Mpc}^{-3} \qquad (7.20)$$

which is consistent with observations based on photometric surveys and standard $(M/L)$ ratios. Further, assuming solar metallicity, we can estimate the local density in metals to be

$$\rho_Z(\text{stars}) \simeq 1.6 \times 10^9 f_* \, \frac{Z}{Z_\odot} \, \Omega_B M_\odot \ \text{Mpc}^{-3} \simeq 7.7 \times 10^6 M_\odot \ \text{Mpc}^{-3}. \quad (7.21)$$

The above discussion illustrates the close interplay between star formation efficiency (which in turn is determined by the poorly known high $z$ IMF), the intensity of extragalactic background radiation and the density of low mass stars in local neighbourhood. Any scenario should be able to account for all these observations.

## 7.3  Intergalactic medium and the Gunn-Peterson effect

We have seen in chapter 6 that formation of neutral matter and the decoupling of radiation occurred at $z \simeq 10^3$. Since it is unlikely that the formation of structures at lower redshifts could have been one hundred percent efficient, we would expect at least a fraction of the neutral material, especially hydrogen, to remain in the intergalactic medium (IGM) with nearly uniform density. This neutral hydrogen could, in principle, be detected by examining the spectrum of a distant source, like quasar. (We will discuss quasars in detail in chapter 9; right now we are only using them as sources of photons produced at high-$z$.) Neutral hydrogen absorbs Lyman-$\alpha$ photons, which are photons of wavelength 1216 Å that corresponds to the energy difference between the $n = 1$ and the $n = 2$ states of the hydrogen atom. Because of the cosmological redshift, the photons that are absorbed will have a shorter wavelength at the source and the signature of the absorption will be seen at longer wavelengths at the observer. We, therefore, expect the spectrum of the quasar to show a dip (known as *Gunn-Peterson effect*) at wavelengths on the blue side (shortwards) of the Lyman-$\alpha$ emission line if neutral hydrogen is present between the source and the observer.

The magnitude of this dip depends on the neutral hydrogen density and can be calculated using the optical depth for such absorption. The general formula for calculating the optical depth $\tau$, of course, is

$$\tau[\nu_0; z] = \int \sigma \, n \, c \, dt = \int_0^z \sigma(\nu) n(z) \left| \frac{c dt}{dz} \right| dz; \qquad \nu = \nu_0(1+z). \quad (7.22)$$

In the case of Lyman-$\alpha$ photon, the absorption cross section is

$$\sigma(\nu) = \frac{\pi e^2}{m_e c} f \delta_D(\nu - \nu_\alpha), \qquad (7.23)$$

where $f = 0.416$ is the oscillator strength and $\nu_\alpha$ is the frequency corresponding to the Lyman-$\alpha$ photon (see Eq. (2.102)). The Dirac delta function in $\sigma(\nu)$ allows us to do the integration in Eq. (7.22) easily and we get

$$\tau_{\mathrm{GP}}(z) = \left( \frac{\pi e^2 f}{m_e c \nu_\alpha} \right) \frac{n_{\mathrm{HI}}(z)}{(1+z)} d_H(z) \qquad (7.24)$$

where

$$d_H(z) = c H_0^{-1} [\Omega_{\mathrm{NR}}(1+z)^3 + \Omega_V]^{-1/2} \qquad (7.25)$$

in a cosmology with $\Omega_{\mathrm{NR}} + \Omega_V = 1$. Here $z$ is related to observing frequency $\nu_o$ by $(1+z) = (\nu_\alpha/\nu_o) = (\lambda/\lambda_\alpha)$. Clearly, all $\nu > \nu_\alpha$ will contribute to the absorption and will lead to an absorption trough in the continuum blueward of the Lyman-$\alpha$ emission line from the source.

To estimate the numbers, let us consider the simple case of $\Omega_{\mathrm{NR}} = 1, \Omega_V = 0$ universe. Then we get

$$\tau_{\mathrm{GP}}(z) = \frac{\pi e^2 f}{m_e H_0 \nu_\alpha} \frac{n_{\mathrm{HI}}(z)}{(1+z)^{3/2}} = 6.6 \times 10^3 h^{-1} \left( \frac{\Omega_B h^2}{0.019} \right) \frac{n_{\mathrm{HI}}}{\bar{n}_H} (1+z)^{3/2} \qquad (7.26)$$

where we have expressed the answer in terms of the mean density of hydrogen nuclei at a redshift $z$, given by

$$\bar{n}_H = \frac{\rho_{\mathrm{crit}}}{m_H} (1-Y) \Omega_B (1+z)^3 = (1.6 \times 10^{-7} \text{ cm}^{-3}) \left( \frac{\Omega_B h^2}{0.02} \right) (1+z)^3 \quad (7.27)$$

(where $Y = 0.22$ is the helium fraction). Equation Eq. (7.26) shows that even if neutral hydrogen is only a small fraction of the total hydrogen density, the optical depth should be fairly high.

No such absorption troughs are seen in the continuum spectra of the quasars! The current upper bound at $z \simeq 5$ is about $\tau_{\mathrm{GP}} < 0.1$, implying that $(n_{\mathrm{HI}}/\bar{n}_H) < 10^{-6}h$. Even if 99 per cent of all the baryons are confined to bound structures and only 1 per cent remained as smoothly distributed component, this result requires the diffuse IGM to be ionized to a level that is better than one part in $10^4$. Assuming that the bound on the optical depth is typically like $\tau \lesssim 0.1$ in the redshift range $0 < z < 5$, we may write the above bound as

$$\Omega_{\mathrm{NR}}(n_H) \lesssim 2 \times 10^{-7} h^{-1}(1+z)^{-3/2}. \qquad (7.28)$$

As we shall see, this poses quite a bit of a problem which is still an open issue. We shall discuss this in some detail in the next section.

## 7.4   Ionisation of IGM

The Gunn-Peterson effect described above shows that IGM has to be predominantly ionized at redshift $z \lesssim 5$. How to do this consistently in a structure formation scenario is one of the key open questions in cosmology. Roughly speaking, we need to form sufficient amount of structures which can emit the required number of UV photons that can ionize the IGM by $z = 5$ or so. This turns out to be not easy.

One can think of collisional ionisation or photoionisation as two possible means of achieving the ionisation. The collisional ionisation requires an IGM temperature of $(10^5 - 10^6)$ K or — equivalently — about 25 eV per atom. The evolution of stellar population, leading to supernova explosions (which reheat the ISM and contaminate it by metals), injects both mechanical energy as well as UV photons into the surroundings. But during the evolution of a typical stellar population, more energy is injected in the form of UV photons than in mechanical form. This is because the energy radiated per baryon (in nuclear burning) which eventually leads to solar metallicity $Z_\odot \simeq 0.02$, is about $0.007\ Z_\odot m_H c^2$ — of which about one-third is radiated as UV light. The massive stars, when they eventually explode a supernova, inject most of the metals as well as about $10^{59}$ ergs of energy per explosion into the surroundings. For Salpeter IMF, there is about one supernova per $150\ M_\odot$ of baryons in the form of stars. The mass fraction in mechanical energy works out to be about $4 \times 10^{-6}$ which is 10 times lower than the energy released in ionizing photons. This suggests that between the two possibilities, the photoionisation could be a major effect in the physics of IGM. Hence we shall ignore collisional ionisation and concentrate on photoionisation.

In the case of photoionisation, the source of photons could be quasars or stars with $M > 10 M_\odot$. We will see below that a high degree of ionisation requires an energy input of about 13.6 $(1 + t/t_{\rm rec})$ eV per hydrogen atom where $t_{\rm rec}$ is the volume averaged hydrogen recombination time scale and $t$ is the cosmic time. The factor $(t/t_{\rm rec})$ is large compared to unity even at $z \simeq 10$ and thus photoionisation requires significantly higher energy than just 13.6 eV per hydrogen atom.

If the mean intensity of ionizing radiation (in units of energy per unit area, time, solid angle, frequency interval) is $J_\nu$, then the photoionisation rate per unit volume is given by

$$\mathcal{R} = n_{\rm HI} \int_{\nu_I}^{\infty} \frac{4\pi J_\nu \sigma_H(\nu)}{h\nu} d\nu \qquad (7.29)$$

where the photoionisation cross section for hydrogen is given by

$$\sigma_I = \sigma_0 \left( \frac{\omega_I}{\omega} \right)^{7/2}, \qquad (7.30)$$

with $\sigma_0 \simeq 8 \times 10^{-18}\,{\rm cm}^2$ and $\omega_I \simeq 2 \times 10^{16}$ Hz (see Eq. (2.106) and note that the dominant effect is from $k_B T \simeq \hbar\omega$). This ionizing flux has to come from a variety of sources at high redshift. We shall see later examples of different ionizing sources and their spectrum. Right now, for the sake of illustration, we will consider a spectrum of the form

$$J_\nu = J_0 \frac{\nu_I}{\nu} \times 10^{-21}\,{\rm erg\ cm}^{-2}\,{\rm s}^{-1}\,{\rm Hz}^{-1}\,{\rm ster}^{-1}, \qquad (7.31)$$

where $\nu_I = 3 \times 10^{15}$ Hz is the ionisation threshold corresponding to $\lambda \simeq 912$ Å. It is assumed that $J_0$ is of order unity. The magnitude of the intensity — with $J_0 \approx 1$ — is typically what arises in most models involving star forming galaxies or quasars as the source which explains the choice of normalization. The $\nu$–dependence is more uncertain and is essentially modeled by the continuum spectrum of quasars (see chapter 9). The rate of ionisation per hydrogen atom due to the flux $J_\nu$ will be

$$\mathcal{R} = \int_{\nu_I}^{\infty} \frac{4\pi J_\nu}{h\nu} \sigma_I d\nu \simeq 3 \times 10^{-12} J_0\,{\rm s}^{-1}. \qquad (7.32)$$

In this case, the equilibrium between photoionisation and recombination will lead to the condition

$$\mathcal{R} n_H = \alpha n_e n_p = \alpha n_p^2, \qquad (7.33)$$

if $n_e = n_p$. In reality we also have to take into account the fact that the plasma may not fill space uniformly but could be clumped. If we assume that a fraction $(1/C)$ of space is uniformly filled by plasma with the rest of the region empty, we need to multiply the right hand side of the Eq. (7.33) by a factor $C$. (Note that, in this case, the mean square density will be $< n^2 > = (n^2/C) = < n >^2 C$). Then, we obtain

$$\frac{n_H}{n_p} = \frac{\alpha C n_p}{\mathcal{R}} \simeq 10^{-6}\frac{C\Omega_{\mathrm{IGM}}h^2(1+z)^3}{J_0}\left(\frac{T}{10^4\,\mathrm{K}}\right)^{-1/2}. \qquad (7.34)$$

Using the Gunn-Peterson bound on $(n_p/n_H)$ we can translate this equation to

$$\Omega_{\mathrm{IGM}} \lesssim 0.4\frac{\Omega_{\mathrm{NR}}^{1/4}J_0^{1/2}}{h^{3/2}C^{1/2}}\left(\frac{T}{10^4\,\mathrm{K}}\right)^{1/4}(1+z)^{-9/4}. \qquad (7.35)$$

While the above bound by itself is not very restrictive, the detailed mechanism for photoionisation as well as the issue of generating the necessary background intensity turns out to be quite non trivial.

What we are really interested in is a statistical description of the ionisation state of IGM. This can be done in terms of a volume filling factor $Q$ for the ionized regions. In the early epochs, $Q$ will be far less than unity and the radiation sources will be randomly distributed. The regions which are fully ionized (usually called *Stromgen spheres*) will be isolated from each other and every UV photon is absorbed somewhere in the IGM leading to a highly inhomogeneous UV radiation field. As time goes on, $Q$ increases and percolation of ionized regions occur around $Q = 1$. Since the mean free path of Lyman continuum radiation is much smaller than the Hubble radius, the filling factor $Q$ can be easily estimated by the following argument. The fact that every ultraviolet photon is either absorbed by a newly ionized hydrogen or by a recombining one, implies that $Q(t)$ must be equal to the total number of ionizing photons emitted per hydrogen atom until that moment $t$, minus the total number of radiative recombinations per atom. That is, we must have

$$Q(t) = \int_0^t dt'\frac{\dot{n}_{\mathrm{ion}}(t')}{n_H} - \int_0^t dt'\frac{Q}{t_{\mathrm{rec}}}. \qquad (7.36)$$

Differentiating, we get

$$\frac{dQ}{dt} = \frac{\dot{n}_{\mathrm{ion}}}{\bar{n}_H} - \frac{Q}{t_{\mathrm{rec}}} \qquad (7.37)$$

which describes statistically the transition of the universe from neutral to ionized form. Given a model for $\dot{n}_{\text{ion}}$, this equation can be integrated to describe the filling factor. Initially, when $Q \ll 1$, recombination can be neglected and the ionized volume increases at a rate $(\dot{n}_{\text{ion}}/n_H)$. At late times, in the limit of a fast recombining IGM, the asymptotic value of $Q$ will be

$$Q \lesssim \frac{\dot{n}_{\text{ion}}}{\bar{n}_H} \, t_{\text{rec}}. \tag{7.38}$$

This shows that the volume filling factor is less than the Lyman continuum photons emitted per hydrogen atom within one recombination time. In other words, of all the photons with $\nu > \nu_I$, only a fraction $(t_{\text{rec}}/t) \ll 1$ is actually used to ionize the new IGM material. The condition for complete reionisation of the IGM will then be

$$\dot{n}_{\text{ion}} t_{\text{rec}} \gtrsim \bar{n}_H \tag{7.39}$$

which says that the rate of emission of UV photons must exceed rate of recombination.

If the matter is clumpy, we need to multiply the right hand side by a clumping factor $C$. Assuming that average clumping factor is a good descriptor and that $t_{rec} \ll t$ we can use Eq. (7.39) to arrive at a critical rate of emission for ionizing photons, $\dot{N}_c$ per unit comoving volume, independent of the previous history. This is given by

$$\dot{N}_c(z) = \frac{\bar{n}_H(0)}{t_{\text{rec}}(z)} = \left(2.51 \times 10^{51} \text{ s}^{-1} \text{ Mpc}^{-3}\right) \left(\frac{C}{10}\right) \left(\frac{1+z}{10}\right)^3 \left(\frac{\Omega_B h^2}{0.02}\right)^2. \tag{7.40}$$

To see the implications of this number, it is useful to determine an effective star formation rate which will provide this rate of photon emission. With usual IMF, for each $1 M_\odot$ star, a fraction 0.08 goes into massive stars with $M > 20 M_\odot$. In these massive stars, at the end of carbon burning, nearly half the initial mass is converted into He and C, with a nuclear energy generation efficiency of 0.007. Further, about 25 percent of the luminosity is in the form of Lyman continuum photons with $E > 20$ eV. Thus, for every $1 M_\odot$ star that is formed per year, we expect an emission rate of

$$N_{\text{photons}} \approx 0.08 \times 0.5 \times 0.007 \times 0.25 \times \left(\frac{M_\odot c^2}{20 \text{ eV yr}}\right) \approx 10^{53} \text{ photons s}^{-1}. \tag{7.41}$$

Using this, we can convert the critical UV photon flux to an equivalent minimum star formation rate per unit comoving volume, $\dot{\rho}_* = (\dot{N}_c/N_{\mathrm{photons}}f_{\mathrm{esc}})$ where $f_{\mathrm{esc}}$ is a factor which determines what fraction of the photons escape out of the star forming region. Numerically, for $\Omega_B h^2 = 0.02$, we have

$$\dot{\rho}_* \approx \left(0.12 M_\odot \ \mathrm{yr}^{-1} \ \mathrm{Mpc}^{-3}\right) \left(\frac{0.5}{f_{\mathrm{esc}}}\right) \left(\frac{C}{10}\right) \left(\frac{1+z}{10}\right)^3. \tag{7.42}$$

This is a fairly high rate of star formation and the energy injected into IGM is comparable to (or even larger than) the energy radiated by quasars at the peak of their activity — which raises the question as to whether it is indeed possible to provide such a high rate of ionizing photons in the universe. It should be stressed that this estimate depends sensitively on the value assumed for $f_{\mathrm{esc}}$ which denotes the fraction of high energy ionizing photons which can escape the star forming regions and reach the IGM. Direct observations of nearby galaxies at the present epoch suggest that this fraction can be very low — possibly only a few percent. If this is indeed true for the high redshift proto galaxies as well, then one would require significantly higher star formation rate in order to make this model work.

A more formal way of doing this calculation is as follows: The galaxy survey data for star forming galaxies at $2 \lesssim z \lesssim 4$ provide direct information about the UV flux of these galaxies, since the rest frame UV continuum at 1500 Å will be redshifted into the visible band for a source at $z = 3$. There has been several detailed studies of ultra violet luminosity functions of Lyman-break galaxies spanning a factor of about 40 in luminosity. Integrating the luminosity function over $L \gtrsim 10^9 L_\odot$ and using the fact that — for a Salpeter mass function — the luminosity at 1500 Å is about 6 times the luminosity at 912 Å, one can determine the comoving emissivity at the ionisation threshold, $\nu = \nu_I$, for these galaxies. The resulting emissivity is about

$$I = \begin{cases} (9 \pm 2) \times 10^{25} h \ \mathrm{erg} \ \mathrm{s}^{-1} \ \mathrm{Hz}^{-1} \ \mathrm{Mpc}^{-3} & (\text{for } z \approx 3) \\ (7 \pm 2) \times 10^{25} h \ \mathrm{erg} \ \mathrm{s}^{-1} \ \mathrm{Hz}^{-1} \ \mathrm{Mpc}^{-3} & (\text{for } z \approx 4). \end{cases} \tag{7.43}$$

There is, however, still the uncertainty as to what fraction of the photons actually escape the galaxy, which need to be addressed. If we assume that about 50 per cent of the photons escape (which is far higher than the fraction which escapes our galaxy today), then one gets about $1.58 \times 10^{51}$ photons $\mathrm{s}^{-1}$ $\mathrm{Mpc}^{-3}$ at $z = 4$ which is just high enough to provide the necessary ionisation. The situation at higher redshifts, of course, is quite uncertain due to lack of direct observational evidence.

## Exercise

**Exercise 7.1**   *Non conventional CMBR:*

Suppose all the energy in CMBR was produced by previous generation of stars and "somehow" thermalised. What would have been the corresponding metallicity of the universe? Does the energy density and metallicity make sense?

**Exercise 7.2**   *Gravity vs nuclear:*

An over dense region in the universe grows by gravitational instability and collapses to form a virialized halo. Baryons cool in this halo forming a galaxy which forms stars that emit radiation. In the entire process, what is the ratio between the nuclear energy which is converted to photons and the gravitational energy that is emitted as radiation?

**Exercise 7.3**   *Metallic shine?*

If the average metallicity of the universe is one-tenth the solar metallicity, estimate the mean bolometric surface brightness of the sky in magnitude arcsec$^{-2}$.

**Exercise 7.4**   *Tinkering with IMF:*

Study of population synthesis models suggest that for the Salpeter IMF the UV luminosity of a galaxy is related to the star formation rate by Eq. (7.13). A drop in the observed value of $L_{\rm UV}$ is what leads to the corresponding drop in SFR at $z > 1.2$. This depends on Eq. (7.13) which in turn depends on Salpeter IMF being valid for the range of masses $(10 - 125)M_\odot$ which is the main source of UV luminosity. Assume that the Salpeter IMF ($\propto M^{-\alpha}$ with $\alpha = 2.35$) gets modified at $z > 1.2$. Suppose the index $\alpha$ varies from $\alpha = 2.35$ at $z = 1.2$ to $\alpha = 2.8$ at $z = 4$. What will happen to the derived value of star formation rate in this case? Can this scenario be constrained in any other manner?

**Exercise 7.5**   *End result:*

Consider the diffuse photoionised intergalactic medium with an over density of $\delta\rho/\rho \approx$ (5 to 10). Long after reionisation is over, this medium can be described by an approximate equation of state with $P \propto \rho^n$. Determine $n$.

**Exercise 7.6**   *Poping once more:*

Astronomers have postulated Pop III stars which have very low core metallicity (say, $Z = 10^{-10}$) and releases energy through CNO cycle. On the other hand, standard Pop II stars has a metallicity of $Z = 10^{-4}$. If a 15 $M_\odot$ Pop II star has $L = 3 \times 10^4 L_\odot, T_{\text{sur}} \approx 4 \times 10^4$ K, determine the corresponding values for a 15 $M_\odot$ Pop III star. What will be its lifetime and emissivity of ionizing photons? Can this help in re-ionizing the universe?

## Notes and References

*Section 7.2:* The fits to star formation rates are based on Steidel *et al.* (1999) and are discussed in detail in chapter 1 of Padmanabhan (2002)

Chapter 8

# Normal Galaxies

The physics of stars, described in chapter 4 was relatively straight forward and one could obtain the basic features of stellar structure starting from the first principles. Ideally, one would have liked to do the same for the galaxies, which constitute the next higher level structures in the universe. This, however, turns out to be quite difficult. As we saw in the last two chapters, the formation of galaxies involves condensation of gaseous structures within the gravitational potential well provided by the dark matter halos. Several aspects of this process are still uncertain preventing us from obtaining the details of galactic physics from first principles. Hence we have to take a more phenomenological approach in this (and the next) chapter. We will describe some of the observed features of galaxies and then try to link up with the theoretical ideas without attempting to "derive" the properties from first principles. As you will see, one can make fair amount of progress by this approach.

## 8.1 Morphological classification of galaxies

The morphology of a galaxy — which is the simplest feature that can be ascertained from observations — will depend on the ratio between the time scale for star formation and the time scale for gravitational collapse.

If most of the early, population II, stars form over a time scale somewhat shorter than the gravitational free fall time, then the protogalactic gas will be converted into a collisionless system of stars before the collapse is well underway. This will lead to spheroidal morphologies with stars moving in orbits with large eccentricity in the mean gravitational field of the galaxy. In this case, each galaxy forms in its own environment through an isolated collapse. One can also think of a variant of this scenario, in which a large galaxy forms through mergers of a number of protogalactic fragments — but in a time scale comparable to the free fall time of the system — accompanied by rapid star formation activity. Since the time scales are comparable, these two scenarios are hardly distinguishable from each other observationally.

It is also possible to imagine another extreme scenario in which the merger of large number of fragments occur over a time scale *comparable* to the Hubble time $t(z) \simeq H(z)^{-1}$ with most of the stars having *already* formed within the merging sub units. This is quite different from the first scenario and will, in general, lead to different final state. The gas, settling down inside a dark matter potential well, will gradually evolve to the lowest energy configuration for a given amount of angular momentum — which will be a centrifugally supported, rotating, thin disk. This occurs *before* most of the gas is converted into stars and hence this scenario — with lower star formation rate — will lead to *disk like* systems. At high redshifts, when merging was more frequent, the formation of massive galactic disks would have been unlikely since they are almost invariably destroyed during encounters. As the merging rate decreases, the gradual assembly of disk shaped galaxies increase.

In reality, you will find that most of the galaxies fall somewhere in between these two extreme configurations. Most galaxies, when observed in visible band, indeed seem to be made of the two components we discussed above, loosely called *bulge* and *disc* with the understanding that there could exist galaxies in which one component may be completely absent. The bulge dominates the central portions of the galaxies, is reasonably spherical in shape and is supported against self-gravity by the orbital motion of the stars in the mean gravitational field. Most of the bulge stars are population II and have a large radial component of orbital velocity. The disc, in contrast, is a flattened structure in which stars move in almost circular orbits and it contains young massive population I stars, interstellar dust and molecular clouds.

Different morphological types of galaxies correspond to combining the bulge and disk components in different proportions. If the galaxy has virtually no disc, it is usually called an *elliptical* and is classified as E0,E1,.... where the notation E$n$ denotes a shape in which the ratio, $(b/a)$, of major to minor axis is given by $(b/a) = (1 - n/10)$. If the bulge-to-disk contribution is of order unity and the disk is reasonably smooth (with no spiral structure *etc.*) the galaxy is called *lenticular* and is denoted by S0. In the same configuration, if the disk shows additional structure in the form of spiral arms, it is called a *spiral* galaxy and is denoted by Sa. If, further, the central region of spiral galaxy contains a prominent bar-like structure, such galaxies are denoted by Sb. Galaxies which do not fall in any of these categories are called — rather unimaginatively — as *irregulars*. This particular classification scheme is called the *Hubble classification* and the E0, E1 *etc.* are said to represent different *Hubble types*. Historically, there was an idea that the sequence E0, E1, ... ,S0, Sa/Sb could represent an evolutionary sequence. One often finds in literature, references to 'early' and 'late' type galaxies with the convention that E0 is the 'earliest' and Sb is the 'latest'. (We will occasionally use this terminology with the understanding that it has *no* direct physical basis.) The ellipticals and the bulges of disk systems currently contain, in their stellar population, about 63 per cent of the baryonic mass that is in the galaxies in the form of stars, stellar remnants or gas. The disks contain only a total of about 21 per cent of the baryonic mass and the gas in the galaxies comprise of about 15 per cent of the total baryonic mass.

The next relevant question is how stars (as determined by starlight) are distributed *within* a galaxy. Here we are faced with the problem that a galaxy does not have as sharp an edge as compared to, say, a star or a planet. The best thing you could do is to describe the stellar distribution by the surface brightness of the galaxy as seen in the plane of the sky. When a galaxy has a bulge and a disk component, we need to specify the surface brightness profiles for each of them. We shall begin with the disk component which also includes, as a limiting case, the disk galaxies.

### 8.1.1 *Properties of disk galaxies*

The surface brightness profile of disks has an exponential fall-off, given by,

$$\Sigma(R) = \Sigma_0 \exp\left(-R/R_d\right) \qquad (8.1)$$

where $R$ is the 2-dimensional radial coordinate in the plane of the disc and $\Sigma_0, R_d$ are constants. Outside the central region, disk galaxies exhibit this exponential luminosity profile though near the center some minor deviations are often observed. When seen edge-on, the luminosity of the disk arises from a very thin sheet well fitted by the function $\text{sech}^2(z/z_0)$ where $z$ is the vertical coordinate normal to the plane of the disk which is assumed to be located at $z = 0$. (For a theoretical motivation for this dependence, see exercise 3.3.) Combining these results, we may take the 3-dimensional luminosity distribution of a disk as

$$L(R, z) = L_0 \exp(-R/R_d) \operatorname{sech}^2(z/z_0). \tag{8.2}$$

The center brightness of many galaxies are typically in the range of about $(22 \pm 1)$ mag arcsec$^{-2}$ and $z_0$ is few hundred parsecs.

Integrating Eq. (8.1) over all $R$ gives the total luminosity to be $L_{\text{tot}} = 2\pi\Sigma_0 R_d^2$. The corresponding surface brightness projected on the sky, $\mu$, is usually measured in V-band mag arcsec$^{-2}$. If the physical luminosity density is $\Sigma$ (measured in $L_{V\odot}$ per square parsec where $L_{V\odot}$ is the V-band luminosity of the Sun), then these two quantities are related by

$$\Sigma = 10^{(2/5)(26.4-\mu)} L_{V\odot} \ \text{pc}^{-2}. \tag{8.3}$$

For example, $\mu = 26.4$ V-mag arcsec$^{-2}$ will correspond to $1L_{V\odot}$ pc$^{-2}$. The easiest way to derive such relations is to note that surface brightness is independent of the distance (when cosmological expansion is ignored) and calculate both $\mu$ and $\Sigma$ at a convenient distance. We choose a distance

$$D = \frac{180}{\pi}(60)^2 \ \text{pc} = 206265 \ \text{pc} \tag{8.4}$$

at which a length of 1 pc will subtend an angle of 1 arcsec. At this distance, the Sun will have the apparent magnitude

$$m_V = M_{V\odot} + 5 \log \frac{D}{10 \ \text{pc}} = 4.83 + 5 \log(20626.5) = 26.4. \tag{8.5}$$

At this convenient distance 1 square arcsec will contain a luminosity $L = \Sigma L_{V\odot} \cdot (1 \ \text{pc}^2)$ and will have a magnitude $m_V = 26.4 - (5/2) \log \Sigma$. Equating this to $\mu$ and solving for $\Sigma$ we get the relation given above in Eq. (8.3).

In terms of $\mu$, which is preferred by observers, Eq. (8.1) becomes

$$\mu(r) = \mu_0 + 1.086 \left(\frac{R}{R_d}\right). \tag{8.6}$$

A wide class of spirals has an exponential form for the surface brightness, with the central surface brightness ranging from 21.5 to 23 mag arcsec$^{-2}$. Taking a value of $\mu_0 \simeq 21$ V-mag arcsec$^{-2}$ and using Eq. (8.3) we get $\Sigma_0 = 170 L_{V\odot}$ pc$^{-2}$. Thus the surface density profile of these spirals can be expressed in the form

$$\Sigma(R) \approx 170 e^{-R/R_d} L_{V\odot} \ \text{pc}^{-2}. \tag{8.7}$$

In our galaxy, the brightness towards the galactic poles above the Sun (which is located at a radial distance of $R = R_0 \simeq 8$ kpc) is about 24 V-mag arcsec$^{-2}$. This implies [through Eq. (8.3)] that $\Sigma(R_0)/2 = 9L_{V\odot}$ pc$^{-2}$ with the factor half arising from the fact that there are two sides to the disk of Milky Way. If the exponential law is applicable to the Milky Way, then we can immediately determine the scale length using the value of $\Sigma(R_0)$; we get $R_d \approx 3.6$ kpc which is consistent with other observations as well.

Let us next consider the central regions of disk galaxies which have been investigated extensively because many of the galactic bulges are suspected to harbor a massive black hole in the center. The radius of influence, $r_H$, of a black hole of mass $M_{\mathrm{BH}}$ on a stellar system with velocity dispersion $\sigma_*^2$ can be obtained by equating the gravitational potential energy per unit mass $(GM_{\mathrm{BH}}/r)$ due to the central black hole to the kinetic energy per unit mass $(\sigma_*^2/2)$. This gives:

$$ r_H \simeq \mathcal{O}(1)\frac{GM_{\mathrm{BH}}}{\sigma_*^2} \simeq 4\left(\frac{M_{\mathrm{BH}}}{10^7 M_\odot}\right)\left(\frac{\sigma_*}{100 \text{ km s}^{-1}}\right)^{-2} \text{ pc.} \qquad (8.8) $$

Stars located at $r < r_H$ will be significantly affected by the black hole. Unfortunately, this region is rather small to be detectable directly in external galaxies. The angle subtended by this radius, in a galaxy located at a distance $D$ from us, is about

$$ \theta \simeq \frac{r_H}{D} \approx 0.1''\left(\frac{M_{\mathrm{BH}}}{10^7 M_\odot}\right)\left(\frac{\sigma_*}{100 \text{ km s}^{-1}}\right)^{-2}\left(\frac{D}{10 \text{ Mpc}}\right)^{-1}. \qquad (8.9) $$

A vast majority of galaxies have $D > 20$ Mpc and $\sigma > 200$ km s$^{-1}$; in such galaxies, we cannot detect the effect of a black hole unless it is very massive.

While confirmed detection of the black hole is difficult, one can indirectly check for the existence of a black hole (of certain mass $M_{\mathrm{BH}}$) by studying the density profile of stars near the origin. The effect of a black hole on the distribution of stars near the central region depends on the gravitational two body relaxation time, $t_R$. Let us begin with the situation in which we are interested in time scales much shorter than $t_R$; i.e., $t \ll t_R$ with $t_R$ given by (see Eq. (1.66))

$$ t_R = 0.34\frac{\sigma^3}{G^2 M \rho \ln \Lambda} \approx 10^{10}\left(\frac{4 \times 10^5 M_\odot \text{ pc}^{-3}}{\rho}\right) \text{ years} \qquad (8.10) $$

where we have taken $\ln \Lambda \simeq \ln N \approx 14$, $\sigma \simeq 150$ km s$^{-1}$ and $M = 1 M_\odot$ for making the estimate. If the black hole forms by slow accumulation of mass over a time scale which is much larger than the orbital time scale $(R_0/\sigma)$ of the stars, then the evolution of the distribution function of the stars can be estimated in the adiabatic approximation. If the initial distribution function is further assumed to be a Maxwellian, then we have

$$f(E) = \frac{n_0}{(2\pi\sigma^2)^{3/2}} e^{-E/\sigma^2} \qquad (8.11)$$

where $n_0$ is the number density of stars and $E = (v^2/2) + \Phi$ is the energy per unit mass. As an example of the application of this result, consider the density contributed by the stars that eventually become bound to the black hole. For these stars, $E \ll \sigma^2$ and we can take $f \approx n_0/(2\pi\sigma^2)^{3/2}$. Then the space density of the stars becomes

$$n(r) = 4\pi \int_0^{\sqrt{2GM_{BH}/r}} v^2 f(E) dv = \frac{4n_0}{3\sqrt{\pi}} \left(\frac{r_H}{r}\right)^{3/2}; \qquad r_H \equiv \frac{GM_{BH}}{\sigma^2}.$$
$$(8.12)$$

Note that the black hole has introduced a weak cusp $(n \propto r^{-3/2})$ in the density profile.

The situation in the case of $t \gg t_R$ is more complicated. Naive application of Maxwellian distribution in energy will give a density $f \propto \exp[-\Phi(r)/\sigma^2] \propto \exp[r_H/r]$ which diverges exponentially near the origin. This result, of course, is unrealistic because any star having a close encounter with the black hole will be tidally shredded and swallowed by the black hole. If these processes lead to a steady state density distribution $n(r) = n_0(r_H/r)^s$, then equilibrium would require energy balance between the potential energy of the stars which are swallowed by the hole and the flux of positive energy that must flow out through the cusp. Since $v^2 \propto r^{-1}$, the relaxation time will vary as $t_R \propto (v^3/n) \propto r^{(s-3/2)}$. The number of stars, $N(r)$ within a radius $r$ scales as $N(r) \propto r^{3-s}$ while the energy of a star at a given radius $r$ goes as $E(r) \propto r^{-1}$. Hence the rate of flow of energy through a radius $r$ scales as $(N(r)E(r)/t_R) \propto r^{(7/2-2s)}$. If this flow has to be independent of radius $r$, then we must have $s = 7/4$ which is a slightly milder cusp $(n \propto r^{-7/4})$ compared to the one obtained in Eq. (8.12). Numerical simulations show that this result is indeed true.

Thus we can distinguish between two possible limiting behaviours in the core region of a galaxy. The first is that of a smooth stellar core with constant density and the luminosity profile of the form like, say, $I(r) =$

$I_0[1 + (r/r_c)^2]^{-1}$ (where $I_0$ is the central brightness and $r_c$ is the core radius). The second is a central region with a cusp in the density so that the intensity goes as $I \propto r^{-\alpha}$ which is indicative of central black hole. HST observations seem to suggest that the distribution function of stars is indeed cuspy near the central regions of many galaxies, indicating the presence of central black holes.

In many spiral galaxies, the inferred mass of the black hole based on the dynamics of stars is found to be well correlated with the absolute blue magnitude of the bulge. A linear relation exists between $\ln M_{BH}$ and the blue magnitude $M_{B,bulge}$ which is equivalent to the relation $M_{BH} \simeq 0.006 m_{bulge}$ where $m_{bulge}$ is the mass of the bulge.

The disk galaxies show fair amount of rotation and the study of rotation velocities as a function of radial distance has been a valuable tool in mapping the density distribution in and around the disc. In all the cases, the rotation curves (which is just the function $v(r)$) are either flat or gently rising at large radii, as far as the observations go. This is one of the key evidences for the existence of dark matter halos around the galaxies. The shape of the rotation curve exhibits some interesting correlations with other properties of the galaxies. For example, the rotation curves tend to rise more rapidly near the center and peak at higher maximum velocity ($v_{max}$) with increasing B-band luminosity. It is possible to understand the origin of such a scaling relation (between the luminosity of the galaxy and $v_{max}$) under certain conditions. Starting from the result $M \propto R v_{max}^2/G$ and assuming that the $(M/L)$ ratio for disk galaxies is approximately constant, we get $L \propto v_{max}^2 R$. Further, if all disk galaxies have approximately the same central surface brightness, we would expect $(L/R^2)$ to be approximately constant. Eliminating $R$ from these two relations, we find that the total luminosity of the disk scales with the rotation velocity as $L \propto v_{max}^4 = C v_{max}^4$ where $C$ is a constant. This is called *Tully-Fisher relation*. (The same result is obtained even if central brightness, $I(0)$, is not a constant, provided $(M/L) \propto I(0)^{-1/2}$.) At a more fundamental level, we can attempt to relate $v$ and $M$ using the results of Sec. 6.4 for a given fluctuation spectrum $\sigma_0(M)$. For $\sigma_0(M) \propto R^{-(n+3)/2} \propto M^{-(n+3)/6}$, we have $v \propto (M/R)^{1/2} \propto M^{(1-n)/12}$. At galactic scales the power spectrum has the approximate slope, $n \approx -2$ giving $M \propto v^4$. If $(M/L)$ is constant, we again get $L \propto v^4$. Taking the logarithms to obtain the absolute magnitude, we get

$$\mathcal{M} = \mathcal{M}_\odot - 2.5 \log\left(\frac{L}{L_\odot}\right) = -10 \log v_{max} + \text{ constant.} \qquad (8.13)$$

Observations show that the best fit relation between absolute blue magnitude $\mathcal{M}_B$ and $v_{max}$ is indeed of the form $\mathcal{M}_B = -\alpha \log v_{max} + \beta$ where the pair $(\alpha, \beta)$ has the values $(9.95, 3.15), (10.2, 2.71), (11.0, 3.31)$ for Sa, Sb and Sc spirals. This is quite close to the relation $L \propto v_{max}^4$ obtained above.

### 8.1.2  Angular momentum of galaxies

Another closely related feature of galactic systems, especially the disk like systems, is their angular momentum. The angular momentum of a galaxy can be conveniently expressed in terms of the dimensionless parameter $\lambda = \omega/\omega_0$ where $\omega$ is the *actual* angular velocity of the system and $\omega_0$ is the *hypothetical* angular velocity that is needed to support the system against gravity purely by rotation. We know that $M\omega_0^2 R \approx GM^2/R^2$ while $\omega^2 = J^2/M^2 R^4$ where $J$ is the actual angular momentum of the galaxy. Eliminating the radius in terms of the energy by $R = GM^2/|E|$ we get

$$\lambda \equiv \frac{\omega}{\omega_0} = \frac{JE^{1/2}}{GM^{5/2}}. \qquad (8.14)$$

It follows that a self-gravitating system with appreciable rotational support has a $\lambda$ comparable to unity. Observations suggest that disk galaxies have $\lambda \simeq 0.4$ to $0.5$.

Of course, we need to understand how the galaxies got this angular momentum, parameterized by observed value of $\lambda$. One possible way of explaining the angular momentum of galaxies is as follows: During the initial collapse of the baryonic structures, tidal forces will be exerted on each protogalaxy by its neighbours. The tidal torque can spin up the protogalaxies thereby providing them with some angular momentum. During the collapse of the gas due to cooling, the binding energy increases while mass and angular momentum remains the same. This can allow $\lambda$ to increase as $\lambda \propto |E|^{1/2}$ and (possibly) reach observed values.

In this scenario, protogalaxies can only acquire initial angular momentum because of tidal torquing due to other collapsing material around it. If the typical mass, comoving radius and density contrast are $M, R, \delta$, then the tidal acceleration of any one of a pair of blobs is $(\delta GM/a^3 R^3)(aR) =$

$(\delta GM/a^2R^2)$ where we have taken the separation between the blobs to be of same order as the size of the blobs and used the expansion factor $a$ to convert the comoving radius to proper radius. If the masses involved are taken to be $M/2$ and the lever arm for the force is $aR/2$, then the net tidal torque is of the order of

$$T \simeq \frac{\delta GM}{a^2R^2} \times \frac{M}{2} \times \frac{aR}{2} = \frac{\delta GM^2}{4Ra}. \tag{8.15}$$

In a $\Omega_{\mathrm{NR}} = 1$ universe, $\delta \propto a$ implying that the tidal torque is a constant and the angular momentum grows linearly with time: $J = Tt$. The total angular momentum acquired until the time of turn around $t_{\mathrm{ta}}$ can now be estimated for a spherical top hat model. Using the relation between energy and virial radius and the details of spherical top hat model, it is easy to show that the angular momentum parameter is about $\lambda \simeq 0.025\,\nu^{-1}$ if the object has an over density which is $\nu$ times fractional density contrast. Once again simple arguments lead to meaningful final results though several factors of order unity have been ignored in the analysis.

Further dissipational collapse of the baryons will increase the value of $\lambda$, especially in the case of disks embedded in spherical halo. In the absence of dark matter halos (and assuming that a protogalaxy was just a self-gravitating cloud of baryonic gas) it turns out that the idea is not viable. But the idea works in the presence of a massive halo. In the presence of a massive dark halo, the initial spin parameter of the system, before collapse of the gas, can be written as $\lambda_i = (J|E|^{1/2}/GM^{5/2})$ where the various quantities, $J$, $E$ and $M$ refer to the *combined* dark matter - gas system, although the contribution from the gas is negligible compared to that of the dark matter. After the collapse, the gas becomes self-gravitating and the spin parameter of the resulting disk galaxy will be $\lambda_d = (J_d|E_d|^{1/2}/GM_d^{5/2})$, where the parameters now refer to the disc. So we find that

$$\frac{\lambda_d}{\lambda_i} = \left(\frac{J_d}{J}\right) \left(\frac{|E_d|}{|E|}\right)^{1/2} \left(\frac{M_d}{M}\right)^{-5/2}. \tag{8.16}$$

The energy of the virialized dark matter - gas system, assuming that the gas has not yet collapsed, can be written as $|E| = k_1(GM^2/R_c)$, while that of the disk is given by $|E_d| = k_2(GM_d^2/r_c)$. Here $R_c$ and $r_c$ are the characteristic radii associated with the combined system and the disk respectively, while $k_1$, $k_2$ are constants of order unity which depend on the precise density profile and geometry of the two systems. The ratio of the binding energy of the collapsed disk to that of the combined system is then

$$\frac{|E_d|}{|E|} = \frac{k_2}{k_1} \left(\frac{M_d}{M}\right)^2 \left(\frac{r_c}{R_c}\right)^{-1}. \tag{8.17}$$

Further, the total angular momentum (per unit mass) acquired by the gas, destined to form the disc, should be the same as that of the dark matter. This is because all the material in the system experiences the same external torques before the gas separates out due to cooling. Assuming that the gas conserves its angular momentum during the collapse, we have $(J_d/M_d) = (J/M)$. Hence,

$$\begin{aligned}
\frac{\lambda_d}{\lambda_i} &= \left(\frac{M_d}{M}\right) \left(\frac{k_2}{k_1}\right)^{1/2} \left(\frac{M_d}{M}\right) \left(\frac{R_c}{r_c}\right)^{1/2} \left(\frac{M}{M_d}\right)^{5/2} \\
&= \left(\frac{k_2}{k_1}\right)^{1/2} \left(\frac{R_c}{r_c}\right)^{1/2} \left(\frac{M}{M_d}\right)^{1/2}
\end{aligned} \tag{8.18}$$

where we have used Eq. (8.17) to simplify Eq. (8.16). The gas originally occupied the same region as the halo before collapsing and so had a pre collapse radius of $R_c$. Hence the collapse factor for the gas is

$$\frac{R_c}{r_c} = \left(\frac{k_1}{k_2}\right) \left(\frac{M_d}{M}\right) \left(\frac{\lambda_d}{\lambda_i}\right)^2. \tag{8.19}$$

We see that the required collapse factor for the gas to attain rotational support has been reduced by a factor $(M_d/M)$, from what would have been required in the absence of a dominant dark halo. For a typical galaxy with a halo which is ten times as massive as the disc, one needs a collapse by only a factor of about 40 or so before the gas can spin up sufficiently to attain rotational support. This is quite easy to achieve. (In the absence of dark matter halo, the factor $(M_d/M) \simeq 10$ is replaced by unity and one requires collapse factors which are ten times larger.)

It is possible to extend these arguments and estimate the characteristic length scale, $R_d$, of an exponential disk profile $\exp(-R/R_d)$ if we assume that there is no exchange of angular momentum between the disk and halo as the disk forms. For an exponential disk, you can easily compute the angular momentum per unit mass and obtain: $J/M = 2R_d v_g$ where $v_g$ is the constant rotational velocity of the stars and gas. Combining this with the virial theorem $|E| = K = (3/2)Mv_h^2$ (where $v_h^2$ is the velocity dispersion of the halo) we can eliminate $J$ and $|E|$ from $\lambda \equiv (JE^{1/2}/GM^{5/2})$ in terms of $M$. Finally, expressing $M$ in terms of the halo rotation velocity by

$GM/R = v_h^2$ and assuming that the density was a factor $f_c$ larger than the background density $\rho_{\text{bg}}(z) = 3\Omega_{\text{NR}}(z)H^2(z)/8\pi G$ at the time of the collapse redshift $z$, we get

$$R_d = \frac{\lambda}{2v_g} \left( \frac{3v_h^2}{2} \right)^{-1/2} v_h^3 \left[ \frac{f_c}{2} \Omega_{\text{NR}}(z) H^2(z) \right]^{-1/2}. \qquad (8.20)$$

If we now use $f_c = 180$ and assume that when the disk settles down, $v_g \simeq v_h$, we get

$$R_d \simeq 6\,h^{-1} \text{ kpc} \left( \frac{v_g}{200 \text{ km s}^{-1}} \right) \frac{H_0}{H(z_f)} \qquad (8.21)$$

which relates the scale length of the disk $R_d$ to the formation redshift $z_f$ of the disc. For reasonable scale lengths, disk could not have formed at redshifts significantly higher than unity. For Milky Way, with $v_g \simeq 220$ km s$^{-1}$, $R_d \simeq 3.6$ kpc we need $H_0/H(z_f) = 0.55$ corresponding to $z_f \simeq 0.5$. The overall picture which arises from these considerations — that the spheroids and bulges form early on while the disks form at a somewhat later epoch — is consistent with the observations. It is, for example, known that metal poor stars in solar neighbourhood have highly eccentric orbits. This will arise naturally if such a spheroidal system of stars have formed early in a time short compared to the collapse time and their orbits ended up mixing through collisionless relaxation.

### 8.1.3  *Properties of elliptical galaxies*

Let us next consider the bulge component which has a more extended distribution. Observations indicate that the bulge can be modeled by taking the fall-off as $\exp[-(R/R_0)^\alpha]$ with $\alpha \simeq 0.25$. Since elliptical galaxies are dominated by a bulge component, this distribution also describes the surface density profiles of ellipticals. The scale length, $R_0$, is related to the size of the elliptical galaxy though defining the size of a galaxy itself is a non trivial task. It is conventional in the description of elliptical galaxies to use the effective radius, $R_e$, defined such that half the light is enclosed by this radius. In terms of $R_e$, the surface brightness of an elliptical galaxy (or, for that matter, the bulge component of a disk galaxy) is well-described by

$$I(R) = I_e \exp \left\{ -7.67 \left[ \left( \frac{R}{R_e} \right)^{1/4} - 1 \right] \right\} \qquad (8.22)$$

which is called the *de Vaucouleurs law*. The corresponding result, in units of magnitude arcsec$^{-2}$, is given by

$$\mu(R) = \mu_e + 8.3268 \left[ \left( \frac{R}{R_e} \right)^{1/4} - 1 \right]. \qquad (8.23)$$

The total luminosity obtained by integration over all $R$ is $L_{\text{tot}} = 22.7 I_e R_e^2$.

In general, a galaxy will have both bulge and disk components. A typical B-band surface brightness profile of a Sa0 galaxy will be a sum of two contributions with different radial dependences adding up to give the observed surface brightness. At large radius the brightness profile is well fitted by the equation of the form:

$$\mu_{\text{disc}}(\theta) = c_0 + c_1 \left( \frac{\theta}{1 \text{ arcsec}} \right) \qquad (8.24)$$

where $c_0$ and $c_1$ are constants. This is characteristic of an exponential disc. Near the origin, the intensity profile is well fitted by the equation

$$\mu_{\text{bulge}}(\theta) = \mu(0) + c_2 \left( \frac{\theta}{1 \text{ arcsec}} \right)^{1/4} \qquad (8.25)$$

with $\mu(0)$ and $c_2$ being two more constants. This is characteristic of a bulge near the center. If we assume this fit to extend all the way to origin, the central brightness due to the bulge will be significantly higher than the extrapolated central brightness due to the disc. Whenever a particular disk galaxy shows a dominant bulge, the intensity distribution *of the bulge* is quite close to de Vaucouleurs law in Eq. (8.22).

In the simplest approximation, an elliptical galaxy is all bulge and no disc. The luminosity profile of the ellipticals are reasonably well fitted by the de Vaucouleurs profile in Eq. (8.22) away from the center but the central region shows a more interesting structure. Current HST observations show that several elliptical galaxies do show a cuspy region with $I \propto r^{-\alpha}$, $\alpha \approx (0.5 - 0.3)$ up to the limit of HST resolution. (We saw in Sec. 8.1.1 that the existence of a central black hole could lead to such a cusp in the density profile.)

One significant systematic trend exhibited by the elliptical galaxies is the existence of what is called a *fundamental plane*. Observations show that the effective radius $R_e$, the central velocity dispersion $\sigma_0$ and the mean surface brightness, $I_e$ (within the effective radius $R_e$) are connected by a relation of the form

$$R_e \propto \sigma_0^{1.49} I_e^{-0.86}. \tag{8.26}$$

That is, if the elliptical galaxies are plotted as points in the $(\ln R_e, \ln \sigma_0, \ln I_e)$ space, they tend to lie on a two dimensional plane rather than scatter all over the three dimensional space. The best-fit equation to the fundamental plane, in terms of the magnitude $\mu_e$, is

$$\log(R_e) = 0.34\mu_e + 1.49\log\sigma_0 - 6.60 \tag{8.27}$$

where $R_e$ is measured in pc, $\mu_e$ in $B-$ mag arcsec$^{-2}$ and $\sigma_0$ in km s$^{-1}$. Since $\sigma_0$ and $\mu_e$ can be measured directly from spectroscopy and photometry, one can use this relation to obtain the physical radius $R_e$. Comparison with the apparent angular radius in the sky will allow us to determine the distance to the galaxy, thereby providing a very valuable observational tool. From virial theorem, one would expect

$$R \propto \frac{M}{\sigma^2} \propto \frac{L}{\sigma^2}\left(\frac{M}{L}\right) \propto \frac{IR^2}{\sigma^2}\left(\frac{M}{L}\right) \tag{8.28}$$

so that $R \propto \sigma^2 I_e^{-1}(M/L)^{-1}$. If the mass-to-light ratio varies as $M/L \propto M^{0.2} \propto L^{0.25}$, one gets $R \propto \sigma^{4/3} I^{-5/6}$ which could approximately account for the fundamental plane of ellipticals.

The existence of the fundamental plane also allows you to derive a simple relationship between diameters and velocity dispersions of these galaxies. Let $D_n$ be the diameter within which the mean surface brightness is $I_n = 20.75B-$ mag arcsec$^{-2}$ and assume that (i) the intensity profile is given by a function of the form $I(R) = I_e f(R/R_e)$; (ii) most of the light interior to $D_n$ comes from radii at which $f(x) \simeq x^{-\alpha}$ with $\alpha \approx 1.2$. It then follows that

$$I_n = 8I_e\left(\frac{R_e}{D_n}\right)^2 \int_0^{D_n/2r_e} dx\, x f(x) \tag{8.29}$$

which, for $f(x) = x^{-\alpha}$, gives $D_n \propto r_e I_e^{1/\alpha} \propto r_e I_e^{0.8}$. Using the Eq. (8.26) for the fundamental plane, we get

$$D_n \propto \sigma^{1.4} I_e^{0.07} \tag{8.30}$$

which is called the *Faber-Jackson relation*. Observations show that this relation holds to a high accuracy. The weak dependence on $I_e$ shows that there will be a tight correlation between $D_n$ and $\sigma$ allowing these relations to be used for estimate of distances.

## 8.2  Models for stellar distribution in a galaxy

For time scales large compared to the age of the universe at the epoch of formation of the galaxy, one can assume that the galaxy to be in a steady state. The stars in such a galaxy will be orbiting in the total gravitational potential generated by all the stars in the system. The distribution of stars in a galaxy can be described by a function $f(\mathbf{x}, \mathbf{v}, t)$ which may be interpreted as the (relative) probability of finding a star in the phase space in the interval $(\mathbf{x}, \mathbf{x} + d^3\mathbf{x}; \mathbf{v}, \mathbf{v} + d^3\mathbf{v})$ at time $t$. The smoothed-out mass density of the stars at any point $\mathbf{x}$ will be

$$\rho(\mathbf{x}, t) = m \int f(\mathbf{x}, \mathbf{v}, t) d^3\mathbf{v}, \qquad (8.31)$$

where $m$ is the average mass of the stars. Such a smoothed-out density will produce a smooth gravitational potential $\phi(\mathbf{x}, t)$ where

$$\nabla^2 \phi = 4\pi G\rho. \qquad (8.32)$$

We may assume that each star moves in this smooth gravitational field along some specific orbit. The conservation of the total number of stars, expressed by the equation $(df/dt) = 0$, will reduce to

$$\begin{aligned}
\frac{df}{dt} &= \frac{\partial f}{\partial \mathbf{x}} + \dot{\mathbf{v}}.\frac{\partial f}{\partial \mathbf{v}} + \dot{\mathbf{x}}.\frac{\partial f}{\partial \mathbf{x}} \\
&= \frac{\partial f}{\partial t} + \mathbf{v}.\frac{\partial f}{\partial \mathbf{x}} - \nabla\phi.\frac{\partial f}{\partial \mathbf{v}} = 0.
\end{aligned} \qquad (8.33)$$

which is usually called *collisionless Boltzmann* equation. Models for galaxies are based on the solution to the coupled equations Eq. (8.32) and Eq. (8.33).

The *actual* gravitational potential at any point $\mathbf{x}$, due to the stars is

$$\phi_{\mathrm{act}}(\mathbf{x}, t) = -\sum_i \frac{Gm}{|\mathbf{x} - \mathbf{x}_i(t)|} \qquad (8.34)$$

where $\mathbf{x}_i(t)$ is the position of the i-th star at time $t$. This will be different from the $\phi$ in Eq. (8.32) which was produced by the *smooth* density. Because of this difference, the actual trajectories of the stars will differ (appreciably) from the orbits of the smooth potential after sufficiently long time intervals.

This time scale is precisely the two-body relaxation time scale derived in Sec. 1.4; since this time scale is large compared to the age of the universe, we can ignore the difference between the exact gravitational potential and the one obtained by averaging the mass density.

While the above equation provides an accurate description of a galaxy in steady state (with $(\partial f/\partial t) = 0$ in Eq. (8.33)), it does not lead to a unique form of stellar distribution. This is because one can obtain a wide variety of solutions to these equations with different symmetry properties. Thus we can, at best, use this equation to describe the properties of a galaxy with a *known* distribution of stars — rather than obtain the result ab initio. We shall discuss a few such models in this section.

When the galaxy is in a steady state, the distribution function does not explicitly depend on time and $f(t, \mathbf{x}, \mathbf{v}) = f(\mathbf{x}, \mathbf{v})$. To produce such a steady state solution we can proceed as follows: Let $C_i = C_i(\mathbf{x}, \mathbf{v})$, $i = 1, 2, \cdots$ be a set of isolating integrals of motion for the stars moving in the potential $\phi$ (which, right now, is not known). It is obvious that any function $f(C_i)$ of the $C_i$s will satisfy the steady-state Boltzmann equation; $(df/dt) = (\partial f/\partial C_i)\, \dot{C}_i = 0$ since $\dot{C}_i$ is identically zero. If we can now determine $\phi$ from $f$ self consistently, and populate the orbits of $\phi$ with stars, we have solved the problem. Let us consider some specific examples to see how this idea works.

In these calculations, it is convenient to shift the origin of $|\phi|$ by defining a new potential $\psi \equiv -\phi + \phi_0$ where $\phi_0$ is a constant. (We will choose the value of $\phi_0$ such that $\psi$ vanishes at the 'boundary' of the galaxy.) The new potential satisfies the equation $\nabla^2 \psi = -4\pi G\rho$ and the boundary condition $\psi \to \phi_0$ as $|\mathbf{x}| \to \infty$. We will also define a "shifted" energy for the stars $\epsilon = -E + \phi_0$; since $\phi_0 = \psi + \phi$, $\epsilon = -E + \psi + \phi = -(1/2)v^2 + \psi$.

The simplest galactic models are the ones in which $f(\mathbf{x}, \mathbf{v})$ depends on $\mathbf{x}$ and $\mathbf{v}$ only through the quantity $\epsilon$ so that $f = f(\epsilon) = f(\psi - \frac{1}{2} v^2)$. The density $\rho(\mathbf{x})$ corresponding to this distribution is

$$\rho(\mathbf{x}) = \int_0^{\sqrt{2\psi}} 4\pi v^2 dv f(\psi - \frac{1}{2}v^2) = \int_0^{\psi} 4\pi d\epsilon f(\epsilon) \sqrt{2(\psi - \epsilon)}. \qquad (8.35)$$

The limits of integration are chosen in such a way to pick only the stars bound in the galaxy's potential. The right hand side is a known function of $\psi$, once $f(\epsilon)$ is specified. The Poisson equation

$$\frac{1}{r^2} \frac{d}{dr}\left( r^2 \frac{d\psi}{dr} \right) = -4\pi G\rho = -16\pi^2 G \int_0^{\psi} d\epsilon f(\epsilon) \sqrt{2(\psi - \epsilon)} \qquad (8.36)$$

can now be solved - with some central value $\psi(0)$ and the boundary condition $\psi'(0) = 0$ — determining $\psi(r)$. Once $\psi(r)$ is known all other variables can be computed.

Three different choices for $f(\epsilon)$ have been extensively used in the literature to describe the galaxies. We will summarize these models briefly.

### 8.2.1  Lane-Emden models

The simplest form of $f(\epsilon)$ is a power law with $f(\epsilon) = A\epsilon^{n-3/2}$ for $\epsilon > 0$ and zero otherwise. Using Eq. (8.35), we see that this corresponds to density distributions of the form $\rho = B\psi^n$ (for $\psi > 0$) with $B = (2\pi)^{3/2}A\Gamma(n-1/2)[\Gamma(n+1)]^{-1}$. [Clearly, we need $n > 1/2$ to obtain finite density]. The Poisson equation now becomes

$$\frac{1}{l^2}\frac{d}{dl}\left(l^2\frac{d\xi}{dl}\right) = \begin{cases} -\xi^n & \xi \geq 0 \\ 0 & \xi \leq 0 \end{cases} \tag{8.37}$$

where we have introduced the variables,

$$L = (4\pi G\psi(0)^{n-1}B)^{-\frac{1}{2}}; \quad l = (r/L); \quad \xi = (\psi/\psi(0)). \tag{8.38}$$

This equation is the Lane-Emden equation which was discussed in Sec. 3.2. The case which is of interest in galactic modeling arises for $n = 5$. In this case, Eq. (8.37) has the simple solution

$$\xi = (1 + \frac{1}{3}l^2)^{-\frac{1}{2}} \tag{8.39}$$

which corresponds to a density profile of $\rho \propto [1 + (1/3)l^2]^{-5/2}$ and a total mass of $M = (\sqrt{3}L\psi(0)/G)$. This model (called the *Plummer model*) provides a reasonable description of a class of elliptical galaxies.

### 8.2.2  Isothermal and King models

The second choice for distribution function is the one of the form

$$f(\epsilon) = \frac{\rho_0}{(2\pi\sigma^2)^{3/2}}\exp\left(\frac{\epsilon}{\sigma^2}\right), \tag{8.40}$$

parameterized by two constants $\rho_0$ and $\sigma$. One can easily verify that the mean square velocity $< v^2 >$ is $3\sigma^2$ and that the density distribution is $\rho(r) = \rho_0\exp(\psi/\sigma^2)$. The central density is $\rho_c = \rho_0\exp(\psi(0)/\sigma^2)$. It is conventional to define a core radius and a set of dimensionless variables by

$$r_0 = \left(\frac{9\sigma^2}{4\pi G\rho_c}\right)^{1/2}; \quad l = \frac{r}{r_0}; \quad \xi = \frac{\rho}{\rho_c}. \tag{8.41}$$

Then the Poisson equation can be rewritten in the form of Eq. (3.34) which is that of isothermal sphere. It was shown Sec. 3.2 that the solution has the asymptotic limit of $\xi \simeq (2/9l^2)$. Hence the exact solution has a mass profile with $M(r) \propto r$ at larger $r$; the model has to be cut off at some radius to provide a finite mass.

For $l \lesssim 2$, the numerical solution is well approximated by the function $\xi(l) = (1 + l^2)^{-\frac{3}{2}}$. The projected, two dimensional, surface density corresponding to this $\xi(l)$ is $S(R) = 2(1 + R^2)^{-1}$, where $R$ is the projected radial distance. In other words, the (true) central surface density $\Sigma(0) \simeq 2\rho_c r_0$ where $\rho_c$ is the central volume density and $r_0$ is the core radius. Since the mass density of the isothermal sphere varies as $r^{-2}$ at large distances, the potential $\phi$ varies as $\ln r$. A disk of test stars embedded in such an isothermal sphere will have constant rotational velocities. A dark matter halo with such an isothermal distribution of particles will thus lead to flat rotation curves. Hence, isothermal sphere is a natural model for the dark matter halos.

An important modification of the isothermal distribution function, which leads to a more realistic stellar distribution, is given by

$$f(\epsilon) = \frac{\rho_c}{(2\pi\sigma^2)^{3/2}} \left(e^{\epsilon/\sigma^2} - 1\right); \quad \epsilon \geq 0 \tag{8.42}$$

and $f(\epsilon) = 0$ for $\epsilon < 0$. (This is called *King model*.) Since $f(\epsilon)$ vanishes for $\epsilon < 0$, this model may be thought of as a truncated isothermal sphere. The density, when integrated from the origin numerically, will vanish at some radius $r_t$ (called the tidal radius). The quantity $c = \log[r_t/r_0]$ is a measure of how concentrated the system is. Bright elliptical galaxies are well described by such a model with $[\psi(0)/\sigma^2] \simeq 10$ and $c \simeq 2.4$.

The above examples show how galactic models of reasonable description tion can be constructed as solutions to the collisionless Boltzmann equation, Eq. (8.33). Unfortunately, this particular approach has no dynamical content and is merely a tool to fit observations. In fact, the situation is worse than that; for a wide class of acceptable $\rho(r)$, it is always possible to obtain an $f(\epsilon)$ which will self consistently reproduce the given density profile. To see this, we begin noting that a given $\rho(r)$ uniquely determines a $\psi(r)$ and hence — on eliminating $r$ between $\rho(r)$ and $\psi(r)$ — the function $\rho(\psi)$. Writing Eq. (8.35) as

$$\frac{1}{\sqrt{8\pi}}\rho(\psi) = 2\int_0^\psi f(\epsilon)\sqrt{\psi - \epsilon}\, d\epsilon \qquad (8.43)$$

and differentiating both sides with respect to $\psi$, we get

$$\frac{1}{\sqrt{8\pi}}\frac{d\rho}{d\psi} = \int_0^\psi \frac{f(\epsilon)d\epsilon}{\sqrt{\psi - \epsilon}}. \qquad (8.44)$$

This equation (called *Abel's integral equation*) has the solution

$$f(\epsilon) = \frac{1}{\sqrt{8}\pi^2}\frac{d}{d\epsilon}\int_0^\epsilon \left(\frac{d\rho}{d\psi}\right)\frac{d\psi}{\sqrt{\epsilon - \psi}}, \qquad (8.45)$$

which determines $f(\epsilon)$. Though this procedure gives an $f(\epsilon)$, there is no assurance that it will be positive definite; obviously, the method works only for those $\rho(r)$ for which $f(\epsilon) > 0$ for all $\epsilon$.

### 8.2.3   Disk models

Similar modeling also works for a disk-like system if the distribution function is taken to be of the form $f = f(\epsilon, J_z)$ where $J_z$ is the conserved $z$−component of the angular momentum. We will mention a few examples: If the distribution function is taken to be

$$f(\epsilon, J_z) = \begin{cases} AJ_z^n \exp(\epsilon/\sigma^2) & \text{(for } J_z \geq 0\text{)} \\ 0 & \text{(for } J_z \leq 0\text{)} \end{cases} \qquad (8.46)$$

then we obtain a surface density for the disk, which is given by:

$$\Sigma(R) = \frac{(n+1)\sigma^2}{2\pi GR} \equiv \frac{\Sigma_0 R_0}{R}. \qquad (8.47)$$

The circular velocity of stars $v_c^2 = -R(\partial\psi/\partial R) = 2\pi G\Sigma_0 R_0$ is a constant for this model, called *Mestel's disk*. These parameters are related to the parameters in the distribution function by

$$n = (v_c^2/\sigma^2) - 1; \quad A = \Sigma_0 R_0 \left[2^{n/2}\sqrt{\pi}\Gamma(n+1/2)\sigma^{n+2}\right]^{-1}. \qquad (8.48)$$

The velocity dispersion in the radial direction $\sigma^2$ is a free parameter characterizing the disk. The parameter $n$ is a measure of the relative values of the steady circular velocity $v_c$ and the velocity dispersion $\sigma$ and quantifies the "coldness" of the disk.

A more complicated set of disk models (called *Kalnajs disk*) can be obtained from the distribution function which has the form

$$f(\epsilon, J_z) = A \left[ (\Omega_0^2 - \Omega^2)a^2 + 2(\epsilon + \Omega J_z) \right]^{-1/2} \qquad (8.49)$$

when the term in square bracket is positive and zero otherwise. This function leads to the following surface density and potential:

$$\Sigma(R) = \Sigma_0 \left( 1 - \frac{R^2}{a^2} \right)^{1/2} ; \quad \phi(R) = \frac{\pi^2 G \Sigma_0}{4a} R^2 \equiv \frac{1}{2} \Omega_0^2 R^2 \qquad (8.50)$$

with $\Sigma_0 = 2\pi A a \sqrt{\Omega_0^2 - \Omega^2}$. The parameter $\Omega$ is free and describes the mean (systematic) rotational velocity of the system: $< v_\phi >= \Omega R$.

## 8.3  Spectral energy distribution of galaxies

Let us next consider the radiation emitted by a galaxy at different frequencies, usually called the spectral energy distribution (SED) of a galaxy. Since the galaxy is made of stars, gas and some amount of dust, it is — in principle — possible to express the total luminosity $L(\lambda)$ of the galaxy at wavelength $\lambda$ at any given time in the form

$$L(\lambda) = \int dV \sum_i \rho_i L_i(\lambda) \exp(-\tau) + L_{\text{others}}(\lambda) \qquad (8.51)$$

where $\rho_i$ is the density of stars of type $i$ having a luminosity $L_i(\lambda)$ and $\tau$ is the optical depth for absorption of light at this wavelength within the galaxy; $L_{\text{others}}(\lambda)$ is the luminosity due to non stellar sources. While stars contribute the dominant amount of energy in a galaxy, there are, of course, other sources which dominate the radiation in different wave bands. For example, the radiation in the x-ray band is possibly due to individual accretion disks in binary stars while the radio luminosity arises from synchrotron radiation from ambient electrons in the galaxy.

Another major component of galaxies is *dust* — a rather disparaging term used by astronomers to describe silicates, carbides, metal oxides and several other components which usually blocks the light from the stars in the visual band. The dust lands up in the interstellar medium through the winds or mass loss from evolving low mass stars and also, of course, from supernova. As the gas moves further away from the stars, it cools and metallic compounds condense out of the gas. The equilibrium temperature of a dust grain at a distance $r$ from a star of luminosity $L$ can be easily estimated by equating the flux of radiation emitted by the dust particle (treated as a blackbody at temperature $T_{\text{dust}}$) to the radiation incident on the particle. This will give $\sigma T_{\text{dust}}^4 = (L/16\pi r^2)(Q_{\text{in}}/Q_{\text{out}})$ where $Q_{\text{in}}, Q_{\text{out}}$ denote the absorption and emission efficiencies of the dust grain and $\sigma = 5.7 \times 10^{-8}$ W m$^{-2}$ K$^{-4}$ is the Stefan's constant. If we take $L = 100 L_\odot$, $r = 700$ AU, then we get $T_{\text{dust}} \approx 30$ K $(Q_{\text{in}}/Q_{\text{out}})^{1/4}$. The radiation from such a dust grain will peak around 100 microns. (In reality, $Q_{\text{out}}$ itself scales as $Q_{\text{out}} \propto a T_{\text{dust}}$ where $a$ is the radius of the dust grain and the actual temperatures of dust grains are somewhat lower.)

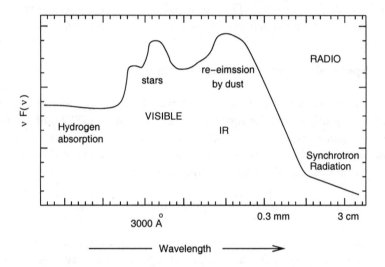

Fig. 8.1  Spectral energy distribution of a typical galaxy. Note that most of the light in the UV and optical band is contributed by the stars, while some amount of far-infrared radiation arises due to reprocessing of starlight by dust. The high energy radiation in x-ray band and radio emission are sub dominant.

Figure 8.1 shows the schematic form of integrated spectral energy distribution of a model galaxy. The UV and optical light comes essentially from stars with different range of masses and corresponding temperatures. If the dominant contribution is from stars in the mass range ($M_1 < M < M_2$) and if the temperature-mass relationship of stars is given by the function $T(M)$, then the energy radiated per logarithmic band of frequency will be

$$\nu I_\nu \propto \int_{M_1}^{M_2} dM \, \frac{\nu^4 n(M)}{exp[h\nu/k_B T(M)] - 1} \tag{8.52}$$

where $n(M)$ is the number density of stars in a mass range $(M, M + dM)$. As the stars evolve, the parameters in this expression will change with time, making $\nu I_\nu$ change. However, two simple limits of this integral can be easily worked out. If the dominant contribution comes from a narrow range of masses, then the integral is dominated by a Planckian at a given temperature and the spectrum will approximately thermal. On the other hand, if a broad range of masses contribute so that the limits $M_1$ and $M_2$ are irrelevant, then the resulting intensity will be an approximate power law if $T \propto M^n$ and $n(M) \propto M^{-\alpha}$. Evaluating the integral, it is easy to see that

$$\nu I_\nu \propto \nu^{4-(\alpha-1)/n} \qquad (8.53)$$

in the $M_1 \to 0, M_2 \to \infty$ limit. These results, however, are applicable only over a limited range. At high frequencies, $(h\nu > 13.6$ eV, $\lambda < 912$ Å$)$, absorption by hydrogen atoms will strongly suppress the intensity (see Fig. 8.1). At low frequencies, it is necessary to take into account re-radiation of energy by dust. A simple model for this re-processing can be based on the assumption that a fraction $f$ of the optical-UV luminosity is absorbed and re emitted in the far infrared so that $L_{\mathrm{FIR}} = fL$. This FIR radiation will be in the form of a modified Planck spectrum with a spectral energy distribution $B_\nu \epsilon_\nu$ where $B_\nu$ is the Planckian intensity for a temperature in the range of $T \approx (45 - 80)$ K and $\epsilon_\nu$ is the dust emissivity which could be taken as a power law: $\epsilon_\nu \propto \nu^x$ where $x = 1$ for $\nu > 3 \times 10^{10}$ Hz $(100 \ \mu m)$ and $x = 2$ for $\nu < 2 \times 10^{11}$ Hz (1 mm) with a linear extrapolation between the two limits. This will lead to the form shown in Fig. 8.1 in the FIR band.

The total integrated light from a galaxy, in a given band, can be used to assign the magnitudes (and colour) in that band and one finds that the colour of a galaxy is broadly correlated with the galaxy type. By and large elliptical, lenticular and early spiral galaxies have absolute blue magnitudes $-22 < M_B < -16$, whereas later type spirals and irregulars have $-18 < M_B < -12$.

This trend suggests, for example, that Sc galaxies will have greater fraction of massive main sequence stars relative to earlier spirals making Sc s bluer than Sa s and Sb s. For successively later type of galaxies, larger part of overall light is emitted in bluer wavelengths implying increasingly greater fraction of younger, more massive, main sequence stars. Irregulars are bluest of the Hubble sequence with $(B - V) \simeq 0.4$. Many irregulars also get bluer towards their centers (unlike early type spirals which become redder) suggesting that irregulars have active star forming regions in the center. These ideas about the composition are confirmed by the $(M/L)$ ratio for spirals — which decreases progressively as one moves to later Hubble type — since upper main sequence stars have lower mass-to-light ratios.

## 8.4   The evolution of galaxies

Estimates of the age of the stars suggest that much of the star formation took place when the universe was about 20 per cent of its current age. If $t_0 \simeq 14$ Gyr, then most ellipticals and bulges of disk galaxies should have formed when the universe was about 3 Gyr old. To form a total stellar mass of about $10^{11} M_\odot$ in a few Gyr, one requires star formation rates of about $10^2 M_\odot$ yr$^{-1}$. For any reasonable initial mass function, one would expect larger number of low mass stars to be formed compared to the high mass stars. But since high mass stars are significantly more luminous than the low mass stars, they dominate the luminosity in the initial stages of the star formation. After the first few million years of continuous star formation at a rate of, say, $10 M_\odot$ yr$^{-1}$ a galaxy would have acquired a stellar mass of $10^7 M_\odot$. Of these, nearly $2 \times 10^5$ are massive stars contributing a total luminosity of about $2 \times 10^{10} L_\odot$. This activity increases the luminosity of the galaxy by about 20 percent and its colour becomes bluer because of the radiation by massive stars. Eventually (after about $10^6$ yr), these high mass stars die out and the luminosity of the galaxy will be dominated by low mass stars.

As the stars evolve over different time scales and move in the H-R diagram, the physical properties of the galaxy will change with time. To study this process in detail, one begins with an initial mass function which characterizes the number of stars in a given mass range (which is produced at any given time) and some assumption regarding the variation of star formation rate in time. The initial mass function is usually taken to be a power law, $(dN/dM) \propto M^{-\alpha}$ with an index $\alpha \approx 2.35$ (see Sec. 4.5.1). As regards the star formation rate, the two extreme limits which will bracket the possibilities are: (i) all the stars have formed in a single burst at sometime in the past or (ii) that the star formation rate is constant over the relevant period of time. Standard stellar evolution theory can then be used to evolve these stars forward in time and to determine the light emitted by the collection of stars at any given instant. This procedure is usually called *population synthesis*.

To illustrate the ideas which are involved, let us consider the case with index $\alpha = 2.35$, metallicity $Z = 0.008$ and a steady star formation rate of $1M_\odot$ yr$^{-1}$. Numerical modeling then shows that the luminosity of such galaxy can be parameterized both at $t = 2$ Myr, and at $t = 900$ Myr by the fitting function:

$$\lambda L_\lambda \propto \left(\frac{\lambda_0}{\lambda}\right)^k \frac{1}{\exp(\lambda_0/\lambda) - 1}. \qquad (8.54)$$

At $t = 2$ Myr, the fitting parameters are given by $\lambda_0 = 3000$Å and $k = 4$ which corresponds to a Planckian distribution with temperature $T = 4.73 \times 10^4$ K. At $t = 900$ Myr, the fit is with $\lambda_0 = 1000$Å and $k = 2.2$ which is fairly distorted from a pure Planckian.

This index $k = 2.2$ can be understood as follows: First, note that, for most of the range of $\lambda \approx (10^3 - 10^4)$Å, the function $(\exp(\lambda_0/\lambda) - 1)^{-1}$ can be approximated as proportional to $\lambda$ making $\lambda L_\lambda \propto \lambda^{-k+1}$. Next, recall (from Eq. (8.53)) that when a broad set of stars contribute to the luminosity, $\nu I_\nu \propto \lambda L_\lambda \propto \lambda^{-4+(\alpha-1)/n}$ giving

$$k = 5 - \frac{1}{n}(\alpha - 1). \qquad (8.55)$$

Finally we saw in chapter 4 that for a wide class of stars, we have $T_{sur} \propto M^{n_L}; L \propto M^{n_L}$ giving $T_{sur} \propto M^{n_L n_{HR}} \propto M^n$ with $n = n_{HR}n_L \approx 0.3-0.6$. If we use $n = 0.5, \alpha = 2.35$, we find from Eq. (8.55) that $k = 2.3$, pretty close to the best fit value. Also note that this relation $L_\lambda \propto \lambda^{-2.2}$ gives, in terms of frequencies,

$$L_\nu \propto \lambda^2 L_\lambda \propto \nu^{0.2} \approx \text{constant.} \qquad (8.56)$$

These galaxies emit almost constant power per unit frequency in the range of $10^3$ Å $< \lambda < 10^4$ Å; putting in all the numbers gives the result that the flux is about $10^{20.6}$ W Hz$^{-1}$ for a SFR of $1M_\odot yr^{-1} Mpc^{-3}$.

In the case of a burst of star formation, we will expect the galaxy to redden as its stars age if the star formation does not continuously replenish the evolution and disappearance of old stars. Simple models for stellar evolution do lead to a conclusion in conformity with this. Figure 8.2 shows the numerical result for colour evolution for a Salpeter initial mass function and a burst of star formation. The colour-colour diagram for main sequence stars is superposed on the figure for comparison.

An Invitation to Astrophysics

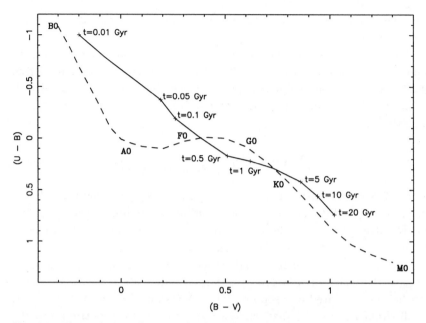

Fig. 8.2 The evolution of a typical galaxy depicted in the color-color diagram. The results for the galaxy, shown by the solid line, is obtained by numerical population synthesis assuming a Salpeter IMF and a single burst of star formation. The dashed curve corresponds to main sequence stars and is shown for comparison.

This result regarding the luminosity evolution of galaxies can also be qualitatively understood from stellar evolution theory. Since the stars with $M \gtrsim 1.25 M_\odot$, evolve in $t < 5$ Gyr, in any stellar population which originated from a burst of star formation, there will be only stars with masses $M \lesssim 1.25 M_\odot$ at $t > 5$ Gyr. Further, these low mass stars emit most of their energy during the giant branch phase. Since the time these stars spend in the giant branch (GB) is fairly short, we can estimate the luminosity of this population by

$$ L \approx \left( E_{\mathrm{GB}} \frac{dN}{dM} \right)_{M_{\mathrm{GB}}} \left| \frac{dM_{\mathrm{GB}}}{dt} \right| \tag{8.57} $$

where $E_{\mathrm{GB}}(M)$ is the total energy emitted by the stars in the giant branch, $(dN/dM)$ is the number of stars in the population with mass in the range $(M, M + dM)$ and $M_{\mathrm{GB}}(t)$ is the age at which a star of mass $M$ turns off the main sequence into the giant branch. Based on our understanding of stellar evolution, we can take

$$\frac{M_{\text{GB}}(t)}{M_{\odot}} \simeq \left(\frac{t}{10 \text{ Gyr}}\right)^{-0.4} ; \quad \frac{dN}{dM} \simeq K \left(\frac{M}{M_{\odot}}\right)^{-\alpha} \tag{8.58}$$

with $\alpha \simeq 2.35$. Combining these, we get

$$L \simeq \frac{K M_{\odot} E_{\text{GB}}(M_{\text{GB}})}{(25 \text{ Gyr})} \left(\frac{M_{\text{GB}}}{M_{\odot}}\right)^{3.5-\alpha} \tag{8.59}$$

or equivalently

$$\frac{d \ln L}{d \ln t} = \left[\frac{d \ln E_{\text{GB}}}{dM_{\text{GB}}} + (3.5 - \alpha)\right] \frac{d \ln M_{\text{GB}}}{d \ln t} = 0.4\alpha - \left(1.4 + 0.4 \frac{d \ln E_{\text{GB}}}{dM_{\text{GB}}}\right). \tag{8.60}$$

Since $(d \ln E_{\text{GB}}/d \ln M_{\text{GB}})$ is in the range $(0, 1)$ and $\alpha \approx 2.35$, we expect $L$ to be a decreasing function of $t$; more detailed models suggest $(d \ln L/d \ln t) \simeq 0.3\alpha - 1.6$ which is in reasonable agreement with the above analysis.

## 8.5 Luminosity function of galaxies

The abundance of galaxies, (or, for that matter, any class of objects) can be characterized in two different ways. The first one provides a count of objects with a given apparent flux (or magnitude) per solid angle in the sky. Since the information is 2-dimensional and one is not measuring the actual distance to the object, this observation is comparatively easy to make and is essentially limited only by: (i) the limiting flux which can be obtained in a given wave band and by (ii) the angular resolution of the instrument. If $n(F_\nu) \equiv (dN/d\Omega dF_\nu)$ denotes the number of sources with the observed flux densities between $F_\nu$ and $F_\nu + dF_\nu$ per solid angle in the sky, we can immediately obtain the total intensity of radiation due to all the sources as

$$I_\nu = \int_0^\infty F_\nu \, n(F_\nu) \, dF_\nu. \tag{8.61}$$

If apparent magnitudes are used instead of fluxes, they can be related by $F_\nu \propto 10^{-0.4 m_\nu}$ with the constant of proportionality depending on the wavelength band in which observations are carried out. In this case, the integrand in Eq. (8.61) is proportional to $10^{-0.4m}(dN/d\Omega dm)$. Multiplying the intensity by $(4\pi/c)$ will give the energy density of photons at the local neighbourhood due to these sources. This has been done, for example, with the source count of galaxies available from Hubble deep fields (HDFs). The HDF survey has detected about 3000 galaxies at $U_{300}, B_{450}, V_{606}, I_{814}$ and 1700 galaxies at $J_{110}, H_{160}$ bands. At still longer wavelengths, there are about 300 galaxies in K-band, 50 at $(6.7 - 15)\mu$m and 9 at $3.2\mu$m. When all the galaxies up to the limiting magnitudes are taken into consideration, the total integrated contribution of light effectively converges and the integration shows that $\nu I_\nu$ varies in the range $(2.9 - 9.0)$ nW m$^{-2}$ sr$^{-1}$ in the wavelength band $(3600 - 16,000)$Å.

A more detailed description of the abundance of the sources is provided by the luminosity function $\Phi(L, z)dL$, which gives the number of sources per unit comoving volume at redshift $z$ in the luminosity interval $(L, L + dL)$. Depending on the nature of the source, $L$ could be either the luminosity in a given wave band or bolometric luminosity. This quantity can be obtained only from a survey which also has redshift information for the sources. Further, in order to be useful, there must exist sufficient number of objects at any given redshift bin.

The luminosity function for galaxies has been investigated extensively for nearby galaxies ($z \approx 0$). We shall concentrate on these results and will briefly mention the status of investigations at higher redshifts. While considering galaxies at $z = 0$, the luminosity function is a function of a single variable, $\Phi(L)dL$, which gives the number of galaxies per unit volume in the luminosity band $L, L+dL$. Observations suggest that this luminosity function is reasonably well described by

$$\Phi(L) = \left(\frac{\Phi^*}{L^*}\right)\left(\frac{L}{L^*}\right)^\alpha \exp\left(-\frac{L}{L^*}\right) \qquad (8.62)$$

which translates to

$$\Phi(M) = (0.4 \ln 10)\, \Phi^*\, 10^{0.4(\alpha+1)(M^*-M)} \exp\left(-10^{0.4(M^*-M)}\right) \qquad (8.63)$$

in terms of magnitudes. The parameter $\Phi^* = (1.6 \pm 0.3) \times 10^{-2}h^3$ Mpc$^{-3}$ sets the overall normalization; $\alpha = -(1.07 \pm 0.07)$ determines the slope of the luminosity function at the faint end; $L_B^* = (1.2 \pm 0.1) \times 10^{10}h^{-2}L_\odot$ (which corresponds to the blue magnitude $M_B^* = (-19.7 \pm 0.1) + 5\log h$) is the characteristic luminosity above which the number of galaxies drops exponentially in the blue band.

This function, called *Schecter luminosity function*, is plotted in Fig. 8.3 by the unbroken line. Since $\alpha \simeq -1$, the integral of $\Phi(L)$ over $L$ can diverge (or become very large) at low $L$ suggesting a copious number of low luminosity galaxies. The total *luminosity density* contributed by all galaxies, however, is finite:

$$l_{\text{tot}} = \int_0^\infty L\,\Phi(L)\,dL = \Phi_0 L^* \Gamma\,(2+\alpha)\,. \tag{8.64}$$

For $\alpha = -1$, $l_{\text{tot}} = \Phi_0 L^*$. There is fair amount of scatter in the observed values of $\Phi^*, M^*$ and $\alpha$ in published literature. Different observations have given $\Phi^*$ varying in the range $[1.4 - 2.6] \times 10^{-2} h^3$ Mpc$^{-3}$, $M^*(B)$ varying in the range $(-18.8$ to $-19.68)$ and $\alpha$ varying in the range $(0.97 - 1.22)$. For example, Sloan Digital Sky Survey (SDSS) gives in R-band, the values $\alpha = -1.16 \pm 0.03$ and $\phi_* = (1.50 \pm 0.13) \times 10^{-2} h^3$ Mpc$^{-3}$. The $l_{\text{tot}}$, however, does not vary all that much and most observations give around $l_{\text{tot}} \simeq 2 \times 10^8 h L_\odot$ Mpc$^{-3}$.

There has been some attempts to study the luminosity function of galaxies at high redshifts in order to ascertain whether there is systematic evolution. At $z \approx 3$, the luminosity function of galaxies can be fitted with $\alpha = -1.60 \pm 0.13, m_* = 24.48 \pm 0.15$ and $\Phi_* = 1.6 \times 10^{-2}$; at $z \approx 4$, the corresponding parameters are about $\alpha \approx -1.60, m_* \approx 24.97$ and $\Phi_* = 1.3 \times 10^{-2}$ in the I-band. It is difficult to make useful comments about galaxy evolution based on these results at present.

Given the luminosity function for galaxies, one can estimate several useful quantities in a relatively straightforward manner. To begin with, ignoring the effects of spacetime curvature, one can relate the redshift $z$ and observed luminosity $f$ of a galaxy to its distance $r$ and intrinsic luminosity $L$ by $z \simeq v/c \simeq (H_0 r/c)$ and $f = (L/4\pi r^2)$. Then the joint distribution function in $z$ and $f$ for galaxies is

$$\frac{dN}{d\Omega dz df} = \frac{1}{4\pi} \int_0^\infty 4\pi r^2 dr\, \Phi\left(\frac{L}{L_*}\right) \delta_D\left(z - \frac{H_0 r}{c}\right) \delta_D\left(f - \frac{L}{4\pi r^2}\right). \tag{8.65}$$

The factor $(1/4\pi)$ in front converts the number density to density per solid angle and the Dirac delta functions select out the required redshift and flux. Using the known form of $\Phi$ and evaluating the integral, we get

$$\frac{dN}{d\Omega dz df} = \frac{4\pi}{L_*}\left(\frac{c}{H_0}\right)^5 z^4 \Phi\left(z^2/z_c^2\right), \quad z_c^2 \equiv \frac{H_0^2 L_*}{4\pi f c^2}. \tag{8.66}$$

The redshift distribution at a fixed apparent luminosity $f$ varies as $z^4 \Phi(z^2/z_c^2)$ which shows the effect of diminishing volume at low $z$ (the $z^4$ factor). The quantity $z_c$ is essentially the mean redshift of galaxies with

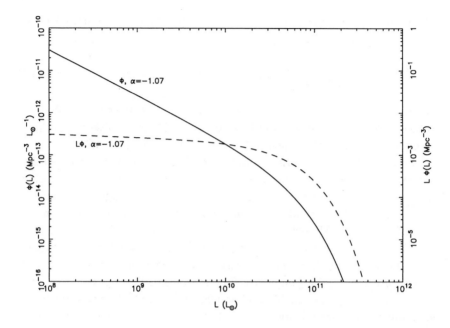

Fig. 8.3 The Schecter luminosity function for galaxies. The solid line gives $\Phi(L)$ and its numerical value is indicated on the left hand side vertical axis. The dashed line gives the number of galaxies per logarithmic band in luminosity and the corresponding numbers are given in the vertical axis on the right hand side.

a flux $f$. The mean luminosity per unit volume is given by

$$j = \int_0^\infty L\Phi\left(\frac{L}{L_*}\right)\frac{dL}{L_*} = \Gamma(2+\alpha)\Phi_*L_* = (0.77 - 1.3)\times 10^8 h\mathrm{L}_\odot\mathrm{Mpc}^{-3}.$$
(8.67)

From this, one can estimate the typical number density of galaxies

$$n_* \equiv \frac{j}{L_*} = \Gamma(2+\alpha)\Phi_* = (6.7 - 14.9)\times 10^{-3}h^3\mathrm{Mpc}^{-3}$$
(8.68)

and the mean separation between the galaxies

$$d_* = n_*^{-1/3} = (4.13 - 5.35)h^{-1}\mathrm{Mpc}.$$
(8.69)

Multiplying the observed mass-to-light ratio of galaxies $\mathcal{R} = (9.8 - 14.65)h(\mathrm{M}_\odot/\mathrm{L}_\odot)$ by the luminosity density $j$ in Eq. (8.67) we get the mass density of galaxies to be

$$\rho = j\mathcal{R} = (0.88 - 1.62)\times 10^9 h^2\,\mathrm{M}_\odot\,\mathrm{Mpc}^{-3} = (5.82 - 10.8)\times 10^{-32}h^2\,\mathrm{g\,cm}^{-3}.$$
(8.70)

The luminosity function also allows us to estimate the gravitational binding energy locked up in the galaxies. If the galaxies are assumed to be embedded in the dark matter halos which are isothermal spheres of mass $M$ and line-of-sight velocity dispersion $\sigma$, then their binding energies will be $U \approx 2M\sigma^2$. Of course $M$ here is the total mass and is related to the galaxy luminosity $L$ by some $(M/L)$ factor. The quantity usually measured in solar units $(M_\odot/L_\odot)$ is about 10 for galaxies. As we saw earlier, (see Eq. (8.13)), the velocity dispersion can be related to the luminosity using the relation: $L/L^* \cong (\sigma/\sigma^*)^4$. Assuming a constant $(M/L)$ ratio allows us to convert the mass to the luminosity and express $U$ in terms of $L$. Integrating over the luminosity function then gives the binding energy per unit volume in galaxies to be

$$U = 2\Phi^* L^* \sigma^{*2} < M/L > \Gamma(5/2 + \alpha) \qquad (8.71)$$

$\sigma^*$ is about 240 km s$^{-1}$ for ellipticals and 170 km s$^{-1}$ for spirals. Taking an average value, the binding energy of the galaxy is given by

$$U \simeq 1300 h^2 \left( \frac{\langle (M/L) \rangle}{10} \right) M_\odot c^2 \text{ Mpc}^{-3}. \qquad (8.72)$$

(Note that $M_\odot c^2$ Mpc$^{-3} = 6.1 \times 10^{20}$ erg cm$^{-3}$.)

Finally, you should appreciate a key observational aspect in any astronomical survey. The faint end of any luminosity function is intrinsically hard to determine. For example, consider a population of sources for which the number of sources in the luminosity bin $(L, L + dL)$ is a power law: $N(L)dL \propto L^{-\alpha}dL$ or, equivalently,

$$\frac{dN}{d\ln L} \propto L^{1-\alpha}. \qquad (8.73)$$

The maximum volume $V(L)$ in which a source of luminosity $L$ could lie and still be detected in a flux limited survey varies as $V \propto L^{3/2}$. It follows that the number of sources observed per logarithmic interval in the luminosity will vary as:

$$n(L) \propto V(L)\frac{dN}{d\ln L} \propto L^{5/2-\alpha}. \qquad (8.74)$$

Suppose you want survey to have dynamic range $y$ (so that $(L_{\min}/L_{\max}) = 10^{-y}$) with about $N_0 = 10^x$ objects in the faintest bin, Then, it is necessary to have $N > 10^x 10^{y((5/2)-\alpha)}$ objects in the survey. The random Poisson fluctuations in the faintest bin will give a fractional error of $N_0^{-1/2} = 10^{-x/2}$. To get 20 per cent accuracy in the lowest (log $L$) bin, say, we need $0.2 \simeq N_0^{-1/2} \simeq 10^{-x/2}$ giving $x \simeq 1.5$. Then the constraint on $N$ translates to about 2500 sources for $\alpha = 1.5, y = 2$ while one needs about $2.5 \times 10^5$ sources for $\alpha = 1.5, y = 4$. If $\alpha = 1$ then one requires $10^4$ sources for $y = 2$ and $10^6$ sources for $y = 4$. These numbers show that accurate determination of luminosity function of any set of faint object, in order to make a census of the universe, will require fairly elaborate surveys.

## 8.6   Distribution of galaxies

Clustering of galaxies into groups is typical in the distribution of the galaxies in the universe. A survey within a size of about 20 Mpc from our galaxy shows that only (10 - 20) percent of the galaxies do *not* belong to any group; they are called 'field' galaxies.

To study the clustering of galaxies quantitatively one needs a good survey of the universe giving the coordinates of galaxies in the sky. Of the 3-coordinates needed to specify the position of the galaxy, the two angular coordinates are easy to obtain. There exist today several galaxy catalogs, containing the angular positions of galaxies in particular regions of the sky, complete up to a chosen depth. If we know the redshift $z$ of these galaxies as well, then we can attribute to it a line-of-sight velocity $v \cong zc$. If we further assume that this velocity is due to cosmic expansion, then we can assign to the galaxy a radial distance of $r \cong H_0^{-1} v$. This will provide us with the galaxy position $(r, \theta, \phi)$ in the sky.

Currently available data points to an interesting pattern in the distribution of galaxies in space. The single most useful function characterizing the galaxy distribution is what is called the *two-point-correlation function*: $\xi_{GG}(r)$. This function is defined via the relation

$$dP = \bar{n}^2(1 + \xi_{GG}(\mathbf{r}_1 - \mathbf{r}_2))d^3\mathbf{r}_1 d^3\mathbf{r}_2, \qquad (8.75)$$

where $dP$ is the probability to find two galaxies simultaneously in the regions $(\mathbf{r}_1, \mathbf{r}_1 + d^3\mathbf{r}_1)$ and $(\mathbf{r}_2, \mathbf{r}_2 + d^3\mathbf{r}_2)$ and $\bar{n}$ is the mean number density of galaxies in space. The homogeneity of the background universe guarantees that $\xi_{GG}(\mathbf{r}_1, \mathbf{r}_2) = \xi_{GG}(\mathbf{r}_1 - \mathbf{r}_2)$ and isotropy will further make $\xi_{GG}(\mathbf{r}) = \xi_{GG}(|\mathbf{r}|)$. From Eq. (8.75) it follows that $\xi_{GG}(r)$ measures the excess probability (over random) of finding a pair of galaxies separated by a distance $\mathbf{r}$; so if $\xi_{GG}(r) > 0$, we may interpret it as a clustering of galaxies over and above the random Poisson distribution. Considerable amount of effort was spent in the past decades in determining $\xi_{GG}(r)$ from observations. These studies show that

$$\xi_{GG}(r) \simeq \left(\frac{r}{r_0}\right)^{-\gamma} ; \qquad r_0 = (5 \pm 0.5)h^{-1} \text{ Mpc}; \qquad \gamma = (1.77 \pm 0.04)$$
(8.76)

in the range $10h^{-1}\text{kpc} \lesssim r \lesssim 10h^{-1}\text{Mpc}$.

Incidentally, it is also possible, using the catalogs of rich clusters (like the Abel catalog which contains 4076 clusters), to compute the correlation function between galaxy clusters. The result turns out to be correlation function

$$\xi_{CC}(r) \simeq \left(\frac{r}{25h^{-1}\text{Mpc}}\right)^{-1.8} .$$
(8.77)

Comparing with the galaxy-galaxy correlation function $\xi_{GG}$ quoted before, we see that clusters are more strongly correlated than the individual galaxies. If firmly established, this result has important implications for the theories of structure formation. Unfortunately, $\xi_{CC}(r)$ is not as well established as $\xi_{GG}$ and hence one has to be cautious in interpreting results which depend on $\xi_{CC}$.

The distribution of matter is reasonably uniform when observed at scales bigger than about $100h^{-1}$ Mpc or so. For comparison, note that the size of the observed universe is about $3000h^{-1}\text{Mpc}$. Thus, one may treat the matter distribution in the universe to be homogeneous while dealing with the phenomena at scales larger than 100 Mpc or so.

## Exercises

**Exercise 8.1** *Dynamical friction near galactic center:*

Assume that the mass density of stars near the center of the galaxy varies as $\rho(r) \propto r^{-7/4}$ with the total mass contributed by the stars within a radius of about 1.8 pc being about $3 \times 10^6 M_\odot$. Near the center, there is a black hole of mass $M_{BH}$ which dominates the dynamics and the stars move in circular Keplerian orbits under the influence of this black hole in the central region. (a) Show that the time scale for dynamical friction of stars scales as $r^{1/4}$. (b) Calculate this time scale for a $10M_\odot$ star at $r = 1pc$. [Ans. $4 \times 10^8$ yr.]

**Exercise 8.2** *Black hole in M 87:*

The giant elliptical galaxy M 87 in the Virgo cluster has a black hole of mass $3 \times 10^9 M_\odot$ at its center surrounded by a diffuse hot intracluster medium with a density of 0.5 cm$^{-3}$ and a sound speed of 500 km s$^{-1}$. Show that an order of magnitude estimate for the accretion rate of this medium on to the black hole is about 0.1 $M_\odot$ yr$^{-1}$. If the efficiency of conversion of energy to radiation is $\eta = 0.1$, what is the resulting luminosity as a fraction of the Eddington luminosity? [Actual observations indicate a luminosity of $10^{42}$ erg s$^{-1}$ suggesting that it is a low efficiency accretion flow.]

### Exercise 8.3  *Rotation curve:*

The rotational velocity of the Sun in the Milky Way galaxy is $v_c(R) = 220$ km s$^{-1}$ at $R = 8$ kpc. What does this imply for the mass contained in the region $r < R$? Radio observations of the 21 cm emission from neutral hydrogen in the direction 30 degrees from the galactic center (on the same side as the direction towards which the Sun is orbiting) show a *maximum* radial velocity of 90 km s$^{-1}$. This could be interpreted as coming from a cloud of neutral hydrogen moving in a circular orbit of radius $r$ in the galaxy and located such that the radial velocity is maximum. The observed radial velocity, of course, is the difference between the true circular velocity $v_c(r)$ and the component of velocity of Sun in the same direction. Use this to estimate the true $v_c(r)$ and the mass contained within this radius. (This is the technique used to obtain the actual rotation curve of Milky Way galaxy.)

### Exercise 8.4  *Practice with surface brightness:*

Consider a typical Sbc galaxy, NGC 5055, located at a distance of $D = 8$ Mpc and seen at an inclination $i = 60°$. (This is the angle between the normal to the plane of the galaxy and the line-of-sight.) R-band photometry along the major axis of this galaxy shows that the surface magnitude can be fitted reasonably well by the relation

$$\mu_R = 19.75 + 0.72 \left( \frac{\theta}{1 \text{ arcmin}} \right). \tag{8.78}$$

Determine the surface luminosity profile in physical units. [Answer: $\Sigma_R(R) = 190e^{-(R/3.5 \text{ kpc})} L_{R\odot}$ pc$^{-2}$. ]

### Exercise 8.5  *Integrated magnitude:*

A spiral galaxy has a surface brightness profile $I(R) = I_0 e^{-R/R_s}$ where $I_0 = 21$ mag arcsec$^{-2}$ and $R_s = 7$ kpc. Calculate the total magnitude of the galaxy if it is at a distance $d$. [Ans. Since magnitudes are logarithmic, one cannot simply add them but one can convert them into fluxes and then add. This should give $I = I_{\text{tot}} = -11.79 + 5\log(d/1 \text{ kpc})$.]

### Exercise 8.6  *Practice with sky background:*

(a) Show that a sky brightness of 21 mag arcsec$^{-2}$ is the same as $1.6 \times 10^5$ photons s$^{-1}$ cm$^{-2}$ Å$^{-1}$ ster$^{-1}$ = $3.78 \times 10^{-6}$ photons s$^{-1}$ cm$^{-2}$ Å$^{-1}$ arcsec$^{-2}$.

(b) Compute how many photons per second will be collected from 1 arcsec$^2$ of the sky by a 1 m telescope with a V-filter of width 1300 Å. (c) Take the B-band Tully-Fisher relation for spiral galaxies to be $M_B = -9.95\log(v_{max}/1 \text{ km s}^{-1}) + 3.15$. Further, let $R_{25}$ be the radius at which the surface brightness of the spiral galaxy falls to 25 mag arcsec$^{-2}$ in B-band. It is also known that $R_{25}$ in kpc obeys the empirical relation $\log R_{25} = -0.249M_B - 4$. Given that a particular galaxy has $v_{max} = 320$ km s$^{-1}$ and has an apparent B = 12.22 mag determine the absolute B magnitude, the distance to the galaxy and the radius $R_{25}$. If one wants to measure the galaxy profile down to $R_{25}$, how much time do you need in a 4m class telescope?

### Exercise 8.7    *Counting the stars:*

The surface brightness of a galaxy at a distance $d$ is $I_0$ mag arcsec$^{-2}$. Assume that the galaxy is made up of stars of absolute magnitude $M$. Show that the number of stars contained within 1 arc second square patch of the image of this galaxy is $N = (d/10 \text{ pc})^2 10^{0.4(M-I_0)}$.

### Exercise 8.8    *From stars to the galaxies:*

One can do a better job on the spectral energy distribution of a galaxy by using our knowledge of stellar luminosities. The results discussed in chapter 4 allow you to relate the surface temperature to the mass of the star. Assuming that each star emits as a blackbody at this temperature and convolving it with an initial mass function, one can obtain the total radiation at any frequency $\nu$ due to the stellar population. One can also compute U-B and B-V indices for the system. This is, of course, valid for a young galaxy in which none of the stars have evolved off the main sequence. Do this and compare your results with those quoted in Sec. 8.4.

### Exercise 8.9    *Poor man's population synthesis:*

The purpose of this exercise is to use a simple model for the evolution of galaxy luminosity to capture the essential physics. To do this, let us assume that there are only 3 types of stars O, A and M with masses (40, 4, 0.5) $M_\odot$, V-band luminosities $(2.5 \times 10^5, 80, 0.06)L_{\odot V}$, B-V colours $(-0.35, 0.0, 1.45)$ and the main sequence lifetime $(1.6 \times 10^6, 5 \times 10^8, 7.9 \times 10^{10})$ yrs. The relative number of each type of star is given by the Salpeter IMF at $t = 0$. (a) When all the stars are still in the main sequence, what is the M/L ratio and the B-V colour of the galaxy? (b) What will be the corresponding values at $10^7$ yrs and at $10^9$ yrs? Note that in evaluating M/L at later times you should take into account the mass contributed by the remnants.

### Exercise 8.10    *How many galaxies can you see?*

Suppose all the light from a galaxy is from stars with $L = 0.2L_\odot$ and that there are about $10^{11}$ stars in the galaxy. Show that the absolute magnitude of such a galaxy is about $-21$. What is the maximum distance at which such a galaxy will be visible to naked eye? Does it make sense? Is the fact that the starlight from a galaxy is spread over a patch in the sky relevant?

**Exercise 8.11**   *Vertical oscillation of stars:*

Assume that our galaxy is a circular disk with radius $R$ and vertical height $H$ with the stars distributed uniformly in the vertical direction (which we take as $z-$axis) and as an exponential disk along $R$. Consider a star which "drops" from a high $z_0$ and oscillates in the vertical direction normal to the plane of the galaxy. Estimate the frequency of oscillations in the vertical direction if the local density of stars near the Sun is about 0.2 $M_\odot$ pc$^{-3}$.

**Exercise 8.12**   *MACHOS:*

If our galaxy has a flat rotation curve with $v_0 = 220$ km s$^{-1}$, estimate the local density of dark matter at $R = 8.5$ kpc. (Assume that dark matter is distributed as a spherical isothermal sphere.) Suppose the dark matter is made of MACHOS (MAssive Compact Halo Objects) each of which has a mass of 0.5 $M_\odot$. (This is *not* true!) Estimate the number density of such objects near solar neighbourhood. How does it compare with the number density of stars?

**Exercise 8.13**   *The galactic potential:*

It is often convenient to have an analytic model describing the gravitational potential of Milky Way. One such model consists of 3 components: disk, spheroid and a halo. The disk and spheroid are assumed to have the potential of the form

$$\Phi(R, z) = \frac{-GM}{[R^2 + (a + \sqrt{z^2 + b^2})^2]^{1/2}} \tag{8.79}$$

with different parameters from the disk and spheroid. For the disk, one takes $a_d = 4.20$ kpc, $b_d = 0.198$ kpc, and $M_d = 8.78 \times 10^{10} M_\odot$ while for the spheroid, it is $a_s = 0, b_s = 0.277$ kpc, and $M_s = 1.12 \times 10^{10} M_\odot$. The halo is described by a spherically symmetric potential

$$\Phi_h(r) = \frac{GM_h}{r_h} \left[ \frac{1}{2} \ln \left( 1 + \frac{r^2}{r_h^2} \right) + \frac{r_h}{r} \tan^{-1} \left( \frac{r}{r_h} \right) \right] \tag{8.80}$$

where $r$ is the spherical polar coordinate radius and $r_h = 6$ kpc, $M_h = 5 \times 10^{10} M_\odot$. (a) Compute the speed of the circular orbit in the galactic plane. What is the value near the Sun? Do you get flat rotation curves at large radius? (b) Compute the density distribution which will give rise to these potentials. What is the mass density in the solar neighbourhood? (c) Consider a star which was moving in a circular orbit at some radius $r_0$. Suppose it was perturbed slightly either in the radial or in the vertical directions. Estimate the radial and vertical oscillation periods.

**Exercise 8.14**  *Sinking to the center:*

(a) Consider a galaxy of mass $3 \times 10^{11} M_\odot$ moving in a cluster containing a large isothermal dark matter halo of density $\rho \approx 10^{-25}$ gm cm$^{-3}$. Compute the time scale for dynamical friction using Eq. (1.66). Also show that the drag force acting on the galaxy makes the angular momentum and the kinetic energy of the galaxy to decay. Work out the time dependence of the orbit if the halo has an isothermal density profile. (b) A satellite galaxy of mass $M$ is orbiting our galaxy which has an isothermal halo of velocity dispersion 220 km s$^{-1}$ and a total mass of $3 \times 10^{12} M_\odot$. Integrate the equations numerically and determine the orbit of the satellite galaxy with and without the dynamical friction. Do you think the orbital decay will have observable effects?

**Exercise 8.15**  *Real life astronomy: Galaxy surveys:*

For this problem, one can choose from one of the many galaxy surveys. Let us concentrate FLASH survey [Kaldare *et al.*, MNRAS, **339**, p. 652 (2003)]. (a) From the survey pictures which goes up to $2 \times 10^4$ km s$^{-1}$, determine the physical distance $d_{\max}$ probed by this survey. The survey is sensitive to galaxies with blue apparent B magnitude brighter than 16.7. What will be the absolute B magnitude of the galaxy at the limit of the survey and the corresponding luminosity? (b) The Milky Way has a blue luminosity of $2 \times 10^{10} L_{B\odot}$. What is the largest redshift at which FLASH would have picked up a galaxy with the luminosity of Milky Way? Comment on the implication. (c) The structures that you see in galaxy survey pictures — which are elongated in the radial direction (dramatically called *fingers of God*) — arise due to velocity dispersion within a cluster. Estimate the velocity dispersion for one of the structures in the picture taking the data from it. (d) If you look at the corresponding figures of a deeper survey (like Las Campanas; look that up) you see less of 'fingers of God'. Why?

**Exercise 8.16**  *Real life astronomy: MACHO lensing:*

Take a look at Alcock *et al.*, Nature, **365**, p. 621 (1993) which has the light curves of the first LMC microlensing event detected by macho experiments. (a) Use these light curves to determine the characteristic time scale $t$ of amplification which may be defined as the time during which $A(t) > 1.34$. (b) Calculate the total mass of Milky Way galaxy up to $R_{\mathrm{LMC}} = 50$ kpc assuming a flat rotation curve at $v_c = 220$ km s$^{-1}$. (c) Using the above results estimate the radius of the Einstein ring $r_E$. [Ans: About 1 AU.] (d) Assuming that the lens is half way between us and the LMC, estimate the lens mass. [Ans: 0.05 $M_\odot$.] Analysis of subsequent events show that a typical mass of a lens is closer to 0.5 $M_\odot$. In principle, this could be a 0.5 $M_\odot$ star or a 0.5 $M_\odot$ white dwarf. Which do you think is more likely? (e) Several years of MACHO experiment allows one to determine the probability for micro lensing. One finds that it is about $p = 1.2 \times 10^{-7}$ per star that is monitored. This allows an estimate of the mass density contributed by the MACHO and thus the total mass of MACHOs up to LMC. Compare this with the result obtained in (b) above.

*An Invitation to Astrophysics*

**Exercise 8.17**   *Real life astronomy: black hole at Milky Way center:*

Take a look at Schodel etal., Nature, **419**, p. 694 (2002) which has a figure giving the orbit of a star (S2) a point near the center of our galaxy (around Sgr A). (a) Assuming that the orbit is an ellipse and that the plane of the orbit is inclined to the plane of the sky, describe how you would calculate the inclination angle from the given data. (b) Assuming that the physical size of the semi major axis is $a = 4.62 \times 10^{-3}$ pc and the angular size is 0.119 arcsec, determine the distance to the galaxy center used in this paper. This distance is known to an accuracy of about 0.5 kpc. What is the error induced on the semi major axis due to this and how does it compare with the measurement uncertainties in the paper? (c) Determine the mass $M$ of Sgr-A and the uncertainty in its estimate.

**Notes and References**

The following textbooks deal with the subject of galaxies in much greater detail: Binney and Tremaine (1987), Elmegreen (1998), Carroll and Ostlie (1996), Binney and Merrifield (1998) and Padmanabhan (2002). Figure 8.2 was adapted from chapter 1 of Padmanabhan (2002).

# Chapter 9

# Active Galaxies

So far we were describing galaxies which are 'normal' in the sense that their energy emission is mostly due to starlight. Another class of extragalactic sources which are of considerable interest but are *not* powered by starlight is called *Active Galactic Nuclei* (AGN, for short). These objects display a variety of exotic features many of which are still not fully understood.

The AGN have spectacular properties. They produce very high luminosities (even $10^4$ times the luminosity of a typical galaxy) from within a tiny volume (less than about 1 pc$^3$). Their spectra extend over a broad range of frequencies and the luminosity per decade in frequency is roughly constant across several orders of magnitude in the frequency. Their line spectrum (in optical and UV) is quite strong and the total flux in lines could be few to tens of percent of the continuum flux. Some of the lines have a width which indicates bulk velocities ranging up to $10^4$ km s$^{-1}$. AGN also show remarkable variety in their radio morphology and strong time variability in different wave bands. They will be the subject of this chapter.

## 9.1  AGN: Basic paradigm and the spectra

One of the simplest characterization of AGN is that they are very compact objects emitting anywhere between $(10^{42} - 10^{48})$ erg s$^{-1}$ of power. Since a typical galaxy has a luminosity of $10^{44}$ erg s$^{-1}$, the AGN have power output which ranges from 1 percent of a typical galaxy to $10^4$ times as large. A power of $10^{47}$ erg s$^{-1}$ is equivalent to the conversion of the rest mass energy at the rate of about 2 solar mass per year. Alternatively, this is equivalent to the energy output of about $10^7$ O type stars confined to a volume of about $10^{-6}$ pc$^3$. You can see that AGN are unusually bright objects!

For nearby AGN, which can be actually seen as embedded in a diffuse host galaxy, the optical image of the nuclear region appears as a bright point source whose flux can exceed the flux from the rest of the host galaxy. This suggests that in most of the AGN, which are too far away for the host galaxy to be detected, we can attribute the emission as arising from a compact nuclear region of the host galaxy, thereby justifying the name active galactic nuclei.

The optical luminosity of most AGN vary at a level of about 10 per cent in a time scale of *few* years, which is quite unlike normal galaxies. The relative change in the amplitude seems to increase at higher frequencies with nearly a factor 2 variation in x-rays. Unlike variable stars, AGN do not show any periodicities or special time scales.

The continuum radiation from the AGN is quite different from the spectra of normal galaxies. Roughly speaking, the luminosity of a typical galaxy arises from the luminosities of the individual stars which radiate as blackbodies. The typical stellar surface temperature varies only by about a factor 10 among the stars that dominate the galactic power output. Since the bulk of the radiation in a blackbody is confined to a range which varies only by about factor 3 in frequency, a typical galaxy emits all its power within no more than one decade in frequency; in fact, it is usually less than a decade. The continuum spectra of AGN, in contrast, is relatively flat in terms of $\nu F_\nu$, all the way from mid-infrared to hard x-rays, spanning a range of $10^5$ in frequencies. That is, the broadband spectrum of an AGN continuum can be characterized by a power law: $F_\nu = C\nu^{-\alpha}$ where $\alpha \lesssim 1$. Such a power-law spectrum has relatively far more energy at high frequencies compared to blackbody spectrum. Compared to normal galaxies, the fraction of energy radiated in radio is about an order of magnitude larger and the fraction that is emitted in x-rays could be even 3 to 4 orders of magnitude larger. There are minor features in the spectra, but no one frequency band dominates the energy output of AGN.

To make a quantitative estimate, let us calculate the total amount of energy density due to AGN using the observed number of AGN (per steradian in the sky) having a flux between $F_\nu$ and $F_\nu + dF_\nu$. The photons we receive from distant cosmic sources sets a (useful) lower bound to the total amount of electromagnetic energy which was radiated in our direction. If the universe is truly homogeneous and isotropic, then the number of photons here, today, is a good indicator of their mean density in the universe. It follows that the local energy density of AGN photons at frequency $\nu$ is just

$$U_\nu = \frac{4\pi}{c} \int dF_\nu \left( \frac{dN}{dF_\nu} \right) F_\nu. \tag{9.1}$$

This quantity can be calculated in any wavelength band (optical, X-ray, ....)in which survey results for AGN searches in the sky are available so that we know $(dN/dF_\nu)$.

Observations show that the number $N(< m_B)$ of quasars (which are a prominent subclass of AGN; in fact we will use the quasars as a prototype of AGN throughout the chapter) per square degree in the sky brighter than $m_B$ in the B band can be expressed in the form $\log N = \alpha m_B + \beta$ with $\alpha = 0.88, \beta = -16.1$ for $m_B < 18.75$ and $\alpha = 0.31$ and $\beta = -5$ for $m_B > 18.75$. Computing $dN/dm_B$, converting $F_\nu$ to $m_B$ and evaluating the integral in Eq. (9.1) in the range, say $m_B = (12, 24)$ we get the total B-band flux (over $4\pi$ steradian) to be about $I_\nu = 40$ Jy. Multiplying by the B-band frequency and dividing by $c$ gives the energy density per logarithmic band $\nu U_\nu \approx 0.8 \times 10^{-17}$ ergs cm$^{-3}$. To find the energy density in all the bands we need to repeat this exercise in each band. However, since good quality data is not available in other bands, we can attempt to estimate this by applying a bolometric correction from, say, the B-band luminosity to the bolometric (i.e. total) luminosity. Because of redshift, the observed frequency $\nu$ will correspond to the frequency $\nu(1 + z)$ at the time of emission and the bolometric correction will, therefore, depend on the redshift. Taking these into account by a factor of about 30, one finds that the local energy density due to AGN radiation is about

$$\epsilon_{rad} \simeq 2 \times 10^{-16} \left( \frac{\mathcal{B}}{30} \right) \text{ erg cm}^{-3} \tag{9.2}$$

where $\mathcal{B}$ is the bolometric and redshift correction. For comparison, note that energy density of starlight is about $10^4$ times larger, CMBR energy density is about $10^3$ times larger and the integrated light from normal galaxies (with $M_B = 20$) is about 10 times larger.

If the AGN radiation was emitted through a process of efficiency $\eta$, then the cosmological mass density used up by all the quasars per unit volume is given by

$$\rho_{\rm acc} = \frac{\epsilon_{\rm rad}}{\eta c^2} = 2 \times 10^{13} \eta^{-1} \ M_\odot \ \text{Gpc}^{-3}. \tag{9.3}$$

This is the mass density which, when converted through a process having efficiency $\eta$, will lead to the observed radiation from the AGN. If the process goes on in a wide class of field galaxies with a luminosity $L \simeq L^*$ (see Sec. 8.5), then we can obtain the amount of mass which each of the galaxies should provide for this process, by dividing this expression by the space density of galaxies with luminosities within, say, one magnitude of $L^*$. The latter is given by $2 \times 10^6 h_{0.75}^3$ Gpc$^{-3}$ (see Sec. 8.5) so that the mass contributed per galaxy is about

$$M_{\text{active}} \simeq 1.6 \times 10^7 M_{\odot} \left(\frac{\mathcal{B}}{30}\right) \left(\frac{\eta}{0.1}\right)^{-1} h_{0.75}^{-3}. \qquad (9.4)$$

The scaling of this expression with respect to $\eta$ is noteworthy. The efficiency of chemical reactions corresponds to about 1 eV per atom so that the energy efficiency per unit rest mass is about $\eta_{\text{chem}} \simeq 10^{-10}$ if the mass of the atom is about $10 m_p$. In the stellar nuclear reactions, the energy released in hydrogen burning is about $(8 \text{ MeV}/m_p)$ corresponding to $\eta_{\text{nucl}} \simeq 8 \times 10^{-3}$. If such processes are to be used for powering AGN radiation, then the active mass which is involved will be $10^{16} M_{\odot}$ in chemical reactions and about $10^9 M_{\odot}$ for nuclear reactions. In contrast, accretion on to a gravitational potential well from the lowest stable circular orbit around a black hole can lead to efficiencies of the order of $\eta_{\text{accr}} \approx 0.1$ which is the scaling used in the above expression. When a body of mass $m$ falls from $r = \infty$ to $r \approx 5 R_s$ where $R_s \simeq (2 G M_{\text{BH}}/c^2)$ is the Schwarzschild radius of a compact object of mass $M_{\text{BH}}$, it radiates the potential energy

$$U = \frac{G M_{\text{BH}} m}{5 R_s} = \frac{G M_{\text{BH}} m}{10 G M_{\text{BH}}/c^2} = 0.1 \ mc^2. \qquad (9.5)$$

This gives an efficiency $\eta$ of 10 percent which is an order of magnitude more than the efficiency achieved in hydrogen fusion. (Only direct particle-anti-particle annihilation can give rise to higher efficiencies.) In other words, among all processes which assemble certain amount of fuel for producing radiation from a compact region, accretion will require the least amount of fuel for a given amount of radiation. Since assembling the fuel in a compact region and converting it into radiation is a difficult process, the problems will be aggravated in any model which uses a process with lower value of $\eta$. This leads to the standard paradigm of AGN being powered by an accretion disk around a massive black hole.

There is another observation which is also suggestive of this paradigm. If the radiation, emitted from a region of size $R$ which is moving with speed

$v$, varies at a time scale $\Delta t$, then we must have

$$R \le \frac{c\Delta t}{\gamma} = c\Delta t \sqrt{1 - \frac{v^2}{c^2}}. \tag{9.6}$$

This constraint arises from the fact that in the rest frame of the emitting region, causal processes can act coherently only over a size $R \le c\Delta\tau$ since the information about the variation in light intensity can only travel a distance $c\Delta\tau$ in a proper time interval $\Delta\tau$. Using $\Delta\tau = \Delta t/\gamma$ leads to the inequality given above. For $\Delta t \approx 1$ hour (which is the time scale of variation observed in AGN) and $\gamma \approx 1$, we get $R \lesssim 10^{14}$ cm which is even smaller than the size of the solar system! Since the total luminosity of the AGN is larger than that of an entire galaxy, we need a very efficient and compact mechanism for power generation. Of all the known radiation mechanisms, accretion on to a massive object fits the bill since it can easily convert rest mass energy into radiation with an efficiency of a few percent from a very compact region.

Let us now examine this process in greater detail. With an efficiency of $\eta \simeq 0.1$, one can achieve a luminosity $L = \eta \dot{M} c^2$ if the accretion rate is $\dot{M}$. That is,

$$\dot{M} = \frac{L}{\eta c^2} = 2 M_\odot \ \mathrm{yr}^{-1} \ \left(\frac{L}{10^{46}}\right) \left(\frac{\eta}{0.1}\right)^{-1} \tag{9.7}$$

which is not difficult to achieve in the dense galactic centers. We, however, know that accretion luminosity is bounded by the Eddington luminosity, $L_E$ (see chapter 2) which is related to the mass $M_{\mathrm{BH}}$ of the compact object by

$$L_E \simeq 1.5 \times 10^{38} \frac{M_{\mathrm{BH}}}{M_\odot} \ \mathrm{erg} \ \mathrm{s}^{-1}. \tag{9.8}$$

To achieve a luminosity $L \simeq 5 \times 10^{46}$ erg s$^{-1}$, we need $M_{\mathrm{BH}} > 3.3 \times 10^8 M_\odot$. Note that the mass corresponding to a black hole with the Schwarzschild radius $R \simeq 10^{14}$ cm (which is the size of the central engine determined from time variability) is indeed $3.3 \times 10^8 M_\odot$. This suggests that one might be able to get a consistent picture involving a massive black hole fueled by accretion.

The real difficulty in accretion models is not with energetics but with: (i) time scales and (ii) in providing a mechanism which will make matter shed its angular momentum. To see the issue of time scales, note that the age of the universe at the redshift of $z \approx 5$ is only about $(5-10) \times 10^8 h_{0.75}^{-1}$ yr. This is not very long compared to the galactic free fall time (about $10^8(r/10 \text{ kpc})$ yr) or the black hole growth time defined by $\tau_{\text{BH}} = (\dot{M}/M_{\text{BH}})^{-1}$. Using $\dot{M} \propto L/\eta$ and $M_{\text{BH}} \propto L_E$, we find that $\tau_{\text{BH}} \approx 4 \times 10^7 \eta_{0.1}(L_E/L)$ yr. If a black hole has to grow from a original mass of about $10M_\odot$ to a final mass of $10^8 M_\odot$, it would take about 16 e-folding times; that is, the time required is more like $16\tau_{\text{BH}} \approx 6 \times 10^8 \eta_{0.1}(L_E/L)$ yr. The universe just may not have sufficient time at high redshifts to allow this. Low efficiency accretion permits more rapid growth of black hole (probably in its early stages) but this will require far more total mass to be accreted in order to provide the same amount of energy.

The situation is made worse by the fact that the angular momentum of the matter needs to be removed efficiently if accretion has to work. For matter in Keplerian orbits in a galaxy, the angular momentum for unit mass is $J/m = (GMr)^{1/2}$ where $M$ is the mass inside a region of radius $r$. If we take $M \approx 10^{11} M_\odot$ and $r \approx 10$ kpc for estimating the initial angular momentum $J_i$ and take $M = M_{\text{BH}} = 10^7 M_\odot$ and $r = 0.01$ pc for estimating the final angular momentum $J_f$, then $J_f \approx 10^{-5} J_i$ showing that almost all the angular momentum needs to be removed by some viscous drag if accretion has to be feasible.

Assuming this can somehow be achieved, an accretion disk around a supermassive ($\sim 10^8 M_\odot$) black hole will provide a nearly thermal radiation in the optical and UV. At first sight this might sound surprising since stellar mass compact objects in HMXB/LMXBs emit in X-rays (see Sec. 5.5) and you might expect supermassive black holes to lead to higher energy radiation. In fact, the increase in the Schwarzschild radius of the black hole works in the opposite direction: If we take the luminosity emitted from regions near the black hole to be proportional to $L \propto (GM_{\text{BH}}\dot{M})/R_s$ where $R_s = (2GM_{\text{BH}}/c^2)$, then $L \propto \dot{M}$ and is independent of $M_{\text{BH}}$. If this energy is emitted from an area $A \propto R_s^2$, the flux will be $F \propto (L/A) \propto (\dot{M}/M_{\text{BH}}^2)$. Assuming the radiation is thermal with temperature $T$, we can write $F \propto T^4$ thereby leading to the estimate of the disk temperature

$$T_{\text{disk}} = \left( \frac{3c^6 \dot{M}}{8\pi\sigma G^2 M_{\text{BH}}^2} \right)^{1/4} \tag{9.9}$$

where all the proportionality constants have been re-introduced. If the total emission is a fraction $f$ of the Eddington luminosity and the efficiency of the process is $\eta$, we also have the relation

$$\eta \dot{M} c^2 = f \frac{4\pi Gc}{\kappa} M_{\mathrm{BH}} \qquad (9.10)$$

(where $\kappa = 0.2(1 + X)$ cm$^2$ gm$^{-1}$ is the electron scattering opacity with $X \approx 0.7$), leading to

$$\dot{M} = \frac{f}{\eta} \frac{4\pi G}{\kappa c} M_{\mathrm{BH}}. \qquad (9.11)$$

Substituting into Eq. (9.9), we get

$$T_{\mathrm{disk}} = \left( \frac{3c^5 f}{2\kappa \sigma G M_{\mathrm{BH}} \eta} \right)^{1/4} \propto M_{\mathrm{BH}}^{-1/4}. \qquad (9.12)$$

This shows that the temperature of the disk actually decreases with increasing mass of the black hole. As an example, consider a situation with $M_{\mathrm{BH}} \simeq 10^8 M_\odot$, $f \simeq 1, \eta \simeq 0.1$ and $L \simeq 1.5 \times 10^{46}$ erg s$^{-1}$. This requires an accretion rate given by Eq. (9.11) which translates to $\dot{M} \simeq 2.6 M_\odot$ yr$^{-1}$. The corresponding temperature obtained from Eq. (9.12) is $T_{\mathrm{disc}} = 7.3 \times 10^5$ K. Such a spectrum will peak around 400 Å in the extreme UV band. This thermal radiation could be responsible for a 'bump' seen in the UV spectra of many AGN.

Further, if the radiation is in equipartition with a magnetic field strength $B_E$, then we obtain a characteristic magnetic field of

$$B_E \simeq 4 \times 10^4 \left( \frac{M}{10^8 M_\odot} \right)^{-1/2} \mathrm{Gauss}. \qquad (9.13)$$

The lifetime of electrons emitting synchrotron radiation in such a magnetic field will be about

$$t_E^{\mathrm{cool}} \simeq \left( \frac{m_e}{m_p} \right) \gamma^{-1} \left( \frac{GM}{c^3} \right) \simeq 0.3 \gamma^{-1} \left( \frac{M}{10^8 M_\odot} \right) \mathrm{s}. \qquad (9.14)$$

These numbers imply that we can expect significant radio luminosity due to synchrotron emission; in fact — if most of the luminosity $L$ is due to synchrotron radiation — then the characteristic frequency will be $\nu \simeq 2 \times 10^{14} (M/10^8 M_\odot)^{-5/14}$ Hz. But the synchrotron lifetime of the electrons will be fairly short and re-acceleration mechanisms are required for sustained emission.

Compared to radio and optical, the origin of high energy spectra is more complex and is ill understood. Pair production processes as well as inverse Compton scattering could contribute to high energy radiation though the details differ from source to source. It is easy to see from the following arguments that pair production has to be be important. The number $n_\gamma$ of photons near the threshold energy for pair production, $h\nu \approx m_e c^2$, can be estimated by dividing the energy density, $U \approx L_\gamma/4\pi R^2 c$, of gamma-ray photons within a region of radius $R$ by the threshold energy $m_e c^2$. The optical depth for pair production is about $\tau_{\gamma-\gamma} \simeq n_\gamma \sigma_T R$ and the pair production will be important if $\tau_{\gamma-\gamma} \gtrsim 1$. This condition can be stated in terms of the *compactness parameter* $\ell \equiv (L_\gamma/R)(\sigma_T/m_e c^3)$ as

$$\ell \equiv \frac{L_\gamma}{R}\sigma_T m_e c^3 \gtrsim 4\pi. \qquad (9.15)$$

Expressing the compactness parameter in terms of Eddington luminosity $L_E$, and Schwarzschild radius $R_s$, as $\ell = 2\pi(L_\gamma/L_E)(R_s/R)(m_p/m_e)$, the condition for pair production becomes

$$\frac{L_\gamma}{L_E} \gtrsim \frac{2m_e}{m_p}\left(\frac{R}{R_s}\right) \approx 1.1 \times 10^{-3}\left(\frac{R}{R_s}\right). \qquad (9.16)$$

Since $R \approx 10R_s$, it is possible to meet this condition in many AGN suggesting that pair production can be an important effect.

Finally, note that the remnant mass in Eq. (9.4) was estimated assuming all bright $(L \sim L_*)$ galaxies host an AGN. Alternatively, if only about 1 per cent of the bright galaxies host the AGN, say, then the remnant mass has to be hundred times larger: $M_{\text{rem}} \approx 2 \times 10^8 h^{-3}\mathcal{B}M_\odot$. Since the total energy radiated by AGN can be expressed in terms of $\dot{M}, M_{\text{BH}}$ and an efficiency factor, one can also estimate the remnant mass as $M_{\text{rem}} \approx 8 \times 10^7 h^{-3}\eta^{-1}(t_Q/t_{\text{dyn}})M_\odot$ where $t_Q$ is the time interval for which the AGN is 'on' and $t_{\text{dyn}}$ is the evolution time scale which will be of the order of the age of the universe. If $t_Q \approx t_{\text{dyn}}$ and $\eta \approx 0.1$, then the remnant masses are large $(M_{\text{rem}} \approx 10^9 h^{-3}\eta_{0.1}M_\odot)$. In this case, only a small fraction of the galaxies harbour AGN and $(L/L_E) \lesssim 0.05h\eta_{0.1}$. On the other hand, if $t_Q \ll t_{\text{dyn}}$, then every bright galaxy could have a AGN with, say, $M_{\text{rem}} \approx 10^7 h^{-3}\eta_{0.1}M_\odot$ if $t_q \approx 0.1t_{\text{dyn}}$. In this case, $(L/L_E) \lesssim 3h\eta_{0.1}$. These arguments illustrate the interdependence of these estimates and highlight the implicit assumptions.

## 9.2  Radio jets and bulk relativistic motion

A sub class of AGN show bright radio lobes and jets connecting the central source to these lobes. Because of the high resolution and sensitivity which can be achieved in radio band, it is often possible to do the imaging of the AGN with milli arc second resolution in this band and study the radio features in detail.

In contrast to the optical/UV emission, the radio spectrum is non-thermal and is most likely to be synchrotron radiation from relativistic electrons. The synchrotron emissivity due to a power law distribution of electrons with $N(E) \propto E^{-x}$ is (see Eq. (2.58))

$$\epsilon_\nu \propto B^{(x+1)/2} \nu^{(1-x)/2} \propto \nu^{-\alpha}. \tag{9.17}$$

Hence a power law spectrum of electron will lead to a power law synchrotron emission with the two indices related by $\alpha = (1/2)(x-1)$. This leads to a non-thermal, power law spectrum in the radio band for the AGN.

For luminous radio jets moving with relativistic speeds, one also needs to consider several effects of special relativity to relate the observed quantities to the intrinsic properties of the source. Some of them can be counter intuitive and deserves a discussion.

The first nontrivial effect due to relativistic motion is that the apparent proper motion of a radio jet in the sky could be at a speed greater than that of light. This arises as follows: Consider a blob of plasma which moves with speed $v$ at an angle $\psi$ to the line of sight. Two signals emitted by the blob in a time interval $dt_e$ in the rest frame of the source will be detected during a time interval $dt_0$ in the rest frame of the observer. There are two effects which make $dt_e \neq dt_0$. First is the cosmological redshift which scales the time intervals by the factor $(1+z)$; it turns out, however, that the really interesting features do not arise from this factor and hence we shall temporarily ignore it. The second effect is due to the fact that the blob moves a distance $vdt_e \cos\psi$ in the direction of the observer between the emission of two signals so that the second signal has to travel less distance compared to the first. This reduction in the distance implies that the signals will arrive in a time interval (in units of $c = 1$)

$$dt_0 = (1 - v\cos\psi)\, dt_e. \tag{9.18}$$

The distance traveled by the blob in the plane of the sky, transverse to the line of sight, is $dl_\perp = vdt_e \sin\psi$ so that the apparent transverse velocity $v_{\text{app}}$ measured by an observer is

$$v_{\text{app}} = \frac{dl_\perp}{dt_0} = (v\sin\psi)\frac{dt_e}{dt_0} = \frac{v\sin\psi}{1 - v\cos\psi}. \tag{9.19}$$

It is obvious that the apparent velocity (indicated with a subscript 'app') can be greater than unity even for large inclination angles provided $\gamma$ is high enough.

The following results can be obtained directly from Eq. (9.19): (i) For a given value of $v$, the apparent velocity $v_{app}$ is a maximum when $\psi = \cos^{-1} v$, with the maximum value being $\gamma v$. (ii) In the limit of $v \to 1$, maximum occurs for $\psi \simeq (1/\gamma)$. (iii) It also follows from this analysis that superluminal motion occurs at some angle provided $v_{app}^{max} > 1$ corresponding to $v > (1/\sqrt{2})$. (iv) The same analysis can be repeated in terms of a fixed observed value for $v_{app}$. The minimum speed $v$ required to produce a given value of $v_{app}$ occurs at $\psi = \cot^{-1}(v_{app})$ with $v_{min} = v/\sqrt{1 + v^2}$. (v) When cosmological redshift is taken into account, Eq. (9.18) needs to modified by a factor $(1 + z)$ on the right hand side which does not change the condition for observing the superluminal motion.

The relativistic motion also affects the flux of the source. If cosmological factors are ignored, then the constancy of $I_\nu/\nu^3$ implies (see Sec. 2.1.2) that the intensity (or surface brightness) of the source will scale as

$$I_\nu(\mathcal{D}\nu_0) = \mathcal{D}^3 I_\nu(\nu_0) \qquad (9.20)$$

where

$$\mathcal{D} = \frac{1}{\gamma(1 - v\cos\psi)} \qquad (9.21)$$

is the standard special relativistic factor (see Sec. 2.3). For an unresolved, optically thin, source or for an optically thick spherically symmetric source, the flux $F(\nu)$ transforms in the same way as the intensity. In particular, for a source with power law spectrum $F(\nu) \propto \nu^{-\alpha}$, we have the result

$$F(\nu) = \mathcal{D}^{3+\alpha} F'(\nu) \qquad (9.22)$$

where the primed quantities are those in the rest frame. It is clear that there is boosting of the observed flux over the rest frame flux. When the cosmological redshift cannot be ignored, the observed frequency $\nu$ and the rest frame $\nu'$ are related by

$$\nu = \frac{\mathcal{D}\nu'}{1 + z} \qquad (9.23)$$

so that the flux transforms to

$$F(\nu) = \left(\frac{\mathcal{D}}{1 + z}\right)^{3+\alpha} F'(\nu). \qquad (9.24)$$

As an application, consider a source which emits two blobs of material moving in two diametrically opposite directions $\psi$ and $\pi + \psi$ with the same speed $v$. If $F_{\text{in}}$ and $F_{\text{out}}$ are the fluxes from the sources moving towards the observer and away from the observer, then we have

$$\frac{F_{\text{in}}}{F_{\text{out}}} = \left(\frac{1 + v\cos\psi}{1 - v\cos\psi}\right)^{3+\alpha} \tag{9.25}$$

which shows that there is a preferential enhancement of fluxes in the component which is moving towards the observer.

The results described above are valid for a single blob of material moving with a given velocity. In the case of a continuous jet formed out of a series of blobs, the number of blobs observed at any given instant itself scales as $\mathcal{D}^{-1}$ and hence the index in the formulas for the flux change from $(3+\alpha)$ to $(2+\alpha)$. So the ratio of the flux arising from advancing jet to the identical receding jet will become

$$\frac{F_{\text{in}}}{F_{\text{out}}} = \left(\frac{1 + v\cos\psi}{1 - v\cos\psi}\right)^{2+\alpha}. \tag{9.26}$$

Such a relativistic beaming effect is often invoked to explain the fact that several observed radio sources show jet like structure only on one side. The idea is that the jet is intrinsically two sided but you only see one of them which has its flux enhanced due to this effect. If the ratio of the fluxes is $R = F_{\text{in}}/F_{\text{out}}$, we can obtain an upper limit on the angle of inclination to be

$$\psi \leq \frac{R^{1/(2+\alpha)} - 1}{R^{1/(2+\alpha)} + 1}. \tag{9.27}$$

The relativistic effects also play a role in determining the time scale of variability of the sources and hence in putting bounds on the size of the source. If the observed variability is in a time scale $\Delta t_{\text{var}}$, then the intrinsic time scale of variability in the rest frame of the source is $\Delta t'_{\text{var}} = \mathcal{D}\Delta t_{\text{var}}$. The minimum source size corresponding to this variability time scale is $R \simeq c\Delta t'_{\text{var}} \simeq c\mathcal{D}\Delta t_{\text{var}}$. The solid angle subtended at the source is $\Omega \simeq (R/r)^2 \simeq (c\mathcal{D}\Delta t_{\text{var}}/r)^2$ where $r$ is the distance to the source. The observed flux is then given by

$$F(\nu) = I(\nu)\left(\frac{c\mathcal{D}\Delta t_{\text{var}}}{r}\right)^2 \tag{9.28}$$

where $I(\nu)$ is the intensity measured in the observer's frame. We can now easily relate the brightness temperature $T'$ corresponding to a given intensity in the rest frame and the brightness temperature $T_{obs}$ as determined in the observer's frame corresponding to the angle $\theta$ subtended by the source where $\Omega \approx \theta^2$. Simple algebra now gives

$$T_{obs} = \mathcal{D}^3 T'. \tag{9.29}$$

The physical origin of the $\mathcal{D}^3$ factor can be understood as follows: The brightness temperature is related to the intensity by $T \propto (I_\nu/\nu^2)\theta^2$. If $\theta$ is treated as a fixed quantity, then $T$ will scale as $(I_\nu/\nu^2) = (I_\nu/\nu^3)\nu$; since $(I_\nu/\nu^3)$ is invariant, this will bring in one factor of $\mathcal{D}$. If we further estimate the size of the source and thus $\theta^2$ by the time variability (rather than treat it as fixed), then we will get an additional $\mathcal{D}^2$ factor from this, thereby producing the $\mathcal{D}^3$ factor. This shows that there is an enhancement of the brightness temperature due to relativistic beaming.

Since relativistic beaming amplifies the observed luminosity of the source, it will also change the luminosity function of the source. Consider, for example, a class of sources all having the same luminosity $l$ but are moving relativistically over randomly distributed directions. The observed luminosity $L$, for a given angle $\psi$, is then given by

$$L = \mathcal{D}^m l, \qquad \mathcal{D} = \frac{1}{\gamma(1 - \beta\cos\psi)} \tag{9.30}$$

where $m = 2 + \alpha$ for a continuous jet and $m = 3 + \alpha$ for a blob. As the angle $\psi$ varies from 0 to $(\pi/2)$, the observed luminosity varies in the range $[(2\gamma)^{-m}l, (2\gamma)^m l]$. For a collection of sources with same $\gamma$ but randomly oriented beaming directions, the probability distribution for $\mathcal{D}$ is given by $P(\mathcal{D})d\mathcal{D} = \sin\psi d\psi$; from this it is easy to work out the conditional probability of observing a luminosity $L$ for a source with luminosity $l$ to be:

$$P(L|l)\,dl = \left(\frac{1}{\beta\gamma m}\right) l^{1/m} L^{-(m+1)/m}. \tag{9.31}$$

If the intrinsic luminosity itself is distributed according to some luminosity function $\Phi_{int}(l)$, then the observed luminosity function is given by

$$\Phi_{obs}(L) = \int dl\, P(L|l)\, \Phi_{int}(l). \tag{9.32}$$

The limits of integration will be fixed by the cut-offs on the intrinsic luminosity function as well as the range of $L$ and $\gamma$. If $\Phi_{int}(l) = Kl^{-q}$ with

$q > 1, l > l_{\min}$, then the observed luminosity function is given by

$$\Phi_{\text{obs}}(L) = \left(\frac{K}{\beta\gamma m}\right) L^{-(m+1)/m} \int_{l_1}^{(2\gamma)^m L} dl\, l^{-q+1/p} \qquad (9.33)$$

where $l_1 = \max\left(l_{\min}, L(2\gamma)^{-m}\right)$. That is,

$$\Phi_{\text{obs}}(L) \propto \begin{cases} L^{-q} & (\text{for } L > (2\gamma)^m l_{\min}) \\ L^{-(m+1)/m} & (\text{for } L < (2\gamma)^m l_{\min}). \end{cases} \qquad (9.34)$$

The physical reason for this scaling is simple: At high luminosities a whole range of $l$ contributes to $\Phi_{\text{obs}}$ thereby maintaining the original shape, while for low $l$, the contribution comes mainly from the sources near $l_{\min}$. The effect of beaming is thus to produce a power law luminosity function with a break which flattens the intrinsic luminosity function at low luminosities.

## 9.3 Quasar luminosity function

We shall hereafter concentrate on a key subclass of AGN, called quasars. The simplest statistical description of the abundance of quasars — or for that matter any other source of a particular kind — is by the source count which measures the number of sources as a function of their brightness. To see the importance of source count, let us consider a survey in which the apparent magnitude of all the sources brighter than a limiting magnitude $m_{\max}$ are measured in a given patch of the sky having a solid angle $\Delta\Omega$. Translating the magnitudes into fluxes using the relation $F = F_0 10^{0.4(m_0 - m)}$ we can determine the number of sources $N(> F)$ brighter than a given flux which exists in the survey. If the sources are distributed in a spherically symmetric manner with a luminosity function $dn/dL$ (which is independent of distance), then it is easy to find an expression for $N(> F)$ in terms of $(dn/dL)$. We note that, for each luminosity $L$ there exists a limiting luminosity distance given by $d_L(L) = (L/4\pi F)^{1/2}$.

Hence, the number of sources which we can see with flux greater than $F$ will be

$$N(> F) = \int dL\, V[d_L(L, F)] \frac{dn}{dL} \qquad (9.35)$$

where $V[d_L(L, F)]$ is the volume within which a source of luminosity $L$ could be seen with a minimum flux of $L$. In a Euclidean universe $V = (\Delta\Omega/3)d_L^3$;

this gives

$$N_{\text{Eucl}}(> F) = \frac{\Delta\Omega}{3(4\pi)^{3/2}} F^{-3/2} \int dL\, L^{3/2} \frac{dn}{dL}. \tag{9.36}$$

Thus, for a spherically symmetric homogeneous population in Euclidean universe, $N(> F) \propto F^{-3/2}$ *independent* of the shape of the luminosity function $(dn/dL)$. An important conclusion is that source counts of such a population in the Euclidean universe cannot be used to obtain any information about the shape of the luminosity function.

Since we expect quasars to evolve over cosmic time scales, the luminosity function of quasars will depend on the redshift as well. Observations suggest that this luminosity function can be expressed in the form

$$\phi(L, z)\, dL = \phi^* \left\{ \left( \frac{L}{L^*(z)} \right)^{-\alpha} + \left( \frac{L}{L^*(z)} \right)^{-\beta} \right\}^{-1} \frac{dL}{L^*(z)}. \tag{9.37}$$

Using the relationships

$$\mathcal{M}_B = \mathcal{M}_B^*(z) - 2.5 \log \left( \frac{L}{L^*(z)} \right);$$

$$\phi(\mathcal{M}_B, z)d\mathcal{M}_B = \phi(L, z) \left| \frac{dL}{d\mathcal{M}_B} \right| d\mathcal{M}_B \tag{9.38}$$

this can be expressed in a more convenient form in terms of the blue magnitudes as follows:

$$\phi(\mathcal{M}_B, z)d\mathcal{M}_B = \frac{\phi^*\, d\mathcal{M}_B}{10^{0.4[\mathcal{M}_B - \mathcal{M}_B^*(z)](\alpha+1)} + 10^{0.4[\mathcal{M}_B - \mathcal{M}_B^*(z)](\beta+1)}} \tag{9.39}$$

where

$$\mathcal{M}_B^*(z) = \begin{cases} A - 2.5k \log(1+z) & (\text{for } z < z_{\max}) \\ \mathcal{M}_B^*(z_{\max}) & (\text{for } z > z_{\max}) \end{cases} \tag{9.40}$$

with

$$\alpha = -3.9, \quad \beta = -1.5, \quad A = -20.9 + 5 \log h, \quad k = 3.45,$$
$$z_{\max} = 1.9, \quad \phi^* = 5.2 \times 10^3 h^3 \, \text{Gpc}^{-3}\text{mag}^{-1}. \tag{9.41}$$

This function gives the number density of quasars with magnitudes between $\mathcal{M}_B$ and $\mathcal{M}_B + d\mathcal{M}_B$ at a redshift $z$ and is plotted in Fig. 9.1 for different redshifts. [The inverse relation is based on $L = L_\odot 10^{0.4[\mathcal{M}_\odot - \mathcal{M}]}$ with $\mathcal{M}_\odot = 5.48$, $L_\odot = 2.36 \times 10^{33}\, \text{erg s}^{-1}$.] In spite of significant amount of efforts to understand the luminosity function of quasars at $z > 3$, the situation remains very inconclusive.

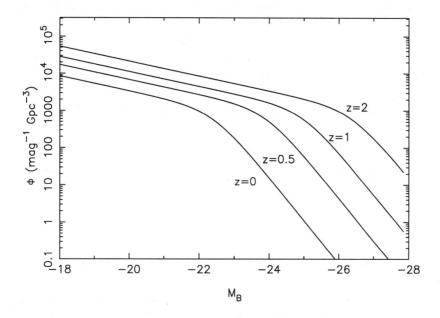

Fig. 9.1   Quasar luminosity function at different redshifts.

One obvious feature of Fig. 9.1 is that the luminosity function of quasars does evolve in time though the actual nature of evolution is not very easy to determine. If the shape of the intrinsic luminosity function of quasars did not change with time but the comoving number density varied with $z$, then the luminosity function will move vertically in Fig. 9.1 (This is called *pure density evolution.*) The other extreme situation corresponds to the assumption that the space density of quasars is constant in time but the luminosity of the population changes with redshift (called *pure luminosity evolution*). In both the cases, one can explain the absence of very high luminosity quasars in the local region: In the case of density evolution, this is because they are spatially very rare; in the case of luminosity evolution, this occurs because the sources are much fainter today than they were, say, at $z = 2$.

There is no physical basis to expect that either extreme assumption should hold and they merely bracket a range of more complicated possibilities. Nevertheless, it is possible to work out some physical consequences of these assumptions which are illuminating. Let us begin with pure density evolution. Since there are fewer quasars now than in the past in this scenario, some of the quasars in the past must have turned off and the dead remnants of quasars must reside at the centers of galaxies, which appear to be normal at present. If we define the 'bright' galaxies to be those with, say, $L \geq 0.1 L^*$, then, from the galaxy luminosity function (see Sec. 8.5) we get the space density of bright galaxies to be

$$n_G(L \geq 0.1\, L^*) = \phi_G^* \int_{0.1}^{\infty} \left(\frac{L}{L^*}\right)^{-1.07} \exp\left(\frac{-L}{L^*}\right) d\left(\frac{L}{L^*}\right)$$
$$\approx 5.7 \times 10^7 h_0^3 \ \mathrm{Gpc}^{-3}. \tag{9.42}$$

The accreted mass per dead quasar can be obtained by dividing the accumulated mass density of Eq. (9.3) by the space density of host galaxies:

$$M_{\mathrm{quasar}} = \frac{2 \times 10^{13} \eta^{-1} M_\odot \ \mathrm{Gpc}^{-3}}{5.7 \times 10^7 h^3 \ \mathrm{Gpc}^{-3}} \approx 3.5 \times 10^5 h^{-3} \eta^{-1} M_\odot. \tag{9.43}$$

While this is not very high, it is at a ball-park value that should lead to observable effects in the nearby galaxies.

The space density of quasars today can be determined from the luminosity function in Eq. (9.37) and is about $1.1 \times 10^6 h^3 \ \mathrm{Gpc}^{-3}$. Comparison with Eq. (9.42) shows that the fraction of bright galaxies containing active quasars today is about $f \simeq 0.019$. If all the bright galaxies contained quasars then the density of quasars would have been higher by a factor 50 or so. If pure density evolution is valid, then a vertical downward translation by factor 50 of the $z = 2$ curve in Fig. 9.1 should make it coincide with the $z = 0$ curve. Such a translation, however, seem to over predict the number of high luminosity quasars at low redshifts. For example, calculating the number density of quasars brighter than $\mathcal{M}_B = -23 + 5 \log h$ at $z = 2$ and rescaling by a factor 50 will lead to a local space density of about $440h^3$ $\mathrm{Gpc}^{-3}$. In a sky survey, extending up to $z = 2$ and covering about 25 per cent of the entire sky, one would expect to find about 100 such objects while only about 8 were found in such surveys. It is obvious that pure density evolution will not work. This is also, of course, clear from the fact that the break point in the luminosity function does evolve with $z$ at least for $z < 1.9$.

Another important application of the observed abundance of quasars is to check whether it is consistent with standard structure formation models. Though the results are not as discriminatory as those based on the abundance of clusters which we discussed in Sec. 6.4 (see Fig. 6.4), it is still interesting from the point of view that it probes much higher redshifts. To obtain such a constraint, we begin by noting that to power the quasars with $z > 4$ having typical luminosities of $10^{47} \text{ergs}^{-1}$, in a universe with $\Omega = 1$ and $h = 0.5$ we require black holes of mass $M_{\text{BH}} \simeq 10^9 M_\odot$ if $L \simeq L_E$. From the luminosity one can also estimate the amount of fuel that must be present to power the quasar for a lifetime $t_Q$. If $\epsilon$ is the efficiency with which the rest mass energy of the fuel is converted into radiation, then the fuel mass is

$$M_f = \frac{L t_Q}{\epsilon} \approx 2 \times 10^9 M_\odot \left( \frac{L}{10^{47} \text{ergs}^{-1}} \right) \left( \frac{t_Q}{10^8 \text{yr}} \right) \left( \frac{\epsilon}{0.1} \right)^{-1}. \qquad (9.44)$$

So if the lifetime is about $10^8 \text{yr}$, and the efficiency is about ten percent, the required fuel mass is comparable to the mass of the central black hole.

The mass estimated above corresponds to that involved in the central engine of the quasar. This mass will, in general, be a small fraction $F$, of the mass of the host galaxy. We can write $F$ as a product of three factors: a fraction $f_b$ of matter in the universe which is baryonic; some fraction $f_{\text{ret}}$ of the baryons originally associated with the galaxy, which was retained when the galaxy was formed (the remaining mass could be expelled via a supernova - driven wind); a fraction $f_{\text{hole}}$ of the baryons retained, which participates in the collapse to form the compact central object. All these factors are model dependent. A very optimistic estimate for $F = f_b f_{\text{ret}} f_{\text{hole}}$ will be $F \simeq 0.01$; a value like $F \simeq 10^{-3}$ is more likely. We then get, for the mass of the host galaxy

$$M_G = 2 \times 10^{11} M_\odot \left( \frac{L}{10^{47} \text{ergs}^{-1}} \right) \left( \frac{t_Q}{10^8 \text{yr}} \right) \left( \frac{\epsilon}{0.1} \right)^{-1} \left( \frac{F}{0.01} \right)^{-1}. \qquad (9.45)$$

For $(F/0.01) \simeq (0.1 - 1)$, with the earlier value being more likely, Eq. (9.45) implies a mass for the host galaxy in the range of $(10^{11} - 10^{12}) M_\odot$, if other dimensionless parameters in Eq. (9.44) are of order unity. Therefore the existence of quasars at $z > 4$ suggests that reasonable number of objects with galactic masses should have formed before this redshift.

An estimate of the fraction of the mass density, $f_{\text{coll}}(> M_G, z)$, in the universe which should have collapsed into objects with mass greater than $M_G$ at redshift $z$ can be made using the luminosity function of quasars. We have

$$\rho(z) f_{\text{coll}}(> M, z) = \int_{-\infty}^{\mathcal{M}(M)} \tau M_{\text{host}}(\mathcal{M}) \phi(\mathcal{M}, z) d\mathcal{M}. \qquad (9.46)$$

In the right hand side, $\mathcal{M}$ is the bolometric magnitude of a quasar situated in a host galaxy with a total mass $M$; $\phi(\mathcal{M}, z)d\mathcal{M}$ gives the number density of quasars in the magnitude interval $d\mathcal{M}$. So the product $M_{\text{host}}(\mathcal{M})\phi d\mathcal{M}$ gives the collapsed mass density around quasars with magnitude in the range $(\mathcal{M}, \mathcal{M} + d\mathcal{M})$. We integrate this expression up to some limiting magnitude $\mathcal{M}$ corresponding to a host mass $M$. Since magnitude decreases with increasing luminosity, this picks out objects with luminosity higher than some particular luminosity or – equivalently – mass higher than $M$. Finally the factor

$$\tau = \max\left(\frac{t(z)}{t_Q}, 1\right) \qquad (9.47)$$

takes into account the possibility that the typical lifetime of a quasar $t_Q$ may be shorter than the age of the universe at the redshift $z$. In that case, typically $(t(z)/t_Q)$ generations of quasars can exist at the particular redshift and we need to enhance the abundance by this factor. Given the luminosity function of the quasars and the relations derived above, we can calculate the fraction by doing the integral numerically. One finds that

$$f_{\text{coll}}\left(> 10^{12}\text{M}_\odot, z \gtrsim 2\right) \simeq 1.4 \times 10^{-3}. \qquad (9.48)$$

To see what this result means, we compare this with the theoretical model developed in Chapter 6 which predicts $f_{\text{coll}}$. A comparison of Eq. (6.76) and Eq. (9.48) shows that at the scales corresponding to the host mass of quasars (which could be above $10^{12}\,\text{M}_\odot$), we must have $\sigma \simeq 1.4$ to 2.6. Any theoretical model for structure formation should produce at least these values of $\sigma$ to be viable.

## 9.4    Neutral hydrogen in the intergalactic medium

Even though quasar spectra do not show any absorption due to smoothly distributed neutral hydrogen in the IGM (see Sec. 7.3), it does exhibit several other absorption features due to hydrogen. We shall first describe some of the broad features seen in a typical quasar spectrum and will then describe the absorption systems in more detail.

Numerous absorption lines are seen in a typical quasar spectrum at wavelengths shortward of the Lyman$-\alpha$ emission line. These can be interpreted as Lyman$-\alpha$ absorption lines arising from clumps of gas along the line-of-sight to the quasar with each absorption line being contributed by a particular region of IGM along the line of sight to the quasar that contains neutral hydrogen. The key feature which allows one to distinguish different Lyman-$\alpha$ absorption systems is the column density of neutral hydrogen present in each of them, which — in turn — can be determined by fitting the observed absorption line to a theoretically known profile (called Voigt profile which we discussed in Sec. 2.5.4). You will recall that Voigt profile is a convolution of natural line width and thermal line width through the formula (see Eq. (2.116))

$$\tau(\lambda) = (1.498 \times 10^{-2}) \left( \frac{Nf\lambda}{b} \right) H(a, u) \qquad (9.49)$$

where

$$H(a, u) = \frac{a}{\pi} \int_{-\infty}^{\infty} \frac{e^{-y^2}}{(u-y)^2 + a^2} \, dy; \quad a = \frac{\lambda \gamma}{4\pi b}; \quad u = -\frac{c}{b} \left[ \left( 1 + \frac{v}{c} \right) - \frac{\lambda}{\lambda_0} \right]. \qquad (9.50)$$

The core of the line is dominated by the Gaussian profile arising from thermal width while the wings are dominated by the Lorentzian profile arising from the natural width. The optical depth at the center of the line is about

$$\tau_0 = 1.497 \times 10^{-15} \frac{N(\text{cm}^{-2}) f \lambda_0(\text{Å})}{b(\text{km s}^{-1})} \qquad (9.51)$$

and essentially determines the nature of the absorption line profile which is observed.

Since the photoionisation cross section for hydrogen is about $\sigma_I \approx 10^{-17}$ cm$^2$, a column density of $N \simeq \sigma_I^{-1}$ provides a natural limit for photoionisation shielding. Lines with column densities between $N \simeq 10^{17}$ cm$^{-2}$ and $N \simeq 10^{20}$ cm$^{-2}$ are usually called *Lyman limit* systems. Systems with low column density — usually below about $N \simeq 10^{17}$ cm$^{-2}$ — are called *Lyman-$\alpha$ forest*. The term 'forest' is used to denote large number of such lines which are typically seen in such absorption systems.

The actual number of absorbing systems seen in the IGM with different column densities of hydrogen, seem to follow a fairly simple power law over a wide range. This result can be expressed in a more useful form in terms of the probability $dP$ that a random line of sight intercepts a cloud with a column density in the range $(\Sigma, \Sigma + d\Sigma)$ at the redshift interval $(z, z + dz)$.

Observations suggest that this probability is given by

$$dP = g\left(\Sigma, z\right) d\Sigma dz \simeq 91.2 \left(\frac{\Sigma}{\Sigma_0}\right)^{-\beta} \frac{d\Sigma}{\Sigma_0} dz, \qquad (9.52)$$

with $\Sigma_0 \simeq 10^{14} \text{cm}^{-2}$, $\beta \cong 1.46$.

### 9.4.1 Lyman forest

We shall now consider the physical properties of the absorption systems beginning from the Lyman-$\alpha$ forest which has the lowest column densities. These low column density systems could arise naturally from the fluctuations in the neutral hydrogen density in the IGM in the following manner. Consider a region of diffuse low density IGM in which the matter is in ionisation equilibrium under the action of an ionizing flux $J_\nu$ (from different sources like quasars *etc.*) and recombination. Since the density contrast in the IGM will be fluctuating and evolving, we will obtain transient absorption systems due to all the low column density neutral hydrogen regions along the line of sight to a quasar. The neutral hydrogen fraction in these systems can be estimated exactly as in Eq. (7.34) and we will get

$$\frac{n_H}{n_p} = \frac{\alpha f n_p}{\mathcal{R}} \simeq 10^{-6} \frac{\Omega_{\text{IGM}} h^2 (1+z)^3}{J_0} \left(\frac{T}{10^4 \, \text{K}}\right)^{-0.76} [1 + \delta(z)]^2. \qquad (9.53)$$

This is same as Eq. (7.34) except for two modifications. We have introduced a factor $[1 + \delta(z)]^2$ to take into account the density variations — $\delta$ is the density contrast compared to the background density — in the IGM; we have changed the exponent of temperature dependence from $-(1/2)$ to a more accurate value of $-0.76$. The neutral hydrogen density in Eq. (9.53) will correspond to an optical depth of

$$\tau(z) = 6.5 \times 10^{-4} \left(\frac{\Omega_B h^2}{0.019}\right)^2 \left(\frac{h}{0.65}\right)^{-1} \frac{(1+z)^6}{H_0} d_H(z) \frac{T_4^{-0.76}}{\mathcal{R}_{12}} [1 + \delta(z)]^2. \qquad (9.54)$$

where $\mathcal{R}_{12} = (\mathcal{R}/10^{-12} \, \text{s}^{-1})$ is the photoionisation rate defined in Eq. (7.32). This equation gives the typical opacity due to Lyman-$\alpha$ forest in diffuse intergalactic medium.

The amount of mass density contributed by these absorption systems is of considerable interest and — fortunately — can be estimated without knowing the size of the neutral hydrogen patch involved in the absorption. Consider a region of IGM from which Lyman-$\alpha$ forest absorption takes place. Let us assume that the typical overdense fluctuation has a size of $l$, neutral hydrogen mass $M_H$, column density in the range $(\Sigma, \Sigma + d\Sigma)$ and let the number density of overdense regions is $n(z)$. Then we have

$$\frac{dN}{d\Sigma dz} d\Sigma \equiv g(\Sigma, z) d\Sigma \simeq \pi l^2 n(z) \frac{cdt}{dz} \qquad (9.55)$$

and

$$M_H \simeq m_p \pi l^2 \Sigma. \qquad (9.56)$$

The total density contributed by neutral hydrogen in these clouds is

$$\rho_H = n(z) M_H = \left[ g d\Sigma \frac{1}{\pi l^2} \frac{dz}{cdt} \right] [m_p \pi l^2 \Sigma] = m_p \Sigma g(\Sigma, z) \left( \frac{dz}{cdt} \right) d\Sigma. \qquad (9.57)$$

Note that this result is independent of the size $l$ of the clouds. The total $\rho_H$ contributed by clouds with column densities in the range $(\Sigma_{\min}, \Sigma_{\max})$ can be obtained by integrating this expression between the two limits. Dividing by $\rho_c(1 + z)^3$ one can find the $\Omega_H$ contributed by the neutral hydrogen in these clouds:

$$\begin{aligned}
\Omega_H(z) &= \frac{m_p}{\rho_c(1+z)^3} \left( \frac{dz}{cdt} \right) \int_{\Sigma_{\min}}^{\Sigma_{\max}} \Sigma g(\Sigma, z) d\Sigma \\
&= \left( \frac{m_p}{\rho_c} \right) \left( \frac{H_0 \Omega}{c} \right)^{1/2} \frac{1}{(1+z)^{1/2}} \int_{\Sigma_{\min}}^{\Sigma_{\max}} \Sigma g(\Sigma, z) d\Sigma. \quad (9.58)
\end{aligned}$$

Taking $g = 91.2(\Sigma/\Sigma_0)^{-\beta} \Sigma_0^{-1}$ with $\Sigma_0 = 10^{14} \, \text{cm}^{-2}$, $\beta = 1.46$, $\Sigma_{\max} \approx 10^{22} \text{cm}^{-2}$, $\Sigma_{\min} \approx 10^{14} \text{cm}^{-2}$, we get, at $z = 3$,

$$\Omega_H \simeq 2 \times 10^{-3} \Omega_{\text{NR}}^{1/2} h^{-1} \left( \frac{\Sigma_{\max}}{10^{22} \, \text{cm}^{-2}} \right)^{2-\beta}. \qquad (9.59)$$

For $\Sigma_{\max} \approx 10^{15} \, \text{cm}^{-2}$, $\Sigma_{\min} \approx 10^{13} \, \text{cm}^{-2}$, which is appropriate for Lyman forest clouds, the corresponding result is

$$\Omega_H(\text{Lyman forest}) \simeq 3 \times 10^{-7} \Omega_{\text{NR}}^{1/2} h^{-1} \qquad (9.60)$$

at $z = 3$.

The $\Omega_{\text{NR}}$ contributed by the *ionized component* in these systems can be much larger. In a cloud with column density $\Sigma$, let the plasma contributes a mass which is $F^{-1}$ larger than that of neutral hydrogen. This fraction $F$

can be determined from Eq. (7.33) rewritten in the form

$$n_p = \left(\frac{\mathcal{R} n_H}{\alpha}\right)^{1/2} = \left(\frac{\mathcal{R} \Sigma}{\alpha l}\right)^{1/2} \qquad (9.61)$$

where $l$ is the characteristic scale of the system. This gives

$$F = \frac{n_H}{n_p} = \frac{\alpha n_p}{\mathcal{R}} = \left(\frac{\alpha \Sigma}{\mathcal{R} l}\right)^{1/2}. \qquad (9.62)$$

So the $d\Omega_{\text{Bary}}$ due to clouds with column density in the range $(\Sigma, \Sigma + d\Sigma)$ will be

$$d\Omega_{\text{Bary}} = \frac{d\Omega_H}{F} = \left(\frac{\mathcal{R} l}{\alpha \Sigma}\right)^{1/2} \left(\frac{d\Omega_H}{d\Sigma}\right) d\Sigma. \qquad (9.63)$$

The result in Eq. (9.58) is now modified to

$$\Omega_{\text{Bary}} = \left(\frac{\mathcal{R} l}{\alpha}\right) \left(\frac{m_p}{\rho_c}\right) \left(\frac{H_0 \Omega_{\text{NR}}^{1/2}}{c}\right) \frac{1}{(1+z)^{1/2}} \int_{\Sigma_{\text{min}}}^{\Sigma_{\text{max}}} \Sigma^{1/2} g(\Sigma, z) d\Sigma. \qquad (9.64)$$

Since the exponent $\beta$ of $g \propto \Sigma^{-\beta}$ is close to 1.5, the integral is close to a logarithm. Taking $\beta = 1.5$ for simplicity, we get, at $z = 3$,

$$\Omega_{\text{Bary}} \approx 6.3 \times 10^{-3} \left(\frac{J_0 l_0 \Omega}{h^3}\right)^{1/2} \left(\frac{T}{10^4\,\text{K}}\right)^{1/4} \ln\left(\frac{\Sigma_{\text{max}}}{\Sigma_{\text{min}}}\right). \qquad (9.65)$$

If we take $\Sigma_{\text{max}} = 10^{15}\text{cm}^{-2}$ and $\Sigma_{\text{min}} = 10^{13}\text{cm}^{-2}$ the logarithm is about 4.6 and we find that

$$\Omega_{\text{Bary}} \approx 3 \times 10^{-2} \left(\frac{J_0 l_0 \Omega}{h^3}\right)^{1/2} \left(\frac{T}{10^4\,\text{K}}\right)^{1/4}. \qquad (9.66)$$

Finally, we note a feature about these systems which is potentially very important. Observations reveal that approximately 50 per cent of the Lyman-$\alpha$ lines with column densities higher than $3 \times 10^{14}$ cm$^{-2}$ and nearly all lines with column densities higher than $10^{15}$ cm$^{-2}$ originate from regions which also show absorption due to carbon. While this shows that stellar nucleosynthesis has already distributed the metals in the IGM at about $z \approx 3$, the actual abundance of metals is difficult to determine. Estimates center around $Z \approx 10^{-2.5} Z_\odot$ with a large — about an order of magnitude — scatter around this value. The metallicity of IGM at high redshifts, when determined more accurately, can impose strong constraints on the structure formation models.

## 9.4.2  Damped Lyman alpha clouds

Historically, damped Lyman-$\alpha$ clouds (DLA, hereafter) were defined[1] as systems with hydrogen column densities higher than $2 \times 10^{20}$ cm$^{-2}$. The high column densities make the optical depth of these clouds sufficiently large for the hydrogen to be mostly neutral. The gas has a temperature $T \lesssim 3 \times 10^3$ K and could even contain molecules. The distribution of column densities follows a slope of $-1.5$ even at these high column densities but there seems to be virtually no evolution in these systems. The number density of these systems per unit redshift interval does not show any significant $z$ dependence and is given by $(dN/dz) \simeq 0.3$ around $z \simeq 2.5$. The cosmic density of neutral hydrogen in DLA at $z \approx 3$ is comparable to that of stars today. All these suggests that DLAs can be thought of as individual clouds (unlike the low column density Lyman-$\alpha$ forest which is most likely to have originated from a fluctuating IGM).

To begin with, $(dN/dz) \equiv B \approx 0.3$, implies that the mean proper distance between the clouds along the line of sight (in a $\Omega_{\rm NR} \simeq 1$ universe) is about

$$L = \frac{c\,dt}{dz}\frac{dz}{dN} = \frac{c}{H_0 B (1+z)^{5/2}} \simeq 10^4 h^{-1}(1+z)^{-5/2}\,{\rm Mpc}$$
$$\simeq 300 h^{-1}\,{\rm Mpc} \qquad (\text{at } z = 3). \tag{9.67}$$

If we take the size of the cloud to be $l = 10 l_0 h^{-1}$ Mpc, then the density of hydrogen atoms and the mass of neutral hydrogen in these systems are given by

$$n_H \simeq \frac{\Sigma}{l} = 0.3 h l_0^{-1} \Sigma_{22}\,{\rm cm}^{-3}$$
$$M_H \approx m_p \Sigma l^2 \approx 10^{10} l_0^2 \Sigma_{22} h^{-2} {\rm M}_\odot \tag{9.68}$$

where $\Sigma_{22}$ is the column density in units of $10^{22}$ cm$^{-2}$. The mass in neutral hydrogen is fairly high allowing these systems to be interpreted as pregenitors of the discs.

---

[1] You may wonder what is "damped" about these systems. The terminology, as usual, has a complicated origin. We saw in Sec. 2.5.4 that the profile of an absorption line will be dominated by the Lorentzian wing at high column densities. This Lorentzian profile, in turn, can be modeled by a damped harmonic oscillator such that the natural line width arises from the damping of the oscillator; see Eq. (2.103). Since these systems with high column densities probe the natural width of the absorption line, the term "damped" has got affixed to them!

The fraction, $f$, of the space filled by these clouds is also of interest and can be estimated as follows. If $n(z)$ is the number density of clouds, then

$$\frac{dN}{dz} = \left(\pi l^2\right) n(z) \frac{cdt}{dz}. \qquad (9.69)$$

The fraction of space filled by clouds in some large volume $V$ is

$$f = \frac{(4\pi/3)l^3 n(z)V}{V} = \frac{4\pi}{3} n(z) l^3. \qquad (9.70)$$

Substituting for $n(z)$ from Eq. (9.69) we get

$$f = \frac{4\pi}{3} l^3 \cdot \left(\frac{cdt}{dz}\right)^{-1} \left(\frac{dN}{dz}\right) \left(\frac{1}{\pi l^2}\right) \approx \left(\frac{dN}{dz}\right) \left(\frac{dz}{cdt}\right) l = \frac{l}{L} \qquad (9.71)$$

where $L$ is the mean distance between the clouds along the line of sight. Numerically, using Eq. (9.67) we get:

$$f = \frac{l}{L} = 3.3 \times 10^{-5} l_0 \qquad \text{(at } z = 3\text{)}. \qquad (9.72)$$

The number density of clouds will be

$$n_{\text{cloud}} = \frac{f}{l^3} = 33 h^3 l_0^{-2} \text{Mpc}^{-3} \qquad (9.73)$$

with a mean inter-cloud separation of $n_{\text{cloud}}^{-1/3} \approx 0.3 h^{-1} l_0^{2/3}$ Mpc. This is nearly 30 times larger than the size of damped Lyman alpha systems if $l_0 \approx 1$.

The high column density of these clouds imply that they will be mostly neutral. The metagalactic flux of photons will be able to ionize only a thin layer of matter on the surface. Near the surface, equilibrium between ionisation and recombination (see Eq. (7.33)) requires

$$\frac{n_H}{n_p} = \frac{\alpha n_p}{\mathcal{R}} \qquad (9.74)$$

leading to a proton density of

$$n_p = \left(\frac{\mathcal{R}}{\alpha} n_H\right)^{1/2} = \left(\frac{\mathcal{R}\,\Sigma}{\alpha\,l}\right)^{1/2}$$
$$\approx 2 \left(\frac{T}{10^4\,\text{K}}\right)^{1/4} \left(\frac{\Sigma}{10^{22}\text{cm}^{-2}}\right)^{1/2} \left(\frac{J_0 h}{l_0}\right)^{1/2} \text{cm}^{-3} \qquad (9.75)$$

which is about six times the neutral hydrogen density of $n_H$ of Eq. (9.68). Further, the penetration length $\lambda$ of photons at such high column densities is low; we have,

$$\lambda \approx \frac{1}{n_H \sigma_{\rm I}} \approx 0.14\, l_0 h^{-1}\, {\rm pc}\, \left(\frac{\Sigma}{10^{22}\, {\rm cm}^{-2}}\right)^{-1}. \qquad (9.76)$$

This length is much smaller than the size of the system. (Note that $(\lambda/l) \approx (\Sigma \sigma_{\rm I})^{-1}$; so $\lambda \approx l$ at column densities of $\Sigma \approx 10^{17}\, {\rm cm}^{-2}$. For $\Sigma \gg 10^{17}\, {\rm cm}^{-2}$, $\lambda \ll l$ and the system is mostly neutral — which is, of course, what you would have expected.) Thus, most of the mass in damped Lyman alpha systems is contributed by the neutral hydrogen:

$$M_{\rm Baryons} \approx M_H \approx 10^{10} l_0^2 h^{-2} {\rm M}_\odot. \qquad (9.77)$$

The $\Omega$ contributed by damped Lyman alpha systems is quite large and comparable to that of luminous matter in the galaxies. We have, from Eq. (9.59),

$$\Omega_H \approx 2 \times 10^{-3} h^{-1} \qquad ({\rm for}\ \Sigma \approx 10^{22} {\rm cm}^{-2}; z \approx 3). \qquad (9.78)$$

The dark matter associated with such a cloud will be higher by a factor $\Omega_B^{-1}$. If $\Omega_B \simeq 0.06$, this factor will be about 18. Thus the total $\Omega_{\rm NR}$ associated with damped Lyman$-\alpha$ systems could be about $(0.05 - 0.1)$ in the redshift range of $z \approx 2 - 3$.

## Exercises

**Exercise 9.1**   *Adding up Planck spectra:*

The analysis in Sec. 3.3.2 (see Eq. (3.54)) showed that the temperature distribution of an accretion disk varies as $T \propto R^{-3/4}$. Assume that the radiation from an AGN arises from such an accretion disk. Then the total intensity which we observe from an accretion disk at any given frequency $\nu$ is an integral over the entire disc. Assuming that each circular patch of the disk emits as a blackbody at the temperature $T(R)$, one can estimate the net intensity at any frequency $\nu$ by a direct integration. Show that the intensity varies as $I_\nu \propto [\nu^2, \nu^{1/3}, \exp(-\nu/\nu_0)]$ for low, intermediate and high frequencies.

**Exercise 9.2**   *Pair production and compactness:*

Radio observations of an AGN at a distance of 200 Mpc shows that at $\nu = 1$ GHz, the brightness is 1 mJy. The source is seen to vary at a time scale of 1 week. (a) Calculate the brightness temperature. (b) Assume that matter and radiation are in thermal equilibrium at this brightness temperature and compute the optical depth for pair production taking the relevant cross section to be $\sigma_{\gamma\gamma} \simeq 10^{-25}$ cm$^2$.

### Exercise 9.3    CMB and the inverse Compton scattering:

Suppose the relativistic electrons are injected into a hot spot of a radio source with a power law spectrum $\gamma^{-p}$. The electrons stay in the hot spot for some fixed time $T$ before being advected away. These electrons will lose energy through inverse Compton scattering off the CMBR. Show that in steady state the spectrum of relativistic electrons will change from $\gamma^{-p}$ to $\gamma^{-(p+1)}$ at a critical $\gamma$ given by $\gamma_c \approx (m_e c^2/\sigma_T c U_{\text{CMB}} T)$ where $U_{\text{CMB}}$ is the energy density of the CMBR. Further show that this will lead to a break in the synchrotron radiation at a frequency $\nu_b$ which varies as $\nu_b \propto (1 + z)^{-9}$.

### Exercise 9.4    Getting the numbers:

(a) Consider a source at redshift $z$, moving across the sky with the angular proper motion $\mu = d\theta/dt_0$ milliarcsec per year, in a $\Omega_{\text{NR}} = 1$ universe. Show that

$$\frac{v_{\text{app}}}{c} = 94.4h^{-1} \left( \frac{\mu}{1 \text{ milliarcsecond yr}^{-1}} \right) \left( 1 - \frac{1}{\sqrt{1+z}} \right). \tag{9.79}$$

(b) Assume that a source emits blobs of matter with speed $v$ in a random direction. Show that the probability that the apparent transverse speed for the source will exceed a value $v_{\text{app}}$ is given by

$$p(> v_{\text{app}}) = \frac{1}{1 + v_{\text{app}}^2} \sqrt{1 - \frac{v_{\text{app}}^2}{\gamma^2 - 1}}. \tag{9.80}$$

### Exercise 9.5    Numerical estimates in variability:

The brightness temperature, $T_b$, for a source with brightness $S$ at wavelength $\lambda$ and subtending an angle $\theta$ is given by $T_b \simeq \lambda^2 S/(2k_B\theta^2)$. Suppose the flux changes by an amount $\Delta S$ (in Jy), in a time scale $\tau$ (in years), show that

$$T_b \geq (2 \times 10^{12} \text{ K}) \lambda_{\text{cm}}^2 \left( \frac{\Delta S}{\tau^2} \right) \tag{9.81}$$

if we ignore relativistic effects. Estimate $T_b$ for the two sources 3C 454.3 ($\Delta S \simeq 5$ Jy, $\tau \simeq 1$ yr, $\lambda \simeq 74$ cm) and AO 0235 ($\Delta S \simeq 1$ Jy, $\tau \simeq 30$ days, $\lambda \simeq 11$ cm). What is the range of Lorentz factor $\gamma$ and direction of motion $\theta$ (with respect to the line of sight) for which relativistic effects can bring down the actual brightness temperature to $10^{12}$ K?

## Exercise 9.6   *Jet scaling:*

Consider a jet with constant opening angle and assume that the surface brightness (luminosity per unit projected surface area) scales as $r^{-2.5}$ where $r$ is the distance along the jet. (a) Determine the scalings for (i) the equipartition pressure in the jet, (ii) equipartition magnetic field strength and (iii) lifetime due to synchrotron cooling. (b) Suppose the jet has a constant bulk Lorentz factor and that the synchrotron radiation is due to relativistic electrons with energy distribution $N(\gamma) \propto \gamma^{-2}$. Show that if the electron gas expanded rapidly and the jet was dominated by a transverse magnetic field, then the surface brightness will scale as $r^{-19/6}$.

[Answer: (a) Constant opening angle $\theta = R/r$ implies $r \propto R$. The volume emissivity is then $\epsilon_\nu \propto I_\nu/R \propto r^{-3.5}$. The volume emissivity in synchrotron radiation scales as $\epsilon_\nu \propto n_e(\gamma^2 B^2/\nu) \simeq p_e(B^2\gamma/\nu)$ where $p_e \simeq n_e\gamma mc^2$ is the pressure of the electron gas. Assuming that most of the radiation is at $\nu \simeq \nu_c \propto \gamma^2 B$, we get $\epsilon_\nu \propto p_e B^{3/2}\nu_c^{-1/2} \propto B_{eq}^{7/2}\nu_c^{-1/2}$ if $p_e \propto B_{eq}^2$ under the assumption of equipartition. Comparing $\epsilon_\nu \propto r^{-3.5}$ with $\nu_c^{1/2}\epsilon_\nu \propto B_{eq}^{3.5}$, we find that the equipartition field $B_{eq} \propto r^{-1}$ and the equipartition pressure scales as $r^{-2}$. The synchrotron cooling time varies as $t_{cool} = E/\dot{E} \propto \gamma/\gamma^2 B^2 \propto \nu^{-1/2}B^{-3/2} \propto r^{3/2}$. (b) For a relativistic gas expanding adiabatically, $p \propto \rho^{4/3} \propto R^{-8/3}$. Equating this to $p_e \propto n_e\gamma \propto \gamma^2(dn/d\gamma)$, we get $\gamma^2 dn/d\gamma \propto R^{-8/3}$. The emissivity varies as $\epsilon_\nu \propto (\gamma^2 dn/d\gamma)B^{3/2}\nu^{-1/2} \propto B^{3/2}\nu^{-1/2}R^{-8/3}$. Using the result $B \propto R^{-1}$, we get $\epsilon_\nu \propto \nu^{-1/2}R^{-25/6}$. The corresponding surface brightness is $I_\nu \propto \epsilon_\nu R \propto \nu^{-1/2}R^{-19/6} \propto r^{-3.1}$. Since the observed distribution scales as $I_\nu \propto r^{-2.5}$, there must be a process of continuous acceleration to restore the energy lost due to adiabatic expansion.]

## Exercise 9.7   *K correction:*

While observing sources like quasars, with redshifts comparable (or larger) to unity, it is necessary to take into account two simple kinematic effects: (i) The frequency at which the radiation is detected is not the same as the frequency at which it is emitted; (ii) the bandwidth over which the observations are carried out will not be the same as the bandwidth over which the radiation is emitted due to redshift. It is conventional to write the apparent magnitude of the object in the form $m_{intrinsic} = m_{observed} - K(z)$, where $K(z)$ is called the K-correction. (a) Show that $K(z) = K_1 + K_2$ where $K_1 = 2.5\log(1 + z)$ and

$$K_2 = 2.5 \log \frac{\displaystyle\int_0^\infty F(\lambda)\,S(\lambda)\,d\lambda}{\displaystyle\int_0^\infty F(\lambda/[1+z])\,S(\lambda)\,d\lambda} \tag{9.82}$$

where $F(\lambda)$ is the spectral energy distribution of the source and $S(\lambda)$ is the filter function determining the wave band over which observations are carried out. (b) Consider a power law source with ($F \propto \nu^{-\alpha}$ and show that, in this case, the total

K-correction is given by

$$K(z) = 2.5(\alpha - 1) \log(1 + z).$$ (9.83)

**Exercise 9.8** *Growing a monster:*

Consider a black hole which grows by accreting matter and radiating at Eddington luminosity with the efficiency $\eta = 0.1$. If it starts at 10 $M_\odot$ at the redshift $z_i$, at what redshift will it grow to $10^9 M_\odot$ ? Alternatively, assume that the black hole should have grown to this value by, say, $z = 5$. What constraint does it put on the initial epoch? Does this sound reasonable?

**Exercise 9.9** *Galactic center:*

Several pieces of evidence suggests that there is a supermassive black hole at the center of our galaxy. Suppose it had been active. How will the spectrum of radiation from this region look like? Assume that it has $M = 10^9 M_\odot$, $L = L_{edd}$ and a radiation spectrum which is $\nu^{-1/3}$ between $\nu = 10^{12}$ and $10^{21}$ Hz. Use this to compute the apparent V-band magnitude as measured from Earth.

**Exercise 9.10** *Iron line:*

The FeK$\alpha$ line representing an $n = 2 \rightarrow 1$ transition in iron is a favourite of x-ray astronomers. Based on the temperature scaling, what is the mass range of the compact objects in which you will expect to see this line — in XRBs or in AGN ?[Hint: Remember that we want $n = 2$ to be reasonably populated but the ionisation fraction should not be too high.]

**Exercise 9.11** *Superluminal motion:*

In the quasar 3C 273 at $z = 0.158$, one component is seen to move away from the nucleus with an angular velocity $2.2 \times 10^{-3}$ arcsec yr$^{-1}$. (a) What is the projected linear velocity of this component? (b) Assuming that the superluminal motion which you found in part (a) above is due to the effect discussed in Sec. 9.2, estimate the angle between the line-of-sight and the direction of ejected matter that will produce the maximum apparent velocity for the ejected component. (c) What is the minimum ejection velocity required to produce the apparent superluminal speed you obtained in (a)?

**Exercise 9.12** *Real life astronomy: Quasars:*

The SDSS web-site has a good navigation tool (http://skyserver.sdss.org) which allows you to explore the objects detected in these surveys. Try, as an example, the coordinates around RA = 176.764° and DEC = 1.412°. There is a quasar in the field which you can click on and look at its spectrum (does it look like a quasar spectrum?) and its redshift. (a) Given the redshift and apparent R magnitude, find the absolute luminosity using the correct luminosity distance formula. Use two cosmological models, one with $\Omega_{NR} = 1$ and the other with $\Omega_{NR} = 0.3$ and $\Omega_V = 0.7$. How much does the estimated luminosity change between these two models? (b) If the quasar is emitting in the Eddington limit, determine the mass of the central black hole for reasonable efficiencies. (c) The spectrum you obtained does not cover all the electromagnetic bands. Often one would like to know whether the same object emits, say, in radio or in x-rays. Try the VLA FIRST survey (check the site http://sundog.stsci.edu) at these coordinates to determine whether this quasar has a radio flux. How will you check its X-ray properties ?

## Notes and References

Some of the specialist textbooks dealing with AGN are: Peterson (1997), Frank King and Raine (1992), Osterbrock (1989), Ferland and Baldwin (1999); also see chapter 8 of Padmanabhan (2002).

# Appendix

# Range of Physical Quantities in Astrophysics

This Appendix summarizes different physical parameters we encounter in the study of astrophysics — like mass, length, time, angular scales, velocity, energy, power, energy density, temperature, pressure and magnetic field. The numbers are approximate and are intended to give an idea of the range. In each of the tables, the values are given in two columns: the first column gives it in a fundamental unit while the second column gives a more appropriate astronomical unit when it iss relevant. The fundamental unit used in each of the tables is indicated in the caption.

## Mass (gm)

| | | |
|---|---|---|
| Electron mass | $9.11 \times 10^{-28}$ | |
| Proton mass | $1.67 \times 10^{-24}$ | |
| Planck mass$((\hbar c/G)^{1/2}$ | $2.2 \times 10^{-5}$ | |
| Mass of Moon | $7.3 \times 10^{25}$ | $(1/80)M_\oplus$ |
| Mass of Earth | $5.98 \times 10^{27}$ | $1M_\oplus$ |
| Mass of Jupiter | $1.9 \times 10^{30}$ | $10^{-3}M_\odot$ |
| Mass of the Sun | $1.99 \times 10^{33}$ | $1\ M_\odot$ |
| Chandrasekhar mass | $2.8 \times 10^{33}$ | $1.4\ M_\odot$ |
| Mass of interstellar clouds | $2 \times 10^{36}$ | $10^3\ M_\odot$ |
| Mass of a globular cluster | $1 \times 10^{39}$ | $5 \times 10^5\ M_\odot$ |
| Mass of Milky Way | $2.6 \times 10^{45}$ | $1.4 \times 10^{12}\ M_\odot$ |
| Virial mass of the Coma cluster | $2.7 \times 10^{48}$ | $1.3 \times 10^{15}\ M_\odot$ |

# Length (cm)

| | | |
|---|---|---|
| Planck length$(G\hbar/c^3)^{\frac{1}{2}}$ | $1.6 \times 10^{-33}$ | |
| Proton Compton wavelength$(h/m_p c)$ | $1.3 \times 10^{-13}$ | |
| Classical electron radius$(e^2/m_e c^2)$ | $2.8 \times 10^{-13}$ | |
| Electron Compton wavelength$(h/m_e c)$ | $2.4 \times 10^{-10}$ | |
| Bohr radius$(\hbar^2/m_e e^2)$ | $5.3 \times 10^{-9}$ | 0.5 Å |
| Molecular mean free path in the atmosphere | $7.0 \times 10^{-6}$ | 700 Å |
| Optical photon wavelength | $5 \times 10^{-5}$ | 5000 Å |
| Neutron star radius | $10^6$ | |
| Radius of the Earth | $6.3 \times 10^8$ | $1 R_\oplus$ |
| Radius of Jupiter | $7.1 \times 10^9$ | $11.3 R_\oplus$ |
| Distance to the Moon | $3.8 \times 10^{10}$ | $60 R_\oplus$ |
| Radius of the Sun | $7.0 \times 10^{10}$ | $1 R_\odot$ |
| Earth-Sun mean distance | $1.50 \times 10^{13}$ | 1 AU |
| Pluto-Sun mean distance | $5.91 \times 10^{14}$ | 39.4 AU |
| 1 light-year (ly) | $9.46 \times 10^{17}$ | |
| 1 parsec (pc) | $3.08 \times 10^{18}$ | 3.26 ly |
| Nearest star to Sun | $4 \times 10^{18}$ | 1.3 pc |
| Radius of Crab Nebula | $4.5 \times 10^{18}$ | 1.46 pc |
| Radius of globular cluster | $1.5 \times 10^{20}$ | 50 pc |
| Scale height of the Milky Way disk | $2 \times 10^{21}$ | 700 pc |
| Distance to the Crab Nebula | $6.6 \times 10^{21}$ | 2.2 kpc |
| Distance from Sun to galactic center | $2.4 \times 10^{22}$ | 8 kpc |
| Distance to the LMC | $1.5 \times 10^{23}$ | 50 kpc |
| Length of a radio jet | $1 \times 10^{24}$ | 300 kpc |
| Distance to Andromeda | $1.9 \times 10^{24}$ | 620 kpc |
| Distance to Coma cluster | $4.1 \times 10^{26}$ | 132 Mpc |

# Time (seconds)

| | | |
|---|---|---|
| Minimum possible spin period for a neutron star | $8 \times 10^{-4}$ | |
| Spin period of pulsar PSR 1957+20 | $1.6 \times 10^{-3}$ | |
| Sun's free fall time scale $(1/\sqrt{G\rho})$ | 2 000 | 33 min |
| Earth rotation period | $8.6 \times 10^4$ | 24 hrs |
| Earth orbital period | $3.2 \times 10^7$ | 1 yr |
| Main sequence lifetime for a 30 $M_\odot$ star | $1.6 \times 10^{14}$ | $5.0 \times 10^6$ yr |

| | | |
|---|---|---|
| Sun's thermal time scale $(GM_\odot/R_\odot^2 L_\odot)$ | $6.3 \times 10^{14}$ | $2.1 \times 10^7$ yr |
| Main sequence lifetime for a 5 $M_\odot$ star | $1.9 \times 10^{15}$ | $5.9 \times 10^7$ yr |
| Orbital time of sun in the galaxy | $7.3 \times 10^{15}$ | $2.3 \times 10^8$ yr |
| Age of Earth and Sun | $1.5 \times 10^{17}$ | 4.7 Gyr |
| Main sequence lifetime for Sun | $3 \times 10^{17}$ | 9.4 Gyr |
| Age of an old globular cluster | $4 \times 10^{17}$ | 12.5 Gyr |
| Age of the universe | $4.1 \times 10^{17}$ | 13 Gyr |

### Angular scales (arcseconds)

| | | |
|---|---|---|
| Resolution of radio telescope (VLBA) | 0.0002 | |
| Accuracy attainable in measuring angular positions of stars | 0.0015 | |
| *Hubble Space Telescope* angular resolution | 0.04 | |
| Atmospheric seeing limit for optical telescope | 0.6 | |
| Gravitational deflection of a light by the Sun | 1.75 | |
| Angular diameter of typical galaxy at $z = 0.5$ | 2.9 | |
| Angular diameter of Jupiter | 36 | |
| Resolution limit of eye | 120 | 2 minutes |
| Angular diameter of the Crab Nebula | 300 | 5 minutes |
| Angular diameter of the Moon/Sun | 1 865 | 30 minutes |
| Angular scale of $d_{eq}$ imprinted on CMBR | 3 100 | 1 degree |
| Angular diameter of Andromeda | 14 400 | 4 degree |

### Velocity (cm s$^{-1}$)

| | | |
|---|---|---|
| Reflex speed induced on Sun by Jupiter | $1.3 \times 10^3$ | |
| Sound speed in air | $3.3 \times 10^4$ | |
| Velocity of molecules in the atmosphere | $4.8 \times 10^4$ | |
| Escape velocity from Moon | $2.4 \times 10^5$ | 2.4 km s$^{-1}$ |
| Escape velocity from Earth | $1.1 \times 10^6$ | 11 km s$^{-1}$ |
| Stellar velocity dispersion in a globular cluster | $2 \times 10^6$ | 20 km s$^{-1}$ |
| Earth's orbital speed around the Sun | $2.9 \times 10^6$ | 29 km s$^{-1}$ |
| Velocity of Sun in Milky Way's center | $2.2 \times 10^7$ | 220 km s$^{-1}$ |
| Escape velocity from Sun | $6.2 \times 10^7$ | 620 km s$^{-1}$ |
| Velocity dispersion of galaxies in a cluster | $8 \times 10^7$ | 800 km s$^{-1}$ |
| Escape velocity from neutron star surface | $2 \times 10^{10}$ | $2 \times 10^5$ km s$^{-1}$ |

## Energy (ergs)

| | | |
|---|---|---|
| Hydrogen hyperfine transition | $9.5 \times 10^{-18}$ | $5.8 \times 10^{-6}$ eV |
| Molecular rotation transition | $4 \times 10^{-15}$ | $2.4 \times 10^{-3}$ eV |
| Thermal energy of atmosphere ($kT_{atmos}$) | $4 \times 10^{-14}$ | $2.4 \times 10^{-2}$ eV |
| Molecular vibration transition | $3 \times 10^{-13}$ | 0.18 eV |
| Hydrogen binding energy | $2.2 \times 10^{-11}$ | 13.6 eV |
| Electron rest mass | $8.18 \times 10^{-7}$ | 0.5 MeV |
| Energy from $4 \times (^1\mathrm{H}) \rightarrow {}^4\mathrm{He}$ fusion | $4.3 \times 10^{-5}$ | 26 MeV |
| Proton rest mass | $1.50 \times 10^{-3}$ | 1 GeV |
| Rotational energy of the Sun | $2.5 \times 10^{42}$ | |
| Gravitational internal binding energy of Jupiter | $3 \times 10^{43}$ | |
| Photon energy from a supernova | $3 \times 10^{51}$ | |
| Gravitational binding energy of a galaxy | $2 \times 10^{60}$ | |
| Gravitational binding energy of a cluster | $5 \times 10^{64}$ | |

## Power (watts)

| | |
|---|---|
| Human being | 150 |
| Luminosity of a white dwarf | $1 \times 10^{23}$ |
| Gravitational wave radiation from PSR 1913+16 | $8 \times 10^{24}$ |
| Solar luminosity | $3.9 \times 10^{26}$ |
| Crab Nebula energy output | $1 \times 10^{31}$ |
| Eddington luminosity of a 1.4 $M_\odot$ neutron star | $2 \times 10^{31}$ |
| Peak photon luminosity of supernova | $5 \times 10^{35}$ |
| Milky Way luminosity | $3 \times 10^{36}$ |
| X-ray luminosity of gas in coma cluster | $1.5 \times 10^{38}$ |
| Quasar luminosity | $1 \times 10^{39}$ |

## Density (gm cm$^{-3}$)

| | |
|---|---|
| Density of the starlight released in a Hubble time | $1 \times 10^{-35}$ |
| Density of the cosmic microwave background radiation | $4.6 \times 10^{-34}$ |
| Baryon density predicted by primordial nucleosynthesis | $2 \times 10^{-31}$ |
| Critical density of the universe ($3H_0^2/8\pi G$) | $4.7 \times 10^{-30}$ |
| Gas density in a cluster of galaxies | $2 \times 10^{-27}$ |
| Gas density in the interstellar medium | $3 \times 10^{-24}$ |
| Density of air | $1.3 \times 10^{-3}$ |
| Mean density of Jupiter | 1.3 |

| | |
|---|---|
| Mean density of the Sun | 1.4 |
| Mean density of the Moon | 3.3 |
| Mean density of the Earth | 5.5 |
| Mean density of a white dwarf | $5 \times 10^4$ |
| Nuclear density | $6 \times 10^{14}$ |
| Central density of a neutron star | $1 \times 10^{15}$ |
| Planck density$(c^5/G^2\hbar)$ | $5 \times 10^{93}$ |

## Temperature (Kelvin)

| | |
|---|---|
| CMBR quadrupole anisotropy | $1.3 \times 10^{-5}$ |
| CMBR dipole anisotropy | $3.3 \times 10^{-3}$ |
| CMBR temperature | 2.726 |
| Surface temperature of Venus | 740 |
| Melting point of iron | 1811 |
| Solar surface temperature | 5770 |
| Center of the Sun | $1.4 \times 10^7$ |
| Gas temperature in a cluster of galaxies | $5 \times 10^7$ |
| Characteristic temperature for$e^+e^-$ pair production | $4 \times 10^8$ |

## Pressure (dynes cm$^{-2}$)

| | |
|---|---|
| Gas pressure in a cluster of galaxies | $7 \times 10^{-12}$ |
| Pressure in the best vacuum achieved on Earth | $5 \times 10^{-11}$ |
| Solar radiation pressure at the Earth | $5 \times 10^{-5}$ |
| Atmospheric pressure | $1 \times 10^6$ |
| Central pressure of Earth | $5 \times 10^{12}$ |
| Central pressure of Jupiter | $4 \times 10^{13}$ |
| Central pressure of the Sun | $2.7 \times 10^{17}$ |
| Central pressure of a white dwarf | $1 \times 10^{24}$ |
| Central pressure of a neutron star | $1 \times 10^{35}$ |

## Magnetic field (gauss)

| | |
|---|---|
| Magnetic field in the local interstellar medium | $5 \times 10^{-6}$ |
| Magnetic field in a radio lobe of a radio galaxy | $1 \times 10^{-5}$ |
| Magnetic field at Earth's surface | $3 \times 10^{-1}$ |
| Magnetic field near Sun's pole | 1 |
| Magnetic field felt by the electron in an $n = 1$ hydrogen atom | $2 \times 10^4$ |
| Pulsar dipole magnetic field | $1 \times 10^{12}$ |

# Bibliography

Arnett D., (1996), *Supernova and Nucleosynthesis*, (Princeton University Press).

Binney J. and S. Tremaine, (1987), *Galactic Dynamics*, (Princeton University Press).

Binney J. and M. Merrifield, (1998), *Galactic Astronomy*, (Princeton University Press).

Blandford R.D. *et al.*, (1993), *Pulsars as Physics Laboratories*, (Oxford University Press).

Bondi H. and M. Weston-Smith, (Eds.), (1998) *The Universe Unfolding*, (Oxford University Press).

Carroll B.W. and D.A. Ostlie, (1996), *Modern Astrophysics*, (Addison-Wesley).

Chandrasekhar S., (1969) *Ellipsoidal figures of equilibrium* (Yale University Press).

Clayton D.D., (1983), *Principles of Stellar Evolution and Nucleosynthesis*, (University of Chicago Press).

Crawford F.S., (1968), *Waves*, Berkeley Physics Course, Volume III, (McGraw-Hill).

de Loore C.W.H. and C. Doom, (1992), *Structure and Evolution of Single and Binary Stars*, (Kluwer).

Dodelson S., (2003), *Modern Cosmology*, (Academic Press).

Eggleton P.P., M.J. Fitchett and C.A. Tout, (1989), Ap. J. **347**, 998.

Elmegreen D. M., (1998), *Galaxies and Galactic Structures*, (Prentice Hall).

Esmailzadeh R., *et al.*, (1991), Ap.J., **378**, 504.

Frank J., A. King and D. Raine, (1992), *Accretion Power in Astrophysics*, (Cambridge University Press).

Ferland G. and J. Baldwin, (1999), *Quasars and Cosmology*, A.S.P. Conf.Series, vol 162.

Hansen C.J. and S.D. Kawaler, (1994), *Stellar Interiors*, (Springer-Verlag).

Henon M., (1983), Les Houches school on *Chaotic behaviour of deterministic systems* (North-Holland publishing co.), p. 55.

Hoyle F., G. Burbidge and J.V. Narlikar, (2000), *A Different Approach to Cosmology : From a Static Universe through the Big Bang towards Reality*, (Cambridge University Press)

Kawaler E., *et al.* (Eds), (1997), *Stellar Remnants*, (Springer Verlag).

Landau L.D. and E.M. Lifshitz, (1975a), *Classical theory of fields*, (Pergamon Press).

Landau L.D. and E.M. Lifshitz, (1975b), *Fluid mechanics*, (Pergamon Press).

Landau L.D. and E.M. Lifshitz, (1975c), *Physical kinetics*, (Pergamon Press).

Lynden-Bell D., (1967), MNRAS, **136**, 101.

Lyne A.G. and F. Graham-Smith, (1998), *Pulsar astronomy*, (2nd edition), (Cambridge University Press).

Marchal C., (1990), *The three-body problem*, (Elsevier).

Meszaros P., (1992), *High energy radiation from magnetized neutron stars*, (Cambridge University Press).

Michel F.C., (1991), *Theory of Neutron Star Magnetospheres*, (Chicago University Press).

Misner C.W., K.S. Thorne and J.A. Wheeler, (1973), *Gravitation*, (Freeman).

Osterbrock D.E., (1989), *Astrophysics of Gaseous Nebulae and Active Galactic Nuclei*, (University Science Books).

Padmanabhan T., (1990), Physics Reports, **285**, 188.

Padmanabhan T., (2000), *Theoretical Astrophysics, Volume I*, (Cambridge University Press).

Padmanabhan T., (2001), *Theoretical Astrophysics, Volume II*, (Cambridge University Press).

Padmanabhan T., (2002), *Theoretical Astrophysics, Volume III*, (Cambridge University Press).

Padmanabhan T., (2003) Physics Reports, **380**, 235 [hep-th/0212290].

Padmanabhan T., (2005), *Understanding Our Universe: Current Status and Open Issues* , [gr-qc/0503107]

Peacock J.A., (1999), *Cosmological Physics*, (Cambridge University Press).

Peterson B.M., (1997), *An Introduction to Active Galactic Nuclei*, (Cambridge University Press).

Rybicki G.B. and A.P. Lightman, (1979), *Radiative Processes in Astrophysics*, (Wiley and Sons).

Seljak U. and M. Zaldarriaga, (1996), Ap.J. **469**, 437.

Shapiro S.L. and S.A. Teukolsky, (1983), *Black holes, White dwarfs and Neutron stars*, (Wiley and Sons)

Shore S. N., (1992), *An Introduction to Astrophysical Hydrodynamics*, (Academic Press).

Shu F.H., (1991), *Physics of Astrophysics* Volume I and II, (University Science Books).

Steidel C. C., *et al.*, (1999), Ap.J., **519**, 1.

Thomson J.J., (1907), in *Electricity and Matter*, Chapter 3, (Archibald constable).

Visser M., (2003), *M. Heuristic approach to the Schwarzschild geometry*; gr-qc/0309072.

# Index

lyman alpha forest, 335, 336

magnitude, 141
main sequence, 168
mass of
    planet, 136
    star, 138
mass-to-light ratio, 21
mean-free-path
    photons, 72
metal production, 266
microlensing, 38
molecular binding energy
    order of magnitude, 78
molecular energy levels, 78
molecular weight
    definition of, 98
monatomic gas, 24

natural width, 82
negative hydrogen ion, 87
neutrino oscillations, 160
neutron decay, 227
neutron star, 173, 194
nucleosynthesis, 222

opacity, 55, 68, 84, 149
optical depth, 55, 57
origin of perturbations
    inflation, 244

p-p reactions, 150
pair production, 323
Pauli exclusion principle, 98
period-luminosity relation, 156
photo-ionisation, 83
    order of magnitude, 84
photon opacity
    order of magnitude, 56
photon-to-baryon ratio, 223
planetary nebula, 170
planets
    tidal friction, 15
plasma dispersion, 127
plasma frequency, 126
Plummer model, 296

Poisson's equation, 2
polarization
    synchrotron radiation, 67
polytropes
    self gravitating systems, 102
polytropic equation of state, 102
Pop I, 153
Pop II, 153
precession, 8
    free, 12
    nutation, 15
    of equinoxes, 12
Press-Schechter formula, 241
pressure
    general expression, 96
principle of equivalence, 2, 30, 31
principle of superposition, 1
protogalaxies, 272
    intergalactic medium, 279
    ionisation, 279
pulsar, 194
    gravitational radiation, 205
    magnetosphere, 195

quasar absorption system
    classification of, 335
    estimates of, 336
    voigt profile, 335
quasar abundance, 330, 333
quasar luminosity function, 329
quasars, 275
    absorption systems, 335
    abundance of, 333, 334
    black hole paradigm, 332
    Gunn-Peterson effect, 273
    K correlation, 343
    luminosity function, 343
    mass of, 333
    scaling, 332

radiation
    beaming, 55
    bremsstrahlung, 59
    classical theory, 44
    dipole, 46
    plane wave, 52